All the World's Air-Ships.

VICE-ADMIRAL SIR PERCY SCOTT'S CONCEPTION OF AERIAL TARGET PRACTICE IN THE FUTURE.

The two war dirigibles are passing each other at varying speeds and altitudes, each towing a special aerial target. The rest of the fleet can be seen lying by in the clouds above.

Frontispiece.

INTRODUCTION

No editor has faced a more daunting task that that which confronted Fred T. Jane when he compiled the original edition of this book. Its predecessor, *Fighting Ships*, had been founded on five thousand years of naval tradition and recorded history when it first appeared in 1898. By contrast, only six aeroplane pilots had made flights longer than one mile by the beginning of 1909, and few people regarded flying as more than a hazardous and pointless sport.

Jane called his new yearbook *All the World's Air-Ships*, not because he regarded aeroplanes as flying ships but because 'airship' seemed likely to become the accepted generic term for both heavier-than-air and lighter-than-air flying machines. However, his nautical background led him to describe the editorial work as "a plunge into unknown and unexplored waters".

There was no worldwide aviation press to keep him informed of attempts to fly by pioneers like Ellehammer in Denmark, Vuia and Breguet in France, Jatho in Germany and Curtiss in the USA. Yet their aircraft were described and illustrated, seeming already more substantial than the "flappers" and "air suckers" designed by some other would-be aviators. Only the French Blériot monoplane and Voisin biplane, and the American Wright Flyer manufactured by the Short brothers in England, could be considered genuine production aeroplanes. The rest were what are known today as homebuilts, but their significance can be appreciated by noting the names of constructors such as A. V. Roe, Henry Farman and Frederick Handley Page.

Fred T. Jane described each aeroplane, so far as possible, under a series of headings, such as maximum length, maximum breadth, supporting surface, total weight, body, wings, motor, speed, propellers, steering and remarks. Specification data were quoted from the start in both metric and Imperial units. Side-view, two-view and even three-view drawings were included, although one comprised no more than six straight lines. The same basic arrangement of information, greatly expanded, has been used in every edition of the book that has been published as *All the World's Aircraft* since 1912. In its seventy-seven years, it has had only four editors and a single printer.

What is known throughout the aviation world simply as "Jane's" is as old as the industry it serves. In the year of its birth, Frederick Handley Page established in Britain the first company to be constituted exclusively for the design and manufacture of aeroplanes; Louis Blériot pointed the way to airlines of the future by making the first international flight over water, from France to England; the first great international flying meeting was held at Rheims in France; and the US Army took delivery of the world's first military aeroplane.

By the time Fred T. Jane, died in 1916, a pioneer airline service had been operated in Florida, and hundreds of warplanes were engaged in deadly combat over the Western Front in Europe. Even so, he could never have imagined that the book he launched in an age of frail stick-and-string flying machines would one day record airliners able to carry a hundred men and women across the Atlantic in armchair comfort at twice the speed of sound, military aircraft and rockets of such terrifying power that their mere existence would impose on Europe an unprecedented period of peace through fear, or spacecraft that would carry men safely to and from the Moon.

By continuing the task of recording factually each stage of the progress of flight, without national or political bias, *Jane's All the World's Aircraft* has gained friends and contributors in every nation that builds aeroplanes. This is, perhaps, the most satisfying legacy of Fred T. Jane in what has proved to be the most barbarous, as well as the most technologically exciting, century in the history of our planet.

John W. R. Taylor
September 1986

ALL THE WORLD'S AIR-SHIPS.

(FLYING ANNUAL.)

FOUNDED AND EDITED BY

FRED T. JANE,

AUTHOR OF "FIGHTING SHIPS," ETC.

With a Special Chapter on "Aerial Engineering," by Charles de Grave Sells, M. INST., C.E.

LONDON:

SAMPSON LOW, MARSTON & CO., LTD.

1909.

PRINTED BY NETHERWOOD, DALTON & CO., PHŒNIX WORKS, RASHCLIFFE, HUDDERSFIELD.

First published by Sampson Low,
Marston and Company Limited in 1909

This edition published
in the United Kingdom in 1986 by
Jane's Publishing Company Limited
238 City Road, London EC1V 2PU

and in the USA by
Jane's Publishing Inc
115 Fifth Avenue
New York NY 10003

ISBN 0 7106 0471 8

Printed and made by Netherwood Dalton & Co. Ltd.,
Huddersfield, England

CONTENTS.

PART I.

	PAGE.
PREFACE	9
GLOSSARY OF TECHNICAL TERMS...	11
AUSTRIAN :—	
Aeroplanes	16
Dirigibles	24
BALKAN STATES	25
BELGIAN :—	
Aeroplanes	26
Dirigibles	30
BRITISH :—	
Aeroplanes	34
Colonial ditto	75
Dirigibles	79
CHINESE :—	
Aeroplanes	85
DANISH :—	
Aeroplanes	86
DUTCH	90
FRENCH :—	
Monoplanes	92
Biplanes	135
Triplanes	149
Helicopters, etc.	153
Dirigibles	159
GERMAN :—	
Aeroplanes	177
Dirigibles	191
GREEK	210
ITALIAN :—	
Aeroplanes	211
Dirigibles	216
JAPANESE :—	
Aeroplanes	219
NORWEGIAN	222
PERUVIAN	223
PORTUGUESE	224

CONTENTS—(*Continued*).

	PAGE.
RUSSIAN :—	
Aeroplanes	225
Dirigibles	235
SOUTH AMERICAN STATES	238
SPANISH :—	
Aeroplanes	239
Dirigibles	245
SWEDISH	247
SWISS	248
TURKISH	250
UNITED STATES :—	
Monoplanes	252
Biplanes	256
Triplanes	280
Miscellaneous	285
Dirigibles	305
SILHOUETTES OF THE WORLD'S DIRIGIBLES ...	313

PART II.

	PAGE.
AERIAL WARFARE (**Admiral Sir Percy Scott**) ...	320
THE FUTURE OF DIRIGIBLES (**"Ripping Panel"**) ...	321
THE POLITICAL ASPECT OF AVIATION (**L. Cecil Jane, M.A.**)	326
AN AERIAL POSSIBILITY (**René Declarges**)	331
AERIAL ENGINEERING (**Charles de Grave Sells, M. Inst. C.E.**)	333
INDICES :—	
(*a*) 'Planes, etc.	*end of book*
(*b*) Dirigibles	*end of book*

PREFACE.

A FEW words may be desirable in introducing this the first annual edition of *All the World's Air-ships*.

Aviation is yet in an early stage, and it may or may not be many years yet before it is on a practical commercial or even military footing. On the other hand the number of dirigibles and aeroplanes of various kinds in existence is already very large and continually increasing, hence the idea that a standard work of reference, giving uniform statistics, should already have a market.

In the first issue of a work of this kind it is impossible to be so all-embracing as might be desired. Equally impossible is it to supply equally full details of every machine dealt with, since in a great number of cases the inventors desire to keep certain details secret. These wishes have been generally respected, and attention mainly directed to securing the complete accuracy of such figures as are furnished.

With the vexed problems of heavier than air or lighter than air, or the relative merits of monoplane or biplane, rigid or non-rigid dirigible, this work has no concern. Nor is it in any sense intended to be a "popular" work on the subject. It is intended to be for the air what *Fighting Ships* is for the sea—a Part I devoted to a record and register, and a Part II dealing with certain general aspects of the flying question, in which all or any expressions of views and opinions are equally welcome.

For the reasons given above, this, the first issue must naturally be a "skeleton." In future issues the kind co-operation of readers is relied upon to assist in the production of a book crammed as far as possible with every possible detail.

At the very outset the arrangement of the book presented certain difficulties. The desirability of arrangement by nationality was obvious enough. Equally obvious was the arrangement of dirigibles into rigid, semi-rigid and non-rigid; and of aeroplanes into monoplanes, biplanes, triplanes, multiplanes and miscellaneous or nondescript. But no sooner was any general arrangement arrived at than the appearance of some new machine upset the whole paging arrangement, and in order to avoid continual chaos it became necessary to adopt the somewhat clumsy method of paging here employed, and abandon all hope of placing all the aeroplanes in strict alphabetical order.

In place of this a considerable system of indexing has been employed, by means of which any particular machine can readily be found.

It was originally intended to place the dirigibles of each nation in a strictly chronological order, but as this would have entailed obscuring immediate appreciation of the progress made in various special types—as for example: *Lebaudy I, Lebaudy II*, etc., etc., the idea was abandoned, and a preliminary table giving this information substituted instead.

Incidentally it may be remarked that not one of the least of the difficulties in compiling this register is the comparatively common practice of changing the name or designation of so many aeroplanes. Where possible these various names have been mentioned so far as possible.

This book being more or less international in character, it has been deemed advisable to give metric as well as English measurements in the vast majority of the statistics. Where engine dimensions are given m/m only have, however, been employed; since to the big majority of those concerned with petrol engines a bore of, say, "$4\frac{3}{8}$ inches," has very little significance, and is probably mentally converted into millimetres before it is really comprehended by the average motorist or aviator.

In the plans of dirigibles a uniform scale has been adopted in order to admit of ready visual reference as to relative dimensions. A scale is also supplied with each.

In the aeroplanes so far as possible a special uniform scale is also employed, but in several cases this has only been approximately possible. When the scale is uniform a scale is given on the plan.

The silhouettes of dirigibles are, perhaps, the most novel feature of the book. The whole of these silhouettes are on a uniform scale with each other. They do not purpose to be exact representations of the various dirigibles, as the characteristic features of each are purposely slightly exaggerated in order to assist recognition. Attention may be drawn to the fact that these special silhouettes are copyright.

Part II requires little in the way of special introduction. Few names are better known in engineering circles than that of Mr. Charles de Grave Sells, and though his name is primarily associated with steam, yet he was among the first to recognise the potentialities of the internal combustion engine. As the motor car has gradually evolved a special type of petrol engine for itself, so it is probable that the flying machine engine of the future will depart more and more from standard motor car practice, and indeed marvellous progress has already been made in that direction.

It is, perhaps, not too much to say that the whole future of aviation rests with the engine and its general reliability.

In conclusion I have to thank many correspondents without whose kind assistance the work could never have been produced. And I venture to express a hope that all readers interested in the subject, who are in a position to supply omissions and corrections, will be kind enough to render that assistance towards the production of the 1910 edition. All information should, if possible, arrive not later than the end of August.

FRED T. JANE.

17, Elphinstone Road,
Southsea.

SPECIAL NOTE.—Only a proportion of the aeroplanes here given have actually flown or arrived at a stage to be photographed. Hence the comparatively large number of non-illustrated pages. A final glance at the proofs seems to render necessary an apology for the many pages that consist of little but blank forms, and for which the only excuse is that there are sure to be some readers able to supply the missing details. I can only plead in extenuation that this, the first edition, is a plunge into unknown and unexplored waters, and draw attention to the fact that however numerous the deficiencies, the residue represents the most strenuous possible labour in a generally unresponsive ocean.—F. T. J.

GLOSSARY OF TECHNICAL TERMS, Etc.

ENGLISH.	FRENCH.	GERMAN.	ITALIAN.	NOTES.
Abaft	Arrière	Hinter	A poppa	
AEROPLANE	Aéroplane	Drackenflieger	Aereoplano	
Aeronaut	Aéronante: Aviateur:	Luftschiffer	Aereonauta	
Aerostat			Aereostato	
Aft	Arrière	Hinten	Addietro	
After (rear)		Hinterer	Poppa	
Air-cooled	Refroidissement par l'air	Mit Lufthüllung	Raffredda ad aria	
Angleiron	Cornière	Eck Schiene	Ferro ad angolo	
Aviation	Aviation	Suftschiffahet	Aviazione	
Balance	Equilibre	Gleichgewicht	Equilibrio	
Ballonet	Ballonet	Ballonet	Palloncino compensatore	
Behind	Arrière	Hinter	Di dietro	
Bevel geared	Engrenage Conique	Konischer Antrieb	Ingranaggio conico	
Biplane	Biplan	Zwei decker	Biplano	
Blades (of propeller)	Branches	Flügel	Pale delt'elica	
Body	Chassis	Körpor	Telaio o chassis	
Box-kite	Cerf-volant	Drachen	Aquilone a celle	
Bracket	Tasseau	Stütze	Sostegno	
Breadth	Largeur	Breite	Larghezza	
Car	Nacelle	Gondel	Navicella	
Centre of Gravity	Centre de Gravité	Schwerpunkt	Centro di gravità	
Chain driven	Transmission par chaine	Ketten antrieb	Trasmissione a catena	
Chassis	Chassis	Motor Rahmen	Chassis	
Circumference	Circonférance	Umfang	Cuconferenza	
Clutch	Embrayage	Kupplung	Innesto	
Coupled	Jumelé	Paarweise	Accoppiato	
DIRIGIBLE	Dirigeable: Aeronat	Motorluftochiff	Dirigibile	
Diameter	Diamètre	Durchmesser	Diametro	
Direct driven	Prise directe	Direkter Antrieb	Presa diretta	
Elevator (horizontal rudder)	Gouvernail de profondeur	Horizontal Steuer	Timone orizzontale	
Engine	Moteur	Motor	Motore	

GLOSSARY OF TECHNICAL TERMS, Etc.

ENGLISH.	FRENCH.	GERMAN.	ITALIAN.	NOTES.
Fan	Ventilateur	Ventilator	Ventilatore	
Flight		Flug	Volo	
Flown	Vole	Geflogen	Volato	
Fore	Avant	Vorderer	Ouvanti	
Forward (in front)	En avant	Vor	Davanti	
Frame	Fuselage	Rahm	Telais	
Framework		Gerüste	Intelaiatura	
Gas bag			Involucro	
Geared to	Multiplié à	Nebersetzt auf	Moltiplicato a	
Gear driven		Gashälter	Trasmissione a ingranaggi	
Girder		durch Zahnrädern getrieben	Longarine	
Glider		Balken	Apparecehio a planaie	
Gondola	Nacelle	Gondel	Navicella	
Helices	Helices	Schrauben	Eliché	
Helicopter	Helicoptre	Schraubenflieger	Elicoplano Elicottero	
Horizontal plane (in a)		Horizontal fläche	Piano orizzontale	
Horizontal rudder (elevator)		Höhensteuer	Timone orizzontale	
Horse power	Puissance en chevaux	Pferdekraft	Forza cavalli	
Hydrogen		Wasserstoff	Idrogens	
Ignition	Allumage	Zündung	Accensione	
Inch	25·39 m/m.	25·39 m/m	Pollice = 25·39 m/m.	
Inclination	Inclination	Schrägstellung	Inclinazione	
Keel	Carène	Kiel	Chiglia	
K.P.M. (kilometres per hour)	Kilometres par heure	Kilometre pro Stunde	Chilometre all'ora	
Kite	Cerf volant	Drachen	Aquilone	
Length	Longueur	Länge	Lunghezza	
Lower (planes)	Inferieur (plans)	Untere Flächen	Piani inferiori	

GLOSSARY OF TECHNICAL TERMS, Etc.

ENGLISH.	FRENCH.	GERMAN.	ITALIAN.	NOTES.
Maximum		Maximum	Massimo	
Middle (plane)		Mittel Deck	Piano medio	
Mile	Mile	Meile	Miglio	
Military	Militaire	Militärische	Militare	
Miscellaneous	General	Verschiedenes	Diversi	
Monoplane	Monoplan	Ein decker	Monoplano	
Motor	Moteur	Motor	Motore	
M.P.H. (miles per hour)	Vitesse		Miglia all'ora	
Multiplane	Multiplan		Multiplano	
Nacelle	Nacelle	Gondel	Navicella	
Non-rigid		Unstare	Non-rigido—flessibile	
Petrol (gasoline)	Essence	Benzin	Benzina	
Pilot (driver)	Flyer : Aviateur	Führer	Aviatore	
Pivot	Pivot	Gewinde Zapfen	Perno	
Planes	Plans	Flächen	Piani	
Pound (lb.)	0·453 kg.	0·453 kg.	Libbra = 0·453 kg.	
Pressure	Pression	Druck	Pressione	
Propeller	Helice	Schraube	Eliche	
Quadruplane		Vier decker	Qudruplani	
Quintuplane		Fürf decker	Quintuplani	
Radiator	Radiateur	Kühler	Radiatore	
Rear (in)	En arrière		Indictro	
Reduction gearing	Engrenage de demultiplication		Ingranaggi di ridugione	
R.P.M. (revolutions per minute)	Tours	Umlauf	Giri al minuto	
Rigid	Rigide	Starr	Rigido	
Rises	S'elève	Hebt sich	Si eleva	
Rubber	Casutchouc	Gummi	Gomma	
Rudder	Governail	Ruder	Timone	
Section	Section		Regione	
Semi-rigid		Halb Starr	Semi-rigido	

ENGLISH.	FRENCH.	GERMAN.	ITALIAN.	NOTES.
Speed	Vitesse	Geschwindigkeit	Velocita	
Stability	Stabilité	Stabilität	Stabilità	
Stabilising fins	Ailerons	Stabilisierungsflächen	Piani stabilizzaton	
Steel	Acier	Stahl	Acciaio	
Steering Gear	Direction	Lenk Übersetzung	Meccanismo di direzione	
Steering Wheel	Volant	Lenkrad	Volante di direzione	
Supporting surface		Tragfläche	Superficio di sostegno	
Surfaces	Surfaces	Flächen	Superfici	
Suspension	Suspension		Sospensioni	
Tail	Queu	Schwanz	Coda	
Total weight	Poids totale	Geramtlast	Peso totale	
Transmission Shaft	Arbre de transmission	Transmissions Welle	Albero di trasmissione	
Trial	Essai	Probe	Prova	
Triplane	Triplan	Drei decker	Triplano	
Universal Joint	Joint universel	Kardan	Guinta universale	
Upper (planes)	Superieur	Obere	Piani superiori	
Useful lift		Nutzlast	Forza utile di elevazione	
Valve	Soupapa	Ventil	Valvola	
Vertical plane (in the)	Plan vertical	(in der) Vertikalfläche	Nel piano verticale	
Vertical rudder	Gouvernail vertical	Seitensteun	Timone verticale	
Volume	Volume	Inhalt	Volume	
Water-cooled	Refroidissement par eau	Wasserkühlung	Raffreddata ad acqua	
Weight	Poid	Gewicht	Peso	
Wheels	Roues	Raeder	Ruote	
Wings	Ailes	Flügeln	Ali	
Wood	Bois	Holz	Legno	
Yard (measure)	0·914 mètres	0·914 meter	Jarda = 0·914 m.	

PART I.

AEROPLANES & DIRIGIBLES.

ARRANGED BY NATIONALITIES IN ALPHABETICAL ORDER.

Aeroplanes first, so far as possible in order, according to whether they are Mono-, Bi-, Triplanes, etc. Then Helicopters and Unclassified Machines. Then Dirigibles in three classes: 1—Non-Rigid; 2—Semi-Rigid; 3—Rigid.

AUSTRIAN.

General Note.—Austria has only just commenced to take much interest in aviation, but a good deal of quiet experiment appears to have been conducted by isolated inventors for many years.

Aerial Societies:—

Upper Austrian Society for Aerial Navigation.
Military Aeronautical Institute.
Wiener Flugtechniseher Verein.
Wiener Aëro Club.

Aerial Journals:—

Weiner Luftshiffer Zeitung.

Flying Grounds:—

Linz.

HIPSSICH. Monoplane.

Maximum length, feet m. **maximum breadth,** 25 feet 7·61 m. **supporting surface,** 400 sq. feet 37 m^2.

Total Weight.—900 lbs. (408 kgs.), *including* aeronaut.

Body.—On 3 wheels, two in front, one behind.

Wings.—2 curved, each with a 25 feet (7·61 m.) spread, placed one behind the other.

Motor.—50 h.p. 4 cylinder. Dutheil-Chalmers. Carried between the planes.

Speed.—

Propellers.—2 of aluminium placed one before and one behind the rear plane. These propellers rotate in opposite directions at 700 r.p.m. A patent clutch connects engine and propeller shaft.

Steering.—Horizontal rudder placed in the back wing.

Remarks.—

WELS & ETRICH. Monoplane.

Photo, C. Malcuit.

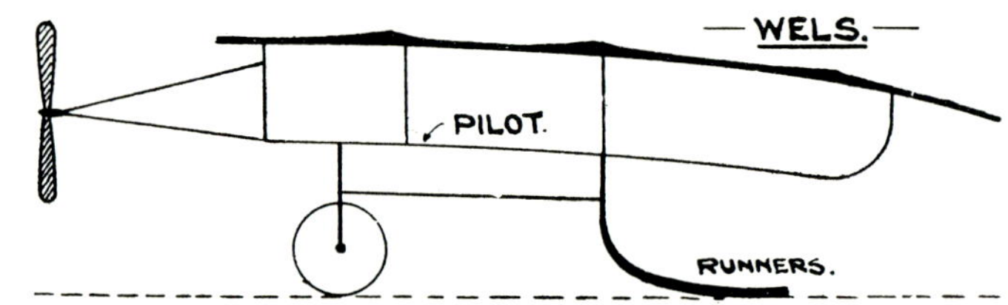

Maximum length, 36 feet / 11 m. **maximum breadth,** feet / m. **supporting surface,** sq. feet / m².

Total weight.—

Body.—On 2 wheels, with two runners just behind them.

Wings.—

Motor.—

Speed.—

Propellers.—1, right forward.

Steering.—

Remarks.—

ZOLYI (Adalar) Biplane.
(*Alias* ZOELGI.)

Building.

Maximum length, 29½ feet / 9 m. **maximum width,** 29½ feet / 9 m. **supporting surface,** 280 sq. feet / 18·5 m^2.

Total weight.—390 lbs. (177 kgs.) *including* pilot.

Body.—On 4 wheels.

Wings.—Span 18 feet (5·48 m.)

Motor.—30 h.p. Dutheil-Chalmers.

Speed.—About 37 m.p.h. (60 km.)

Propellers.—One; direct drive.

Steering.—Like a Wright.

Remarks.—Generally of the *Voisin* type; but with the Wright system of warping the wings.

BURCHARD. Triplane.

Maximum length, feet / m. **maximum breadth,** feet / m. **supporting surface,** sq. feet / m^2.

Total weight.—

Body.—On two bicycle wheels, with a couple of small supporting wheels under the main plane ends.

Wings.—Curved as in the Wright machines. A main section, with extensions on either side, slightly lower than the central planes which act as stabilisers. Distance between the planes: 4 feet (1·20 m.)

Motor.—*Not yet settled.*

Speed.—

Propellers.—

Steering.—Horizontal rudder of two planes in front, with two semi-circular balancers. There is a celluloid tail behind, consisting of two horizontal planes and two vertical rudders. Steering is by means of an ordinary motor car wheel.

STOLFA. Miscellaneous.

Building.

Maximum length, feet / m. **maximum breadth,** feet / m. **supporting surface,** sq. feet / m².

Total Weight.—

Body.—

Wings.—Machine consists of cylindrical boxes and is somewhat suggestive of the *Givaudin* (see France).

Motor.—

Speed.—

Propellers.—

Steering.—

Remarks.—Building in Vienna.

URBANEK.

Completing.

Maximum length, 33 feet / 10 m. **maximum breadth,** 38 feet / 11·50 m. **supporting surface,** sq. feet / m².

Total weight.—

Body.—

Wings.—Span, 38 feet (11·50 m.) The special feature is that the dimensions of the wings can be altered at will.

Motor.—

Speed.—

Propellers.—One.

Steering.—Single elevator in front.

Remarks.—This machine is said to have been 14 years building. It is not yet complete. Constructed at Prague.

NEMETHY. Monoplane.

Maximum length, feet / m. **maximum breadth,** feet / m. **supporting surface,** sq. feet / m².

Total weight.—Only 66 lbs. (30 kgs.)

Body.—Tricycle with 2 small runners in place of hind wheel.

Wings.—

Motor.—¾ h.p. single cylinder.

Speed.—

Propeller.—1 forward.

Steering.—Vertical rudder in front of propeller.

Remarks.—

HENRI FARMAN I. Biplane (now converted to a Triplane).

(Voisin type) *q.v.*

Maximum length, 38 feet / 11·50 m. **maximum span,** 42⅔ feet / 13 m. **supporting surface,** 645 sq. feet / 60 m². including tail, etc.

Total weight.—1234 lbs. (560 kgs.) *including* driver.

Body.—On 3 wheels, two in front, one behind. Continental material. Maximum breadth of body, 6½ feet (2m.)

Wings.—42⅔ feet (13 m.) × 5¾ feet (1·70 m.) Planes are 5 feet (1·50 m.) apart. (Third plane now added above).

Motor.—50 h.p. 8-cylinder Antoinette, air-cooled: 110 × 105 m/m. 1100 revs. Weight, 209 lbs. (95 kgs.) = 4·1 lbs. per h.p. (1·9 kgs. per h.p.)

Speed.—34 m.p.h. (55 km.) *Now slower.*

Propellers.—One 7 feet (2·10 m.) in diameter.

Steering.—2 wooden horizontal rudders in front, one on either side of the prow. Vertical rudder inside box tail.

Remarks.—This was the first machine to turn a corner. It first flew in 1907. It has done 16¾ miles (27 km.) Chalons S. Marne to Reims in 20 minutes. Won the Deutsch.-Archdeacon Grand Prix, 13th Jan., 1908. Time, 1 min. 28 sec. Also *l'Auto* Prix de la Hauteur, 30th Oct., 1908. Built by Voisin Fréres. Subsequently converted into a triplane. Bought by an Austrian syndicate. The syndicate failed to secure the results they expected. They dissolved the syndicate and presented the machine to the Austrian army. Being turned into a biplane again.

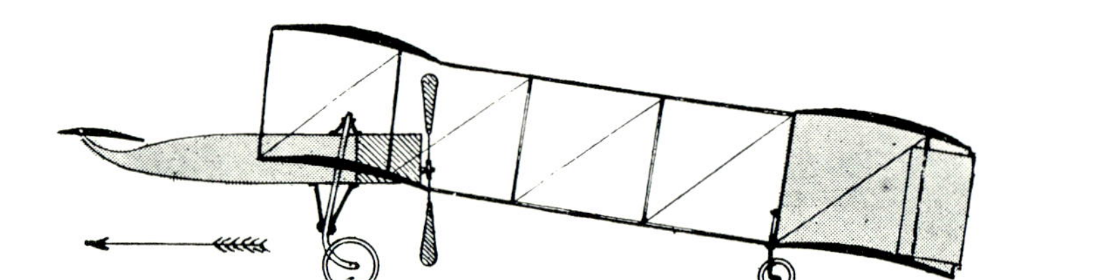

WRIGHT BROS. Biplane. *(see U.S.A.)*

Concessionaire :—

Of this type there are in Austria, built, building, or on order :—

Model 1908-9

1 Wels and Etrich (*purchased*)
2 ———
3 ———
4 (?)

WRIGHT type.

PARSEVAL Type.

Approx. maximum length, 148 feet (45 m.) **approx. diameter,** 25 feet (7·60 m.) **volume,** 63,570 c. feet (1,800 m^3.)

Total lift.—4,211 lbs. (1,910 kgs.) **Useful lift,** lbs. (kgs.)

Gas bags.—Continental rubbered fabric.

Motor.—Probably one 50 h.p.

Speed.—

Propellers.—Parseval type.

Steering.—

Remarks.—The designs of all contemporary classes of air-ships were considered by the Austrian Military Authorities, and this was one of the two types selected for construction.

BALKAN STATES.

BULGARIA. MONTENEGRO. ROUMANIA. SERVIA.

General Note.—Captain Goliescu is building a monoplane of his own design in Roumania. No machines appear to be on order for any of these States. Some time ago a machine of novel design was reported building in Bulgaria, but no trace of its existence can be discovered.

BELGIAN.

General Note.—The Belgian, M. Adhemar de la Hault, editor of *La Conquête de l'Air*, was one of the pioneers of aviation, and public interest in flying is great.

Aerial Societies:—

Aero Club Belgique.
Aero Club of Hainaut.
La Ligue Nationale Belgique.

Aerial Journals:—

La Conquête de l'Air.

Flying Grounds:—

Mons (Aero Club).

MARCHAND. Monoplane.

Completing.

Maximum length, feet / m. **maximum breadth,** $39\frac{1}{2}$ feet / 12 m. **supporting surface,** 533 sq. feet / 50 m^2.

Total weight.—

Body.—Dimensions, $26\frac{1}{4} \times 3\frac{1}{4}$ feet (8×1 m.) Mounted on 2 fixed wheels in front, single wheel, connected with vertical rudder, behind.

Wings.—Dimensions, $18 \times 14\frac{3}{4}$ feet ($5{\cdot}50 \times 4{\cdot}50$ m.) Ailerons fitted.

Tail.—

Motor.—40 h.p., 4-cylinder, water-cooled, carried in front of aviator.

Speed.—

Propellers.—2 wooden, placed forward, running up to 600 r.p.m. Diameter, $8\frac{1}{2}$ feet (2·65 m.) Pitch, variable, but normally *about* 8·2 feet (2·50 m.) Propellers are $9\frac{1}{4}$ feet (2·80 m.) apart from centre to centre.

Steering.—By tiller working on universal joint. Elevator aft. Consists of two planes, with vertical rudder placed between. Area of planes, also of rudder, $10\frac{3}{4}$ sq. feet (1 m^2.) in each case.

Remarks.—Built by M. Marchand, at Casteau, near Mons. The driver sits about the middle of the machine.

VOISIN Biplanes. (*see France*).

Belgian Concessionaire :—

Of this type there are owned by Belgians :—

Model 1908-9.

1 Baron de Caters

2

LOUIS DE LAMINNE (Chevalier). Biplane.
(Wright Pattern).

Maximum length, feet / m. **maximum breadth,** feet / m. **supporting surface,** 430 sq. feet / 40 m^2.

Total weight.—1,102 lbs. (500 kgs.), *including* aviator.

Body.—

Wings.—2 planes, Wright pattern.

Motor.—60 h.p. Vivinius.

Speed.—

Propellers.—Two.

Steering.—Mainly as *Wright*.

Remarks.—Completed for trials August, 1909.

LEPOUSSE. Aero-torpille.

Building.

Maximum length, 16½ feet / 5 m. **maximum breadth,** feet / m. **supporting surface,** sq. feet / m^2.

Total weight.—Said to be only 165½ lbs. (75 kgs.)

Body.—

Wings.—Span 10 feet (3 m.)

Motor.—Gas turbine.

Speed.—

Propellers.—

Steering.—

Remarks.—

DE LA HAULT. Flapper.

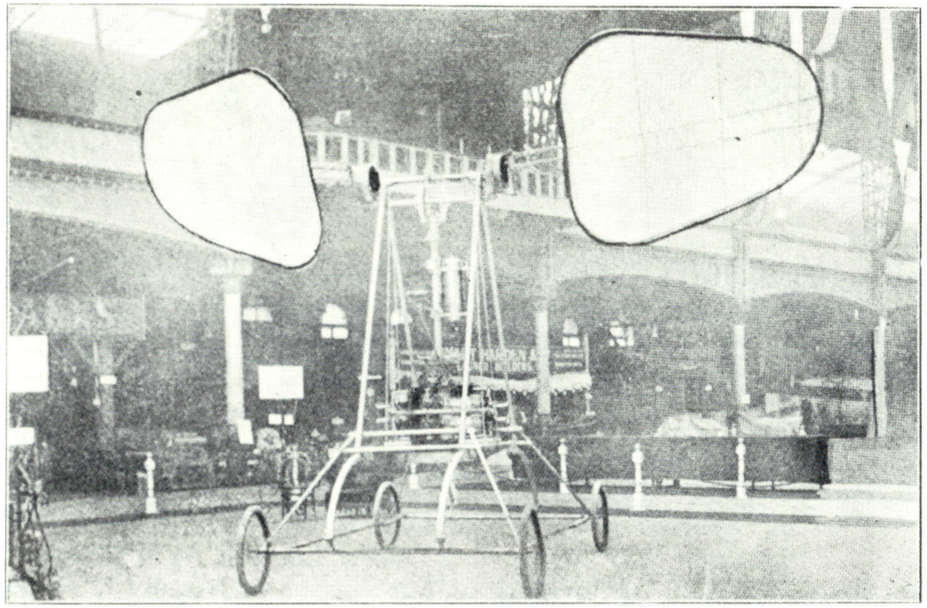

Maximum length, 13 feet. / 4 m. **maximum breadth,** feet. / m. **supporting surface,** 72 sq. feet. / 6·5 m².

Total weight.—400 lbs. (181·50 kgs.)

Body.—Rectangular tube frame on 4 wheels, supporting a double triangle at the apex of which the wings are mounted.

Wings.—2 wings which in turning trace a figure "8." Mounted on trunnions, actuated by a pair of spur wheels through ball and socket connections. Spur wheels geared to the engine.

Motor.—100 h.p. 8-cylinder horizontal opposed Miesse. Bore, 130 m/m; stroke, 140 m/m. Air-cooled. Vertical crank shaft.

Speed.—

Steering.—Entirely by altering the inclination of the wings.

Remarks.—Belongs to the Miesse Co.

VLEMINCKZ. Biplane.

Altering.

Maximum length, feet / m. **maximum breadth,** feet / m. **supporting surface,** *about* 300 sq. feet / 28 m^2.

Total weight.—

Body.—Entirely of tubular steel. On 4 wheels; one forward, two amidships, one aft.

Wings.—Small spread. 2 vertical surfaces between the planes.

Motor.—Gnome rotary, air-cooled.

Speed.—

Propellers.—Two.

Steering.—Elevator and vertical rudder both in front.

Remarks.—Has made two attempts to rise without success.

VANDENBERGH. Flapper.

Building.

Maximum length, feet / m. **maximum breadth,** feet / m. **supporting surface,** sq. feet / m^2.

Total weight.—1,320 lbs. (599 kgs.)

Body.—

Wings.—2 large flappers, with bird action.

Motor.—

Speed.—

Propellers.—

Steering.—Only 1 rudder, carried right aft.

Remarks.—

LA BELGIQUE.

Length, 180 feet / 54·8 m.; **maximum diameter,** 32 feet / 9·75 m.; **capacity,** 29,064 cubic feet / 2,700 m³.; **total lift,** 6,614 lbs. / 3,000 kgs.

Gas bag.—Rubber proofed Continental fabric. Ballonet, filled by a separate motor giving 7·5 inches of water pressure.

Motors.—2, each of 55 h.p.

Propellers.—2, one in front, and the other in rear of the car. 285 revolutions per minute. Wood and steel construction.

Speed.—25 miles per hour. 40 km. per hour.

Planes.—Horizontal: a gas tube bent horizontally round the tail.
Vertical: vertical fins on the tail, and a long vertical keel under the gas bag.

Car.—A long girder, square in section, tapered at both ends.

Miscellaneous.—Built by L. Godard, France, 1909. Crew, 3 men. Accommodation for 1 passenger. Fuel for 10 hours. Greatest height attainable, 3,280 feet / 1,000 m.

Table of weights.—

	lbs.	kgs.
Gas bag, complete with ballonet, valves, planes, suspension, etc.	1,951	885
Propellers (2)	275½	125
Blower	33	15
Motors (2) complete with gearing and shafting ...	1,410	640
Car...	496	225
Fuel for 10 hours	738½	335
Ballast	826¾	375
1 passenger (or ballast)	154	70
Crew (3)	463	210
Guide ropes, etc.	132¼	60
Miscellaneous	88	40
	6,610	3,000

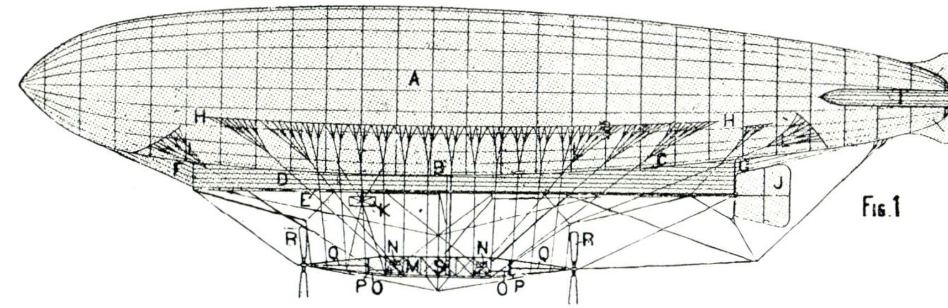

Fig. 1

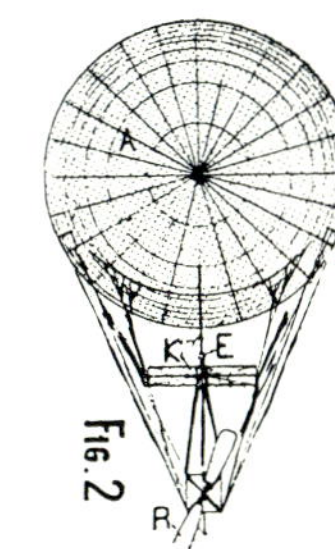

Fig. 2

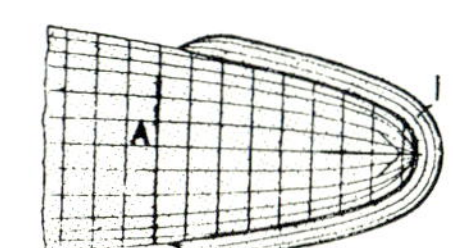

Fig. 3

KEY TO PLAN.

A = Gas bag.
B = Ballonet.
C =
D = Vertical surface.
E = Keel beam.
F & G = Front and rear continuations of D.
HH = Suspension strip.
I = Horizontal tail fin.
J = Vertical rudder.
K = Horizontal rudder or aeroplane.
L = End suspensions.
M & N = Motors.
O = Motor attendant's post.
P = Reduction gearing.
Q = Propeller shaft.
R = Propeller.
S = Captain's post.

SAINT-MARCQ.
(Military).

Building

Maximum length, feet / m. **maximum diameter,** feet / m. **volume,** 141,270 c. feet / 4,000 m³.

Total lift.— lbs. (kgs.) **Useful lift.**— lbs. (kgs.)

Gas bags.—Reported to be divided into a number of gas-tight compartments.

Motors.—Two 100 h.p. Pipe motors.

Speed.—37 m.p.h. (60 km.)

Propellers.—

Steering.—

Remarks.—No details are known concerning this vessel, whose name has been spoken of for a very long time; the date of her completion is unknown.

LA FLANDRE.

(Astra type.)

(*See France*).

Maximum length, 246 feet (75 m.) **maximum diameter,** 48¼ feet (14·7 m.) **volume,** 69,914 c. feet (6,500 m^3.)

Total lift.—15,763 lbs. (7,150 kgs.) **Useful lift,** lbs. (kgs.)

Gas bags.—Continental rubbered fabric, yellow. Ballonet, 16,146 c. feet (1,500 m^3.)

Motors.—2 Pipe motors of 110 h.p. each, placed in line with each other in the fore and aft line, and with clutches and the necessary gearing in between them.

Speed.—35 m.p.h.

Propellers.—3, namely, one at the fore end, driven by the two motors when coupled together, and two placed above and on either side of the centre of the car, for use when only one motor is running. Made of wood, and of the "Integrale" type.

Steering.—Vertical steering by means of a large double aeroplane fixed above the car, about a third from the front. Horizontal steering by means of a double vertical rudder above the rear end of the car. Stability is secured by the usual Astra pear shaped stabilising gas bags, with fins of rubbered cloth spread between the inner edges of these shapes.

Remarks.—The distinctive feature of this ship is the arrangement of the propellers. Both motors can be coupled either on to the front propeller or on to the two rear propellers, or on to all three together, but they are actually intended only to drive the front one. On stopping either motor the other is connected to the two rear propellers, which are designed for a slower speed of translation than the front one, with the result that the running motor does not find itself overloaded as it would if the same propeller had to serve both for one and for two motors.

LAMBERT.

The *Lambert* is Belgian owned. It is a pioneer dirigible, rather like an early Zodiac.

It was sent to the U.S.A. for the St. Louis Exhibition.

Its whereabouts at the present time cannot be accurately ascertained.

BRITISH.

General Note.—Till the last few months the subject of flying, apart from mere ballooning, attracted neither interest nor attention save from a few isolated experimenters. A decade or more ago Sir Hiram Maxim was a pioneer in aeroplanes. There was also the pioneer dirigible with aeroplanes of Dr. Barton, another dirigible at Cosham, Portsmouth (burned about 1902), and at a later date the dirigible of Mr. Willans, of Cardiff.

The close of 1908 saw a marked change, and the number of aeroplanes under construction here recorded indicates the vast amount of attention now being bestowed upon the subject. Quite a large number of machines are projected in addition to those here mentioned.

The extreme tendency of the British inventor to isolate himself and work in secret is responsible for the meagre details about some machines and the possible omission of some others.

Aerial Societies:—

- Aerial League.
- Aeronautical Society.
- Aero Club.
- Aeroplane Club (with local branches).
- Institute of Flight.
- Association of British Aerocraft.

Aerial Journals:—

- *Aeronautical Journal.*
- *Aeronautics.*
- *Aerocraft.*
- *The Aero.*
- *Flight.*

Flying Grounds (with hangars built or to be built):—

- Dagenham (Aeronautical Society).
- Shellness, Sheppey (Aero Club).
- Wembley Park (Aeroplane Club).
- Salisbury Plain
- Fambridge
- Rye
- Lea Marshes.

ANTOINETTE Monoplanes. *(See France).*

British Concessionaire:—
Of this type there are in England, built, building, or to order: –

Model 1908-9.	
1	Order
2	„
3	
4	
5	
6	
7	
8	
9	

ANTOINETTE type.

BLERIOT Monoplanes. ***(XI or Cross Channel Model.)***

British Agency: Bleriot, Ltd., Long Acre, London.
Built, building, or to order in England:—

Model 1908-9.		Model 1908-9.
1	Parkinson (Newcastle), *built*	15
2	Ballin Hinde	16
3	Percy Richardson	17
4	Sheffield Simplex Motor Co.	18
5	G. H. Cox (XII)	19
6	Captain Windham	20
7	Grahame White	21
8		22
9		23
10		24
11		25
12		26
13		27
14		

In addition: Humber's Ltd. are building 50 at Coventry.

HANDLEY PAGE. Monoplane.
(Woolwich.)

Maximum length, feet / m. **maximum breadth,** feet / m. **supporting surface,** sq. feet / m^2.

Total weight.—

Body.—The wheels remain behind when the machine rises, and it alights on fixed runners.

Wings.—

Motor.—25 h.p. New Engine Co. motor.

Speed.—

Propellers.—

Steering.—

Remarks.—Standard special type.

Built, building, or ordered :—

1 Handley Page, demonstration.

2 Bridgewater Motor Co., demonstration.

MAC FIE. Monoplane.

Building.

Maximum length, feet / m. **maximum breadth,** feet / m. **supporting surface,** sq. feet / m^2.

Total weight.—
Body.—
Wings.—
Motor.—
Speed.—
Propellers.—1, Clark propeller.
Steering.—

Remarks.—Building at Fambridge. All details confidential.

LING. Monoplane.

Building.

Maximum length, feet / m. **maximum breadth,** feet / m. **supporting surface,** sq. feet / m^2.

Total weight.—
Body.—
Wings.—Fixed.
Motor.—
Speed.—
Propellers.—
Steering.—

Remarks.—Building at Hull. The special feature is an automatic device for maintaining equilibrium.

NEALE. Monoplane.

Maximum length, feet / m. **maximum breadth,** feet / m. **supporting surface,** sq. feet / m².

Total weight.—

Body.—On 3 wheels; two forward, one under the tail. Brake fitted to this. Triangular spruce girder, stayed with steel piano wire. Engine carried in front, in strong wooden supports.

Wings.—Main planes at a slight dihedral angle. Large ailerons at tips. Mainplane 18 × 5 feet (5·48 × 1·52 m.) Vertical wooden fins at ends. Behind them, flaps for steering and stabilising.

Tail.—2 planes, vertical rudder between them.

Motor.—12 h.p. 2-cylinder J.A.P. mounted high in front. Simms magneto.

Speed.—

Tractor.—1 large, of wood, 6½ feet (2 m.) in diameter, placed right forward, geared down to ⅓ engine revolutions.

Steering.—By warping flaps. Vertical rudder in tail.

Remarks.—Completed July, 1909. First trial, August, 1909.

GROSE AND FEARY. Monoplane.

Building.

Maximum length, feet / m. **maximum breadth,** feet / m. **supporting surface,** sq. feet / m².

Total weight.—

Body.—

Wings.—

Motor.—

Speed.—

Propellers.—

Steering.—

Remarks.—Fitted with a patented special device for maintaining lateral stability.

BARBER. Monoplane.

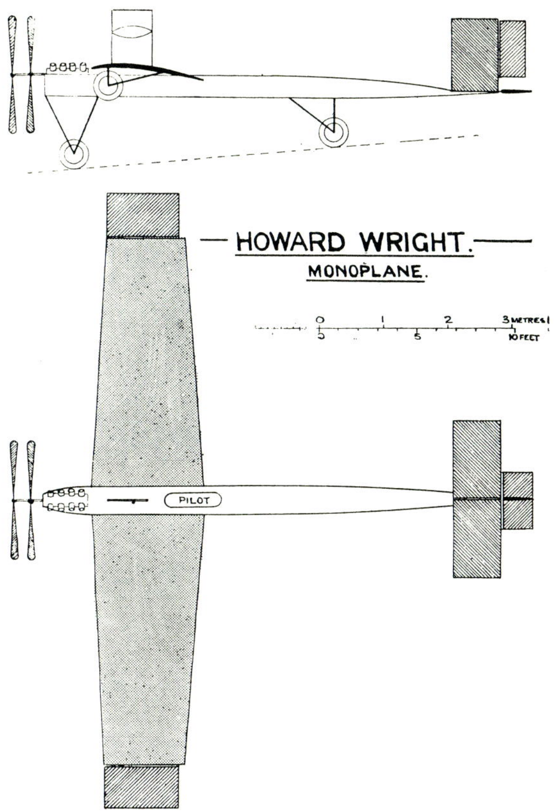

Maximum length, 27 feet (8·23 m.) **maximum breadth,** 32 feet (7·75 m.) **supporting surface,** 200 sq. feet ($18\frac{1}{2}$ m².)

Total weight.—1,000 lbs. (453 kgs.)

Body.—Enclosed. 2 seater. Mounted on 3 wheels, spring forks on all.

Wings.—Span, 32 feet (9·75 m.) Barber's patent automatic lateral stability.

Motor.—50 h.p. Antoinette, water-cooled.

Speed.—

Propellers.—2, revolving in opposite directions.

Steering.—Horizontal and vertical rudder at rear.

Remarks.—8 gallons petrol carried. Built for C. H. Barber by Howard Wright.

WEISS (José). Monoplane.

Completing or on trial.

Maximum length, feet, m. **Maximum breadth,** 34 feet, 10·36 m. **supporting surface,** 150 sq. feet, 14 m².

Total weight.—500 lbs. (227 kgs.), *including* aviator.

Body.—On 3 wheels which are left behind when the machine rises. A novel feature of the chassis is that it automatically faces the wind.

Wings.—Span, 34 feet (10·36 m.) The rear edge of planes is flexible and fitted with flaps for horizontal steering. The tips have a double curve to secure automatic stability.

Motor.—12 h.p. 3-cylinder, Anzani, air-cooled.

Speed.— m.p.h. (62 km.)

Propellers.—2, chain driven through gearing. Reduction ratio, 3 to 1. The propellers are placed one each side behind the wings. Diameter, 6 feet (1·95 m.)

Steering.—By 2 small planes at back of main planes.

Remarks.—This machine is based upon a long series of well-known experiments by Mr. Weiss.

WILSON (Edgar). Monoplane.

Building.

Maximum length, feet, m. **maximum breadth,** feet, m. **supporting surface,** sq. feet, m².

Total weight.—

Body.—On three wheels.

Wings.—

Motor.—25 h.p., by David Smith & Co., London.

Speed.—

Propellers.—Two, abaft the wings.

Steering.—Flexible outer edges to the wings.

Remarks.—

WORSWICK. Monoplane.

Building.

Maximum length, feet / m. **maximum breadth,** 33 feet / 10 m. **supporting surface,** 240 sq. feet / 22 m^2.

Total weight.—600 lbs. (272 kgs.) including aeronaut.

Body.—Cigar-shaped wooden girder (ash) covered with aluminium plate. It is 14 feet long by 2½ feet maximum diameter (4·26 × 0·75 m.) and weighs 40 lbs. (18 kgs.)

Wings.—

Motor.—

Speed.—

Propellers.—

Steering.—Horizontal rudder 40 sq. feet (3·71 m^2). Vertical 10 sq. feet (0·92 m^2).

Remarks.—

F

DUNNE I.

Building at Sheppey.

Maximum length, feet / m. **maximum breadth,** feet / m. **supporting surface,** sq. feet / m².

Total weight.—

Body.—

Wings.—

Motor.—40-50 h.p. Wolseley-Siddeley.

Speed.—

Propellers.—

Steering.—

Remarks.—It generally resembles *Dunne II.* in principles, but differs in appearance.

DUNNE II. Monoplane.

(*Alias* **HUNTINGDON.**)

Completing.

Maximum length, *about* 50 feet / 15 m. **maximum breadth,** *about* 59 feet / 18 m. **supporting surface,** sq. feet / m².

Total weight.—

Body.—Driver sits in the extreme bow.

Wings.—The main plane, elevated well clear of body, is of enormous size, *very roughly*, 59 × 10 feet (18 × 3 m.), very possibly *more*. Ailerons in rear of plane edges. The 2 end planes are of equal width and also fairly large.

Motor.—Placed forward of main plane.

Speed.—

Propellers.—2, chain and gear driven, placed a long way behind motor, abaft main plane, one on either side of the body.

Steering.—

Remarks.—Very great secrecy has been observed about these machines, and the utmost that can be gleaned about them is recorded above. They are of enormous size compared to other monoplanes, and the (approximate) measurements given are probably well under the actual rather than over it.

Models on a large scale have been built and flown well.

The diagram is merely to give a general idea of the extremely interesting and novel type.

Built for and the property of Professor Huntingdon.

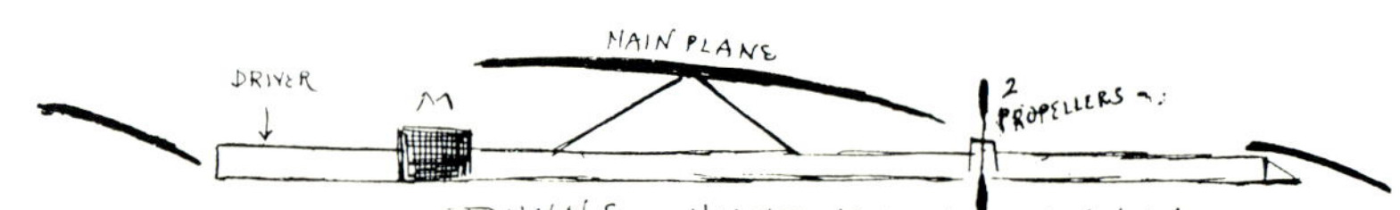

DUNNE – HUNTINGDON – Sketch plan.

CODY (British Government) Biplane.

Max. length, about 40 feet / 12 m. **Max. width,** 7½ feet / 2·28 m. **Planes,** 9 feet apart / 2·74 m. **supporting surface,** sq. feet / m².

Total weight.—

Body.—

Planes.—

Motor.—8-cylinder light Antoinette, 50 h.p., placed immediately in front of the lower surface. A 50 h.p. Simms ***V*** type since ordered. Weight, 4⅓ lbs. per h.p.—Weight, 217 lbs. (98·4 kgs.)

Speed.—

Propellers.—2 amidships, immediately in front of pilot.

Steering.—Horizontal plane rudder in front of the main surfaces. Vertical rudder, carried on 2 bamboo supports, 12 feet (3·65 m.) in rear (originally had a big fantail).

Remarks.—A large governor 20 feet (6 m.) in span is carried about 12 feet (3·65 m.) to the front. On each side are 2 small auxiliary planes, which assist in the steering. This machine was originally somewhat different, but altered to the details here given after a smash.

BADEN-POWELL. Biplane.

Maximum length, feet / m. **maximum breadth,** feet / m. **supporting surface,** sq. feet / m^2.

Total weight.—

Body. —

Wings.—Practically 3 biplanes in series—tail and elevator being nearly as large as the main planes.

Motor.—

Speed.—3-cylinder 12 h.p.

Propellers.—

Steering.—Biplane elevator in front. Vertical rudder in tail.

Remarks.—Departs considerably from usual practice.

BARNWELL. Biplane.

Maximum length, feet / m. **maximum breadth,** 48 feet / 14·63 m. **supporting surface,** sq. feet / m².

Total weight.—1,568 lbs. (711 kgs.), *not including* aviator.

Body.—Mounted on 2 good-sized wheels in the centre line. 2 smaller wheels in line with the forward wheel, placed about midway from it and the ends of the lower plane.

Wings.—Main planes are 48×8 feet (14·63×2·43 m.) They are slightly concave—about 3 inch (76 m/m) drop. Distance between planes, 7½ feet (2·30 m.)

Motor.—Ordinary motor car Humber engine. Bore, 100 m/m. Stroke, 150 m/m. Weight complete, 460 lbs. (210 kgs.)

Speed.—

Propellers.—In rear of planes. 2 yellow pine, 2-bladed, chain driven. Diameter, 10 feet (3 m.) Pitch, 10 feet (3 m.) Thrust, *about* 300 lbs. per propeller.

Steering.—Combined vertical and horizontal rudder in front. Small stabilising planes at the extremities between the main planes.

Remarks.—First flight, 28th July, 1909. Designed and built by Harold and Frank Barnwell, Grampian Motor and Engineering Co., Stirling, N.B.

WINDHAM. Tandem Monoplane I.

Maximum length, 50 feet / 15·23 m. **maximum breadth,** 24 feet / 7·31 m. **supporting surface,** sq. feet / m^2.

Total weight.—

Body.—Bamboo. Weighs *about* 125 lbs. (56·50 kgs.)

Wings.—These form a square, placed point forward, each wing being triangular in shape. They are supported by large bamboo booms. The wings are non-rigid.

Motor.—

Speed.—

Propellers.—1, right forward.

Steering.—Both vertical and horizontal rudders are placed right aft.

Remarks.—This machine embodies several novel and interesting features.

WINDHAM. Monoplane II.

Building.

Maximum length, 25 feet / 7·61 m. **maximum breadth,** *about* 30 feet / 9 m. **supporting surface,** sq. feet / m^2.

Total weight.—

Body.—Main frame V shaped, 25 feet (7·61 m.) long. Built of poplar, stanchions fitted into aluminium sockets. 2 special shock absorbing chassis, one in front on two wheels, one on one wheel in rear.

Wings.—Main plane span *about* 30 feet (9 m.) Depth, 6 feet (1·80 m.) One fixed plane aft.

Motor.—35 h.p. Dutheil-Chalmers.

Speed.—

Propellers.—Special Windham design. 1, in front.

Steering.—2 elevators and vertical rudder in tail.

Remarks.—Building at the Clapham Junction Works.

FAY-WILKINSON. Tandem Biplane.

Building.

Maximum length, feet / m. **maximum breadth,** feet / m. **supporting surface,** sq. feet / m^2.

Total weight.—

Body.—On 8 small wheels in pairs; two pairs under front planes, two pairs under the rear planes.

Wings.—Each plane is divided into 3 sections, the outer sections being hinged and capable of any adjustment. Tips of outer planes are curved away to the rear and warp automatically to any very strong pressure.

Tail.—1 plane in triplicate, which duplicates the front planes, working in unison with them. Central portion rigidly connected to front central planes by girders.

Motor.—Special 70 h.p., 6-cylinder, water-cooled. Weight, 300 lbs. (136 kgs.)

Speed.—

Propellers.—Four 4-bladed, two placed in front of main planes, two immediately abaft front planes. Chain driven. Mounted 2 on a shaft.

Steering.—By altering outer planes. Also by patent semi-automatic stabiliser. The principle of this is a weight which follows to an exaggerated degree any movement of the driver's body.

Remarks.—Great longitudinal stability is secured by the form of tail adopted. Building at Withernsea, near Hull.

H. FARMAN Biplanes. (*see France*).

British Concessionaire :—

Of this type there are in England, built, building or on order :—

Model 1908-9.

1 G. B. Cockburn, Gloucester (*built*)
2 E. Ballin Hinde
3
4

Farman type.

HUMPHREY. *Biplane.*

Maximum length, 13 feet (4 m.) **maximum breadth,** 45 feet (13·71 m.) **supporting surface,** sq. feet (m²)

Total weight.—

Body.—Boat-shaped, designed for rising from water. Hydroplane side floats. Also mounted on 4 wheels fixed at the end of bamboos for land use. These wheels can be folded out of the way when the machine descends on water.

Wings.—Span, 40 feet (12 m.) The planes in front have a very decided dip in the centre, and roughly, form a couple of semi-circles. They straighten out to a single slight semi-circle in rear. A triangular keel on top placed vertically in the centre line, to assist stability. Ailerons at the wing tips.

Tail.—Rigid for about half its length. Its feathers are flexible, in order to assist longitudinal stability and damp oscillations.

Motor.—8-cylinder air-cooled J.A.P. Bosch magneto. Engine is mounted on a chassis athwart the frame, and placed moderately high up.

Speed.—

Propellers.—2 steel, bevel-geared, placed immediately in rear of planes. Diameter of propellers, 8 feet (2·44 m.) Propellers revolve in opposite directions at 400 r.p.m. They can be declutched.

Steering.—Wedge-shaped elevator in front. 2 side wings. Tail.

Remarks.—This machine is extremely novel in very many details; but it is the fruit of many years of experiment. The special features are the wedge-shaped elevator, keel, tail and curved planes. Main idea: great lateral stability and constant centre of effort.

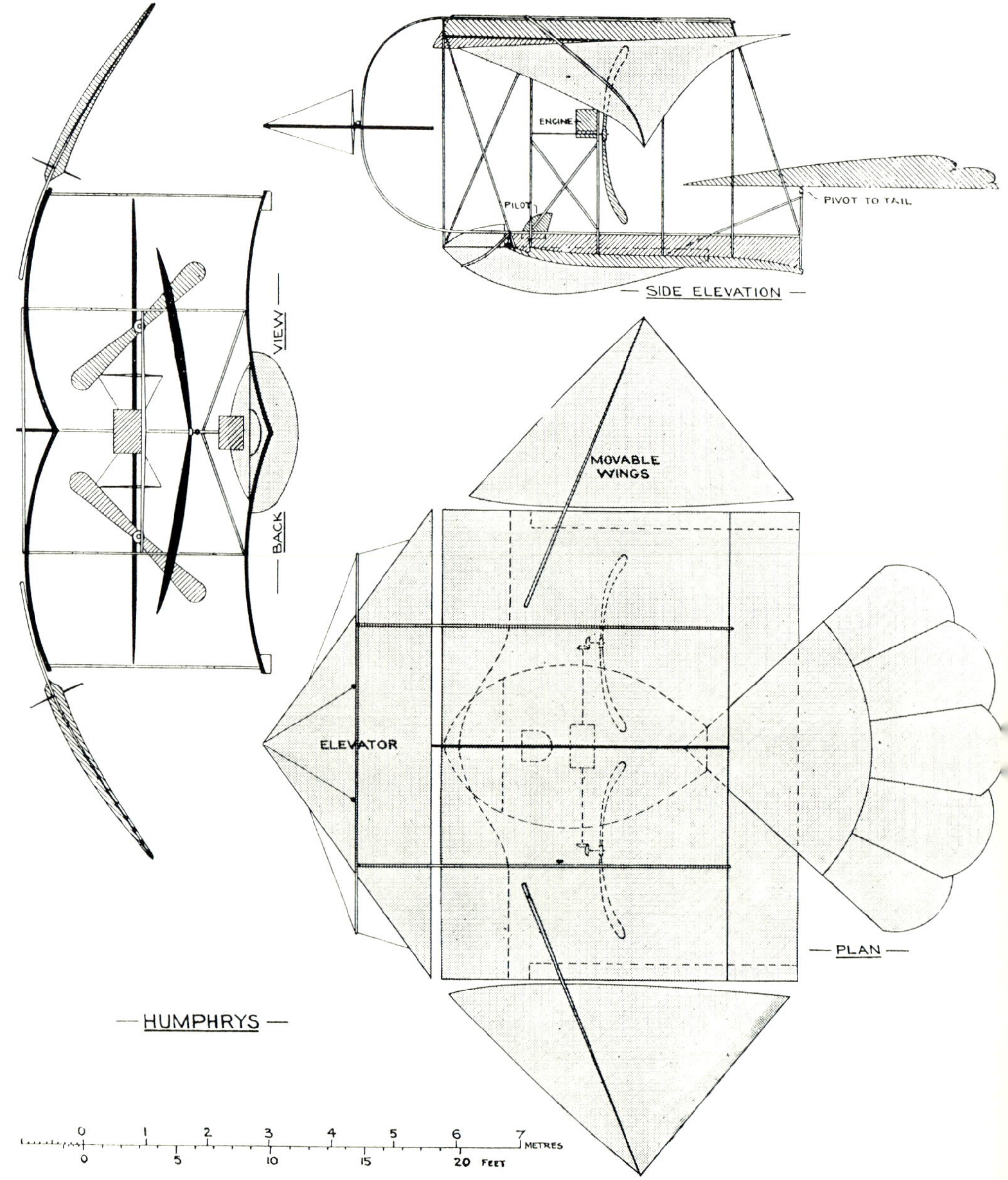

PIFFARD. Biplane.

Not yet photographed.
Completing.

Maximum length, 34½ feet / 10·50 m. **maximum breadth** *of planes,* 6½ feet / 2 m. **supporting surface,** 510 sq. feet / 46½ m².

Total weight.—

Body.—American white wood and ash. Cane and steel landing skids. 3 large wheels which can be shifted out of the way for landing.

Wings.—Curved planes of the usual pattern. 2 fixed longitudinal stabilisers: one, a curved lifting surface, forward of and above elevator; one, of two convex surfaces, aft. These are about 1 foot (305 m/m.) above upper plane. Ailerons equidistant between the planes at ends. Midway between planes, between ends and centre, are fixed vertical planes.

Motor.—40 h.p 8-cylinder E.N.V., mounted on a sliding bed. It automatically shifts by the action of the propeller's speed, thus altering centre of gravity. There is also a lever to control this action at will.

Speed.—

Propeller.—1, direct driven.

Steering.—Elevator in front. Vertical rudder in front.

Remarks.—Several novelties are embodied in it. It is the result of a number of experiments with models.

HAMMOND.

Building at Brooklands.
No details.

Maximum length, feet / m. **maximum breadth,** feet / m. **supporting surface,** sq. feet / m².

Total weight.—

Body.—

Wings.—

Motor.—

Speed.—

Propellers.—

Steering.—

Remarks.—

PHILLIPS (A). Miscellaneous.

Preparing to build.

Maximum length, feet / m. **maximum breadth,** feet / m. **supporting surface,** sq. feet / m^2.

Total weight.—

Body.—

Wings.—Two adjustable biplanes, one on either side: flat in front and curved downwards in rear.

Motor.—Carried on a platform low down.

Speed.—

Propellers.—Two 6-bladed placed side by side, chain driven off a bevel gearing which permits the propellers to work in a horizontal, vertical or any intermediate plane. The blades can be feathered, and either propeller feathered independently of the other. They can be used for steering and lifting as well as propulsion, also as brakes. For starting the propellers are set vertically with blades feathered, and when the engine has been speeded up, gradually feathered to give the required lift.

Steering.—The planes tilt, and are used as horizontal rudders. The propellers act as vertical rudder, and the machine is steered in the horizontal plane by feathering one or other propeller.

Remarks.—

FRISWELL.

Building.

Maximum length, feet / m. **maximum breadth,** feet / m. **supporting surface,** sq. feet / m^2.

Total weight.—

Body.—

Wings.—

Motor.—

Speed.—

Propellers.—

Steering.—

Remarks.—

PORTE AND PIRIE. Biplane.

FRONT VIEW PORTE & PIRIE

Maximum length, feet / m. **maximum breadth,** 26 feet / *about* 8 m. **supporting surface,** sq. feet / m^2.

Total weight.—

Body.—On runners. Designed to rise like *Wrights*.

Wings.—Convex planes. The upper plane is advanced considerably, so that its rear edge coincides with the front edge of the lower plane. Stabilising fins and ailerons at extremities between planes.

Motor.—J.A.P

Speed.—

Propellers.—

Steering.—Single elevator forward. Second elevator aft, with vertical rudder above it.

Remarks.—Designed by Lieutenants Porte and Pirie, R.N., at Gosport, Hants. Preliminary trials as a glider at Portsmouth, August, 1909. Damaged its tail in rising. Can carry 2.

SHORT BROS. Biplane (for two).

Maximum length, feet / m. **maximum breadth,** feet / m. **supporting surface,** sq. feet / m^2.

Total weight.—

Body.—Spruce throughout.

Wings.—Continental fabric.

Motor.—40 h.p.

Speed.—

Propellers.—Two.

Steering.—Elevator forward, vertical rudder aft.

Remarks.—Design embodies a number of small special features.

SHORT BROS. Biplane (for one).

Maximum length, feet / m. **maximum breadth,** feet / m. **supporting surface,** sq. feet / m^2.

Total weight.—

Body.—

Wings.—

Motor.—12 h.p.

Speed.—

Propellers.—

Steering.—

Remarks.—Entirely a smaller edition of the other.

STOCKS AND COCHRANE. Biplane.

(Gosport).

Maximum length, feet / m. **maximum breadth,** feet / m. **supporting surface,** sq. feet / m^2.

Total weight.—

Body.—Special steel, wooden ribs.

Wings.—Main planes, 40 feet (12m.) wide; with stabilising fins attached to the extremities.

Motor.—40 h.p.

Speed.—

Propellers.—2, chain driven.

Steering.—Elevator in front. Another behind; vertical rudder being just in front of it.

Remarks.—Built by two naval officers. On trials the chain drive to propellers was a continual source of trouble. Completing.

SANDARS. Biplane.

Building.

Maximum length, feet / m. **maximum breadth,** 50 feet / 15·23 m. **supporting surface,** sq. feet / m^2.

Total weight.—

Body.—

Wings.—Span 50 feet (15·23 m.) Dihedral planes.

Motor.—

Speed.—

Propellers.—2, placed right in front.

Steering.—

Remarks.—Believed to be "Wright" type, except for points noted. Not yet flown.

TINLINE.

Building.

Maximum length, feet / m. **maximum breadth,** feet / m. **supporting surface,** sq. feet / m^2.

Total weight.—

Body.—

Wings.—

Motor.—

Speed.—

Propellers.—

Steering.—

Remarks.—No details can be obtained.

VOISIN Biplanes. *(see France.)*

British Concessionaires:—Aero Motors, Ltd.

Similar to the usual Voisin type, but all after No. 3 will be fitted with the 50 h.p. 6-cylinder V type Simms engine.

Of this make there are in England, built, building, or on order:—

Model 1908-9.

1 J.T.C. Moore Brabazon's *Bird of Passage* (to be fitted with a 50-60 Green engine), sold to A. George (wrecked Aug., 1907).

2 Ernest Pitman (put together by Short Bros.) (E.N.V. Engine)

3 F. R. Simms (Simms engine)

4 — Moering

5

6

7

8

9

VOISIN type. *Photo, C. Malcuit.*

WILSON Biplane.
(Wright Type.)

Building.

Maximum length, feet / m. **maximum breadth,** feet / m. **supporting surface,** sq. feet / m².

Total weight.—

Body.—

Wings.—

Motor.—

Speed.—

Propellers.—

Steering.—

Remarks.—

WILKES. Biplane.

Building.

Maximum length, feet / m **maximum breadth,** feet / m. **supporting surface,** 533 sq. feet / 50 m².

Total weight.—

Body.—

Wings.—

Motor.—50 h.p. Gladiator.

Speed.—

Propellers.—

Steering.—

Remarks.—

WINDHAM. Biplane.

Maximum length, feet / m. **maximum breadth,** feet / m. **supporting surface,** sq. feet / m².

Total weight.

Body.—On 6 wheels, two of which are at the ends of the lower deck.

Wings.—Ailerons between the planes at the extremities. Automatic or hand control.

Motor.—Duthiel-Chalmers, various horse-powers may be fitted.

Speed.—

Propellers.—

Steering.—Biplane elevator placed forward. Vertical rudder aft in a biplane tail.

Remarks.—This is a Pischoff-Koechlin biplane, considerably altered by Captain Windham, and to some extent his design. *Special features*: The elevator and tail are placed further away from the main planes than is usual. The object of this is to secure extra stability.

WRIGHT BROS. Biplanes. (*see U.S.A.*)

British Concessionaires:—Short Bros.

Of this type, (besides privately built copies) there are in England, built, building, or on order:—

Model 1908-9.

1 Hon. C. S. Rolls
2 Alexander Ogilvy
3 ——— unnamed
4 ——— „
5
6
7

WRIGHT type. *Photo, C. Malcuit.*

BULL'S EYE or ROE I. Triplane.

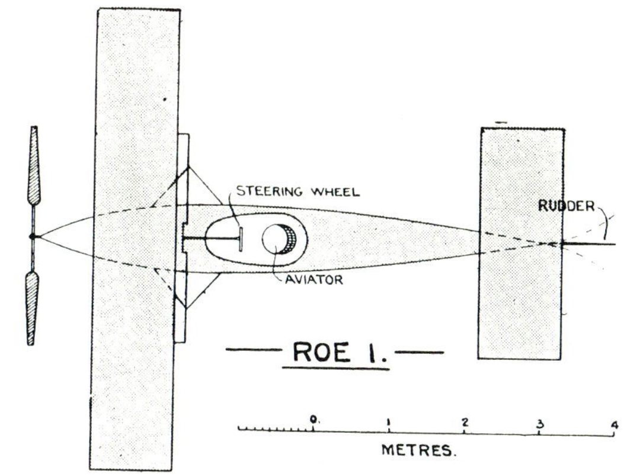

Maximum length, 23 feet. / 7 m. **maximum breadth,** 20 feet. / 6·09 m. **supporting surface,** 320 sq. feet. / 29·7 m².

Total weight.—200 lbs. (90·7 kg.), not including the aeronaut.

Body.—Triangular section fusiform deal girder, covered with cotton-oiled paper backed with muslin. The motor is in the fore part and the aeronaut sits in a well about 4 feet ( ) abaft the main planes.

Wings.—The 3 superposed main planes, each 20 feet by 3 feet 7 inches (6·09 × 1·09 m.), are set at an angle of 5°. The 3 superposed tail planes measure 10 feet by 3 feet 7 inches (3·04 × ·96 m.)

Motor.—10 h.p. J.A.P.

Speed.—

Propellers.—

Steering.—Vertical steering is effected by altering the inclination of the main planes bodily, the tail remaining stationary. Horizontal steering is effected by means of a rudder abaft the tail, and by twisting the rear edges of the main planes at the same time. All these movements are controlled by a single lever which is pushed forward or pulled back for descending or rising, and moved from to either side for turning.

Remarks.—A vertical plane is fitted on each side, both between the main planes and between the tail planes. The motor being air cooled channels are cut in the front of the body to concentrate the air on the motor. The outer 6 feet of the main planes are made to fold back for housing or transport.

ROE II. Triplane.

Building.

Maximum length, 32 feet 9·75 m. **maximum breadth,** feet m. **supporting surface,** 620 sq. feet 57·59 m^2.

Total weight.—650 lbs. (295 kgs.) *not including* aviator.

Body.—

Wings.—Planes are $32 \times 5\frac{1}{3}$ feet ($9·75 \times 1·62$ m.), otherwise as *Roe I.*

Motor.—35 h.p. 8-cylinder J.A.P.

Speed.—

Propellers.—

Steering.—As in *Roe I.*

Remarks.—This machine is entirely a larger replica of *Roe I.* Both machines are the property of A. V. Roe.

SAUL I.

(Type: Miscellaneous.)

With wings folded for transit.

Maximum length, 20 feet / 6 m. **maximum breadth,** 20 feet / 6 m. **supporting surface,** 210 sq. feet / 22 m².

Total weight.—320 lbs. Made up as follows:—

Motor	... ...	50 lbs.
Propeller	... ...	2 „
Frame	... ...	50 „
Wings	... ...	60 „
Aeronaut	... ...	158 „
		320 „

Body.—Formed of lengths of hickory wood of small section bent to shape, stayed with crossbars of hickory and adjustable wires. It is mounted on 2 sledge-shaped runners formed by the underside of the frame. Dimensions: 19 feet (5·79 m.) long, 3 feet (0·91 m.) wide, 5 feet (1·52 m.) high.

Wings.—The 4 planes, each 20 feet (6 m.) broad by 3 feet (0·91 m.) wide, are detachable from the frame. Built of hickory bars with white wood and hickory ribs, covered with light varnished fabric. The rear edge is flexible, becoming more so towards the tips which are slightly narrowed. The planes of each "pair" are 3 feet (0·91) apart vertically, the rear edge of the upper one being just above the front edge of the lower. The 2 lower planes are 5 feet (1·52 m.) apart.

Motor.—8 h.p. Air-cooled. 2,000 r.p.m.

Speed.—30 m.p h. (18 km. p.h.)

Propellers.—2-bladed. 4 feet (1·21 m.) diameter at the back of the body. Direct coupled. Starting thrust 60 lbs

Steering.—Is obtained solely by flexing the front planes, the 2 tips together for vertical movement, or separately for turning. Inherent automatic stability is relied upon to preserve the balance at all times.

Remarks.—This machine is conspicuous for lightness (160 lbs. without the aeronaut) and portability (the wings fold up and pack on top of the frame, the total housing space required being 19 feet × 3 feet × 6 feet 8 inches (5·79 × 0·91 × 2 m.), and cheapness, the cost with an 8-12 h.p. motor being about £250.

SAUL II.

Maximum length, feet / m. **maximum breadth,** feet / m. **supporting surface,** sq. feet / m².

Total weight.—

Body.—

Wings.—

Motor.—

Speed.—

Propellers.—

Steering.—

Remarks.—

PRIDE. Monoplane.

No photo procurable.

Maximum length, $32\frac{2}{3}$ feet / 10 m. **maximum breadth,** feet / m. **supporting surface,** sq. feet / m².

Total weight.—684 lbs. (310 kgs.)

Body.—

Wings.—

Motor.—50 h.p.

Speed.—

Propellers.—

Steering.—

Remarks.—Completed 1900. First flight up to 1,000 yards. Designed and built by Christopher Pride, of Fishponds, Bristol.

HOWARD WRIGHT. Biplane.

Max. length, 43 feet / 10·10 m. **Max. breadth,** 40 feet / 12·19 m. **Height,** $10\frac{1}{2}$ feet / 3·19 m. **supporting surface,** 520 sq. feet / 48 m².

Total weight.—1,100 lbs. (499 kgs.)

Body.—Entirely enclosed. 2 seats. 15 gallons petrol carried. Mounted on 2 wheels in centre line, spring forks. Wheels at the wing tips.

Wings.—Main planes are 40 feet (12 m.) wide by $6\frac{1}{2}$ feet (2 m.) deep. Vertical stability planes. Ailerons. Box tail.

Motor.—26 h.p. Metallurgique (water cooled).

Speed.—

Propellers.—2, revolving in opposite directions.

Steering.—Vertical rudder in the tail; biplane elevator with vertical side planes in front. Righting tips in rear edges of the main planes

Remarks.—The special feature of this machine is the care taken to diminish skin friction. The body is covered with plane fabric, carefully smoothed and varnished, and its shape is also a matter of careful design. Everything else is done to minimise wind resistance.

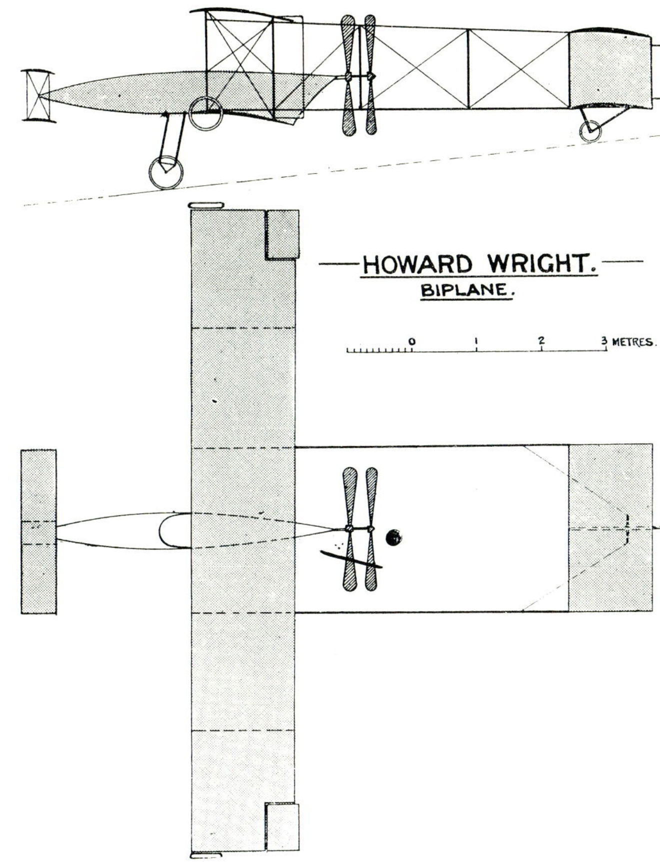

Machines of this type in England, built, building or on order:—

1 Howard Wright

2 Malcolm Seton-Karr

LAMPLOUGH. "Lifter Plane Glider."

Completing.

Maximum length, feet / m. **maximum breadth,** feet / m. **supporting surface,** sq. feet / m^2.

Total weight.—

Body.—No wheels (see Wings).

Wings.—Cross biplanes for gliding, fore and aft biplanes for lifting. The action of these last, which are mechanically operated, is a series of short reciprocating glides to which a feathering motion is imparted, in order to give the necessary lift. When the machine is up these lifters go to reinforce the gliding planes.

Motor.—25-30 h.p.

Speed.—

Propellers.—

Steering.

Remarks.—

MUMFORD. Helicopter.

Completing.

Maximum length, 60 feet / 18·28 m. **maximum breadth,** 41 feet / 12·49 m. **supporting surface,** sq. feet / m².

Total weight.—

Body.—

Wings.—6 helices, each 25 feet (7·62 m.) in diameter, revolving at 40 r.p.m. Mounted on shafts inclining slightly forward and with a slight outward set to ensure stability.

Motor.—30 h.p.

Speed.—

Propellers.—The 6 lifting helices are bevel geared to longitudinal side shafts, chain driven from engine. The propellers are 2-bladed and when the machine is up give a forward drive by the whole machine being tilted.

Steering.—By one triangular rudder aft; universally hinged for use as both horizontal and vertical rudder.

Remarks.—Built at Denny Bros. Shipbuilding Yard, Dumbarton, Scotland.

H. PHILLIPS.

Maximum length, feet } m. } **maximum breadth,** feet } m. } **supporting surface,** sq. feet } m^2. }

Total weight.—
Body.—
Wings.—
Motors.—
Speed.—
Propellers.—
Steering.—
Remarks.—

NYBORG.

Maximum length, feet / m. **maximum breadth,** feet / m. **supporting surface,** sq. feet / m^2.

Total weight.—

Body.—

Wings.—

Motors.—

Speed.—

Propellers.—

Steering.—

Remarks.—

McCLEAN. Biplane.

Building.

Maximum length, feet / m. **maximum breadth,** feet / m. **supporting surface,** sq. feet / m².

Total weight.—

Body.—Howard Wright style of body.

Wings.—Fixed planes with flexible extensions in rear of wing tips.

Motor.—

Speed.—

Propellers.—Two.

Steering.—Elevator in front. Vertical rudder in box tail behind.

Remarks.—Generally of the Voisin or Howard Wright types, but embodying a large number of special features.

SMITH AND MATEKA. Biplane.

Building.

Maximum length, feet / m. **maximum breadth,** feet / m. **supporting surface,** sq. feet / m².

Total weight.—

Body.—

Wings.—Upper plane straight. Lower plane curved up to meet it. Ailerons in front of upper plane. No tail.

Motor.—

Speed.—

Propellers.—

Steering.—Biplane elevator (like *Wright*).

Remarks.—

PISCHOFF-KOECHLIN Biplanes. *(see France.)*

Of this make there are in England, built or building:—

Model 1908-9.

1 Captain Windham

2 Reported on order for an unnamed aviator

At Yarmouth.

Building.

Maximum length, feet / m. **maximum breadth,** feet / m. **supporting surface,** sq. feet / m².

Total weight.—

Body.—

Wings.—

Motor.—

Speed.—

Propellers.—

Steering.—

Remarks.—Being built with extreme secrecy.

WOKINGHAM FLYER. Miscellaneous.
(Telescopic Cigar-shaped).

Building.

Maximum length, 140 feet / 42·67 m. **maximum breadth,** 20 feet / 6 m. **maximum height,** 31 feet / 9·44 m.

Total weight.—

Body.—Constructed of ⅛ inch three-ply boarding, screwed together and covered with stout canvas. Cigar-shaped extending from central section.

Wings.—*Nothing known.*

Motor.—80 h.p. Weight 616 lbs. (279·41 kgs.) = about 5½ lbs. per h.p. placed amidships.

Speed.—

Propeller.—One spiral rotascope forward, cone-shaped; direct drive, 1,200 r.p.m.

Steering.—One vertical rudder aft. Believed to have two horizontal rudders, one on either side of the vertical rudder.

Remarks.—The machine is telescopic and can close in to about 60 × 14 × 16 feet (18 × 4 × 4·87 metres). It is claimed that a model flew successfully. Still building.

McCURD. Multiplane.

Building.

Maximum length, feet / m. **maximum breadth,** feet / m **supporting surface,** sq. feet / m².

Total weight.—

Body.—

Wings.—Adapted to be set at different angles of inclination. One half arranged so as to move in an opposite direction to the other half.

Motors.—

Speed.—

Propellers.—

Steering.—

Remarks.—Building at North Finchley.

RUCKER. Biplane.

Building.

Maximum length, feet / m. **maximum breadth,** feet / m. **supporting surface,** sq. feet / m^2.

Total weight.—

Body.—

Wings.—

Motor.—50 h.p. 6-cylinder Simms.

Speed.—

Propellers.—Special.

Steering.—

Remarks.—Nothing more known about this machine. It is possibly a *Voisin* with a few special minor fittings.

ROLLS. Biplane.

Building.

Maximum length, feet / m. **maximum breadth,** feet / m. **supporting surface,** sq. feet / m^2.

Total weight.—

Body.—

Wings.—

Motor.—

Speed.—

Propellers.—Two.

Steering.—

Remarks.—This is not the Wright Bros. machine owned by Mr. Rolls, but a species of Voisin type embodying a number of his own ideas.

(LATE) AERIAL EXPERIMENT ASSOCIATION.

(4) ***SILVER DART. Biplane. (now British Colonial).***

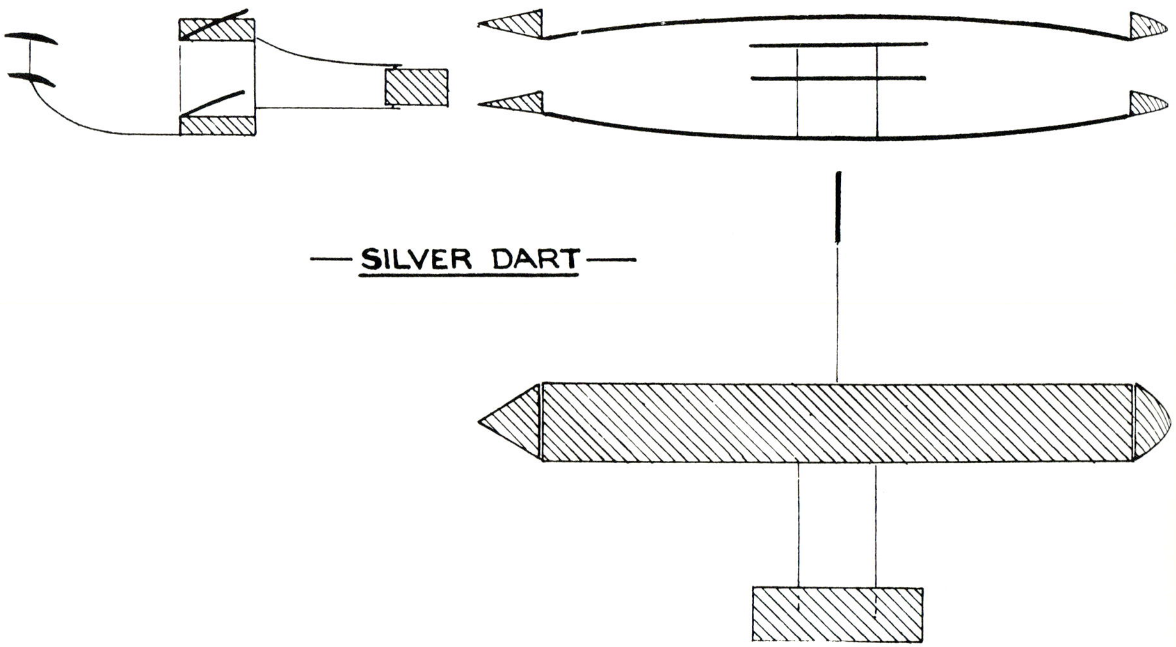

Maximum length, feet / m. **maximum breadth,** $36\frac{1}{4}$ feet / 11 m. **supporting surface,** sq. feet / m^2.

Total weight.—600 lbs. (272 kgs.)

Body.—Wood. On 3 wheels; one in front, two behind.

Wings.—30×4 feet (9·14×1·22 m.) *Special feature* is that the planes are curved towards each other. Stabilising triangular planes at wing tips.

Motor.—

Speed.—

Propellers.—1; diameter, $5\frac{1}{2}$ feet (1·67 m.)

Steering.—Elevator in front; wing tips also warp. Single vertical rudder in rear (this is the *special feature* of *Silver Dart* compared to its predecessors).

Remarks.—Special automatic stability. Designed by Dr. Graham Bell. Has flown over 2,000 miles.

(LATE) AERIAL EXPERIMENT ASSOCIATION.

BADDECK. Biplane. (Nos. 1, 2 and 3.)

***(Alias* No. 5 AERIAL EXPERIMENT ASSOCIATION; *alias* also CANADIAN AERODROME CO.; also CYGNET I, LOON III, SILVER DART II.)**

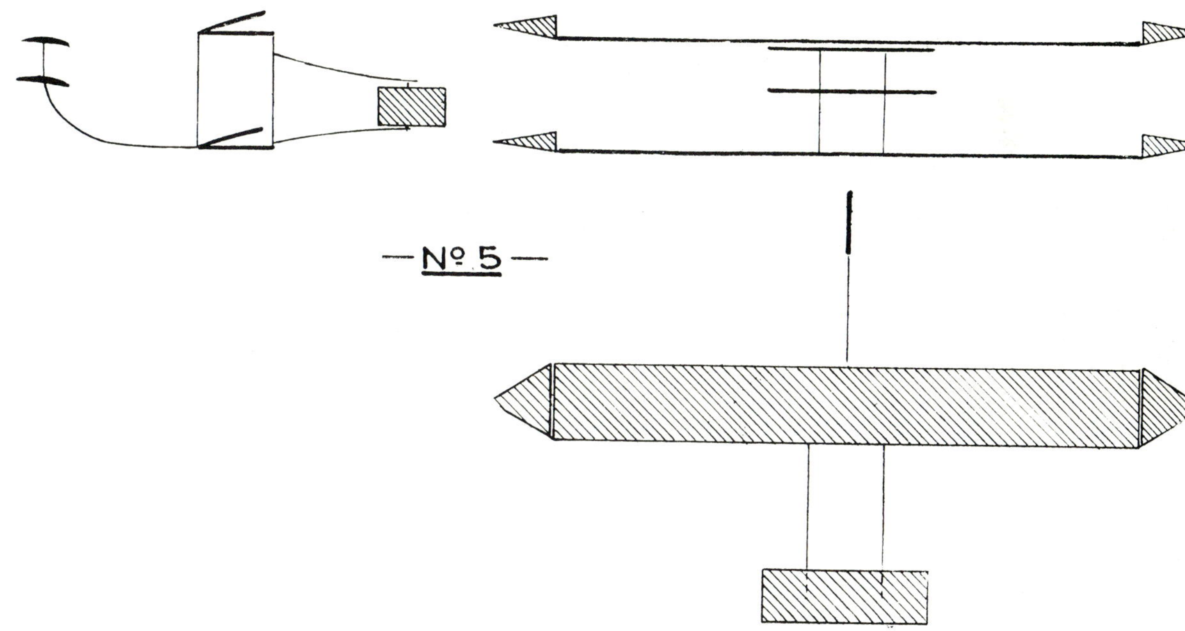

Maximum length, feet / m. **maximum breadth,** feet / m. **supporting surface,** sq. feet / m².

Total weight.—*About* 600 lbs. (272 kgs.)

Body.—Wood. On 3 wheels.

Wings.—Straight planes, 30 × 4 feet (9·14 × 1·22 m.) Triangular stabilising planes at wing tips.

Motor.—

Speed.—

Propellers.—1, $5\frac{1}{2}$ feet (1·67 m.) diameter.

Steering.—Elevator in front. Single vertical rudder in rear (as in *Silver Dart*).

Remarks.—This Association is now dissolved. These 5 machines form an interesting group, indicating as they do the results of the experimental flights. The *Herring-Curtis* is the U. S. result.

CYGNET II.

Designed by Dr. Graham Bell, the designer of *Silver Dart.*

Maximum length, 13 ft. 4$\frac{3}{4}$ in. **maximum width** (on top), 5 ft. 7 in. (at bottom), 40 ft. 2$\frac{1}{2}$ in.
4 m. 16 m. 12 m.

Total weight.—950 lbs. (431 kgs.) with one man.

Body.—

Wings.—

Motor.—50 h.p.

Speed.—

Propellers.—1 in rear, 10 feet (3 m.) diameter, 13 feet (3·96 m.) pitch, driven at three-fourths the motor speed.

Steering.—2 plane horizontal rudder, and a vertical rudder in front.

Remarks.—3,610 Tetrahedral cells. The cells are cut away in the lower centre to make room for motor and aeronaut. Mounted on 3 runners or skates.

GIBSON. Monoplane.

Building.

Maximum length, 65 feet / 19·81 m. **maximum breadth,** feet / m. **supporting surface,** sq. feet / m^2.

Total weight.—

Body.—

Wings.—

Motor.—60 h.p.

Speed.—130 m.p.h. claimed.

Propellers.—

Steering.—

Remarks.—Building by William Gibson, of Victoria, British Columbia.

ELLIS (Peter). Biplane.

Building.

Maximum length, feet / m. **maximum breadth,** feet / m. **supporting surface,** sq. feet / m^2.

Total weight.—

Body.—

Wings.—

Motor.—

Speed.—

Propellers.—

Steering.—

Remarks.—Being built to inventor's designs at Wellington, N.Z. No details have transpired.

2, for experimental purposes.

6,458 c. feet (600 m^3.) and 1,238 c. feet (115 m^3.) respectively.

The models are not intended to carry men.

They have been delivered to a private company, who propose controlling them in the air by means of wireless telegraphy, and so gaining valuable experience.

They were built by the Societée Astra.

CLEMENT BAYARD II. Will be bought by Great Britain.

(Clement Bayard class.)

Maximum length, 300 feet / 91 m. **maximum diameter,** 51 feet / 15·51 m. **volume,** 75,350 c. feet / 7,000 m³.

Total lift.—16,975 lbs. (7,700 kgs.) **Useful lift,** lbs. (kgs.)

Gas bags.—Continental rubbered fabric.

Motors.—2 Clement motors, 4-cylinder, of 130 h.p. each. 2 motors of 220 h.p. each have also been provided in case the speed should prove disappointing. The 2 motors are placed on either side of the motor space.

Speed.—35 m.p.h.

Propellers.—2, of wood; type, "Integrale." Placed one on either side of the motors, well above the level.

Steering.—Vertical steering by means of a treble horizontal rudder over the rear end of the car. Horizontal steering by means of 2 vertical rudders placed one on each side of the rear horizontal rudder.

Remarks.—The feature of these two ships which distinguishes them from the Astra ships of about the same size is the arrangement of the propellers and the use of a 2 speed gear in connection with these. Normally each motor drives its own propeller through two sets of gearing connected by a Cardan shaft. On stopping one motor, the stopped motor is unclutched from its propeller shaft, which is then connected up by chain drive to the opposite shaft. The running motor is then put on to a "low gear," so that it can make the revolutions necessary for obtaining full power, while the propellers run slower than before. The ratio of "low gear" to "high" is 2 to 1, so that a single motor will be running under its best conditions when well throttled down.

LEBAUDY.

(MORNING POST.)

[Generally follows the lines of the *Liberté* (France) *q.v.*]

Note.—There are one or two other dirigibles building in England, including one driven by a steam engine; but no details are procurable.

None appear to be of any importance, and one, of which a good deal has been heard, seems to be entirely mythical.

NAVAL AIR-SHIP I.

(Also known as VICKERS.)

Maximum length, 510 feet / 155 m. **maximum diameter,** feet / m. **volume,** c. feet / m³.

Total lift.—20 tons (kgs.) **Useful lift,** lbs. (kgs.)

Gas bags.—Of special Continental fabric, as used in the latest Zeppelin ships. Under construction in England by Messrs. Short Brothers, Aeronautical Engineers.

Motors.—Believed:—Two 8-cylinder, 200 h.p. Wolseley motors.

Speed.—45 miles per hour (72 km.)

Propellers.—

Steering.—As in Zeppelin class.

Remarks.—The utmost secrecy has been preserved concerning this vessel. It has been officially announced that she is of the Zeppelin type, and therefore, of aluminium. Also that she is being built by Messrs. Vickers, Sons & Maxim, whose allied Wolseley company exhibited sets of 8-cylinder engines of 200 h.p., designed for some air-ship, at the Olympia.

At the time of printing, September, 1909, only the preliminary work in connection with the air-ship shed appears to be in hand.

BRITISH ARMY II.

Maximum length, feet / m. **maximum diameter,** feet / m. **volume,** 26,857 c. feet / 2,500 m³.

Total lift.—6,062 lbs. (2,750 kgs.) **Useful lift,** lbs. (kgs.)

Gas bags.—Of Continental rubbered fabric, made to the designs of the British Military Authorities, by the Astra Company.

Motors.—2 Green motors, of 50 h.p.

Speed.—

Propellers.—

Steering.—

Remarks.—The gas bag is being made in Paris, and the car and all other parts in England.

ARMY DIRIGIBLE, No. III.

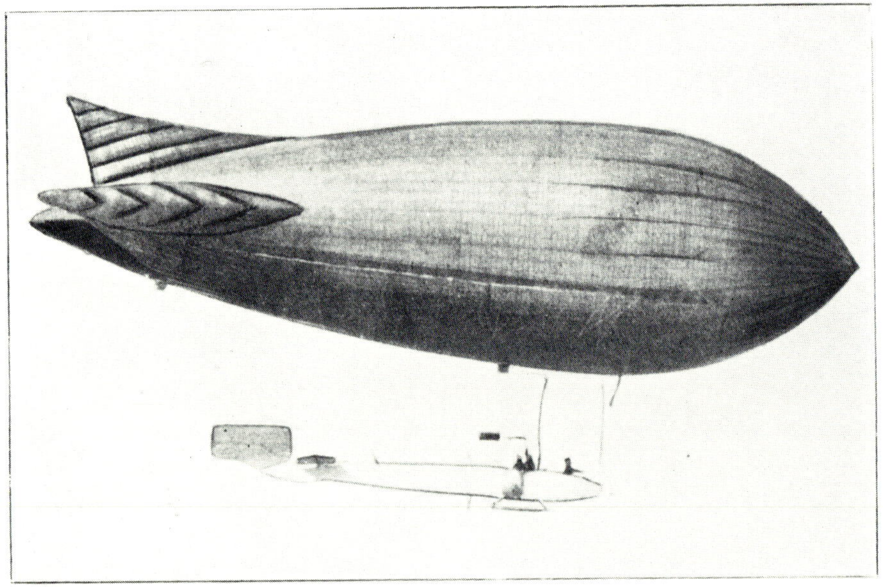

Maximum length, 84 feet / 25·60 m. **maximum diameter,** feet / m. **volume,** 21,000 c. feet / 594 m³.

Total lift.—1,470 lbs. (667 kgs.). **Useful lift,** lbs. (kgs.)

Gas bags.—Gold beater skin. 3 gas filled tail fins are fixed to the rear end. Ballonet, 4,200 c. feet (170 m³.)

Motors.—Two 8 h.p. 3 cylinder Buchet.

Speed.—18 m.p.h. (29 k.m.)

Propellers.—6 feet (1·82 m.) diameter.

Steering.—A vertical rudder at the rear of the car. 2 horizontal rudders, one each side of the car, near the rear end.

Car.—Built of hickory wood and steel tubes, covered with silk. Fitted with 2 landing runners underneath.

Remarks.—Although this vessel is purely experimental, and much too small even for private pleasure uses, she is extremely interesting in several ways. Gold beater skin is used, and for the first time it seems to have been found possible to support the car from it without having bands or nets right over the gas bag. The shape of the gas bag very good considering the necessity of preserving a large average diameter to get the required lift. The shape and disposition of the fins is not only novel but possibly superior to any other at present known. The use of 2 motors in such a small vessel is unique, and suggests that however future ships develop they will certainly not be left to the mercy of a single motor again.

WILLOWS I.
(Cardiff, Wales).

Maximum length, 74 feet (m.) **maximum diameter,** 18 feet (m.) **capacity,** 12,000 c. feet (m³.)

Total weight.—650 lbs. (kgs.)

Frame.—Triangular section. Length, 45 feet (m.) Weight of frame, 450 lbs. (kgs.)

Gas bags.—

Motor.—7 h.p. 2-cylinder Peugot, placed amidships; direct drive.

Propellers.—1 aft, 10 feet (m.) diameter. In front, 1 of special design for control.

Steering.—By patent front propeller entirely.

Remarks.—Balanced to be about 40 lbs. (kgs.) heavier than air. 2 subsequent frames purchased by the British Government. Privately owned by E. T. Willows, of Cardiff, Wales.

WILLOWS II.

Building.

Maximum length, feet / m. **maximum diameter,** feet / m. **capacity,** 14,000 sq. feet / m^2.

Total weight.—

Frame.—

Gas bags.—

Motor.—20 h.p. Curtiss.

Propellers.—2, both amidships.

Steering.—

Remarks.—Envelope by Spencer Bros.

CHINESE.

General Note.—So far as is known, no machines, but it should not be forgotten that the Chinese, with their thousands of years of experience of kite flying, are likely to be apt pupils in all that pertains to the science of wind resistances, etc.

An aeroplane reported building by Professor Chatley at Tientsin Imperial Engineering College, turns out to be a glider for experimental and research work.

DANISH.

General Note.—There are 3 aeroplanes now flying in Denmark, and rumours of another constructing near Copenhagen.

No dirigibles, but a naval officer has been sent to study the *Lebaudy* machines in France.

Aerial Societies:—

Flying Grounds:—

Klampenburg, Copenhagen.

Aerial Journals:—

None.

ELLEHAMMER. Biplane.

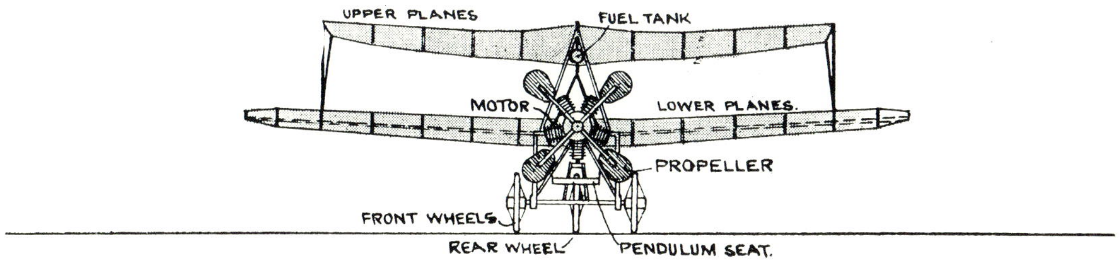

Maximum length, feet / m. **maximum breadth,** 39·37 feet / 12 m. **supporting surface,** 398 sq. feet / 37 m².

Total weight. –286½ lbs. (130 kgs.)

Body.—Mounted on 3 wheels, the rear one being carried by the vertical rudder, giving good steering while on the ground.

Wings.—Superposed, tapered towards the ends. Upper wing 32·8 feet (10 m.) wide, lower wing 39·3 feet (12 m.) wide,

Motor.—Ellehammer. 5 radial revolving cylinders, like the "Gnome" motor. 30 h.p. at 900 r.p.m. Weight, 75 lbs. (34 kgs.) Air cooled.

Speed.—42 m.p.h. (67·5 k.m.)

Propellers.—One 4-bladed, fixed on the revolving motor.

Steering.—Vertical and horizontal rudders in rear, carried by longitudinal wooden bars.

Remarks.—The main feature of this aeroplane is that the aeronaut sits on a swinging seat, hung as a pendulum; this pendulum is connected to the horizontal rudder in such a manner that a forward swing gives up helm and vice versa.

H. FARMAN Biplanes. *(see France.)*

Danish Concessionaire:—

Of this type there are in Denmark, built, building, or on order:—

Model
1908-9

1 Dr. Folmes-Hansen

Farman type.

WRIGHT BROS. Biplanes. *(see U.S.A.)*

Danish Concessionaire:—

Of this type there are in Denmark, built, building, or on order:—

Model 1908-9.

1 Leon?Delagrange

2 *Reported on order* (?)

WRIGHT type.

DUTCH.

General Note.—Little or no interest is yet taken in flying. Two Dutch inventors are rumoured to have aeroplanes in hand or projected, but no details or confirmation are procurable.

A society has been formed at Sourabaya, Java, for the purchase of air-ships.

Aerial Societies:—

Aero Club of Holland.
Java Aerial Society.

Aerial Journals:—

None.

Flying Grounds:—

None.

WRIGHT BROS. Biplane. *(See U.S.A.)*

Concessionaire :—

Of this type there are in Holland, built, building, or to order :—

Model 1908-9.

1 Lefevre
2 (?) *for Java*
3 G. P. Küller

WRIGHT type.

FRENCH.

General Note.—While keeping pace with Germany in the matter of dirigibles, France has been the pioneer of heavier-than-air machines, and at present occupies the premier position in this field.

Though every type of machine finds some support there, the most essentially French development is the monoplane, a type that has generally speaking been neglected in all other countries till quite recently.

The bulk of aeroplanes that have actually flown are of French design.

Aerial Societies:—

Aero Club de France.
Academie Aeronautique de France.
Aeronautique Club de France.
Société des Aëronautes du Siège.
Aero Club du Sud Ouest.
Aero Club du Rhone.
Aero Club du Nord.
Club Aeronautique de l'Aube.
La Ligue Aerienne du Sud.
Société Francaise de Navigation Aérienne.
Union des Aviateurs de la Seine.
Société d'encouragement à l'Aviation.
Société Aeriel.

Aerial Journals:—

L'Aerophile.
L'Aero.
L'Aeronaute.
Aerostat (Bulletin Aeronautique).
Aerostat (Academie d'Aerostation).
Revue de l'Aerostation.
Le Ballon.
L'Aerostation.
L'Aeronautique.
Bulletin Aeronautique.
Encyclopediede l'Aviation.
La Ligue Nationale Aerienne.
Revue de l'Aviation.
L'Aeromécanique.

Flying Grounds (with *hangars*):

Juvissy.
Beauce.
Châlons.
Issy les Moulineaux.
Moisson.
Douai.
La Brayelle.
Reims.
Croix d'Hins (Bordeaux) (Aer. Lig. du Sud.)
Grand Camp, Lyons.
Buoy.
St. Cyr.
Le Mans.

La Compagnie Transaérienne (branch of the Astra Cie), M. M. Surcouf and Kapferer, directors, intend running dirigible lines as follows:—

(1) Issy les Molineaux—Beauval—Reims—Nancy.
(2) Juvissy—Orleans.
(3) Paris—Bordeaux—Pau.
(4) Paris—Brussels (in conjunction with the Belgian *Avia* Society.

It has or is building its own hangars at Issy-les-Molineaux (double), Meaux-Beauval (2), Reims, Nancy, Juvissy, Orleans, Sartrouville-Montesson, Pau, Brussels.

ANTOINETTE Monoplanes.

Special features of Antoinette monoplanes, which are developed through *III*, from the original *Gastambide et Mengin*, are (1) fixed tapering wings with ailerons in rear; (2) the characteristic tail.

The 1910 model has a 100 h.p. engine, 16-cylinder. **Warping wings:** Length, $39\frac{1}{3}$ feet (12 m.) Span, $49\frac{1}{4}$ feet (15 m.) Lower vertical fin done away with.

Of this kind (models *IV* or *VII* or *V* and *VI*) there are in existence, built, building, or to order:—

Model 1908-9.		Model 1908-9.	
1	? (III.)	9	M——
2	Rouchonnet (IV.)	10	Juvissy Aerodrome
3	Réne Demanest (V.)	11	Lieut. Charry
4	Capitaine Burgeat (VI.)	12	Capt. Windham
5	Hubert Latham (VII.)	13	
6	Cie Aérienne	14	
7	,,	15	
8	,,		

ANTOINETTE type.

ANTOINETTE III. Monoplane.

Maximum length, $42\frac{3}{4}$ feet / 14 m. **maximum width,** 41 feet / 12·50 m. **supporting surface,** 430 sq. feet / 40 m^2.

Total weight.—1,146 lbs. (519 kgs.) *including* driver.

Body.—On 1 wheel forward; sledge runners aft.

Wings.—Span 32·5 feet (10 m.)

Motor.—50 h.p. Antoinette, 8 cylinder. 110 × 105 m/m. 1,100 revs. Weight, 209 lbs. (95 kgs.) = 4·1 lbs. (1·9 kgs.) per h.p.

Speed.—$46\frac{1}{2}$ m.p.h. (75 k.m.)

Propeller.—1 forward.

Steering.—Defective—*see Remarks.*

Remarks.—Steered with difficulty. Flew short distances.

ANTOINETTE IV. Monoplane.

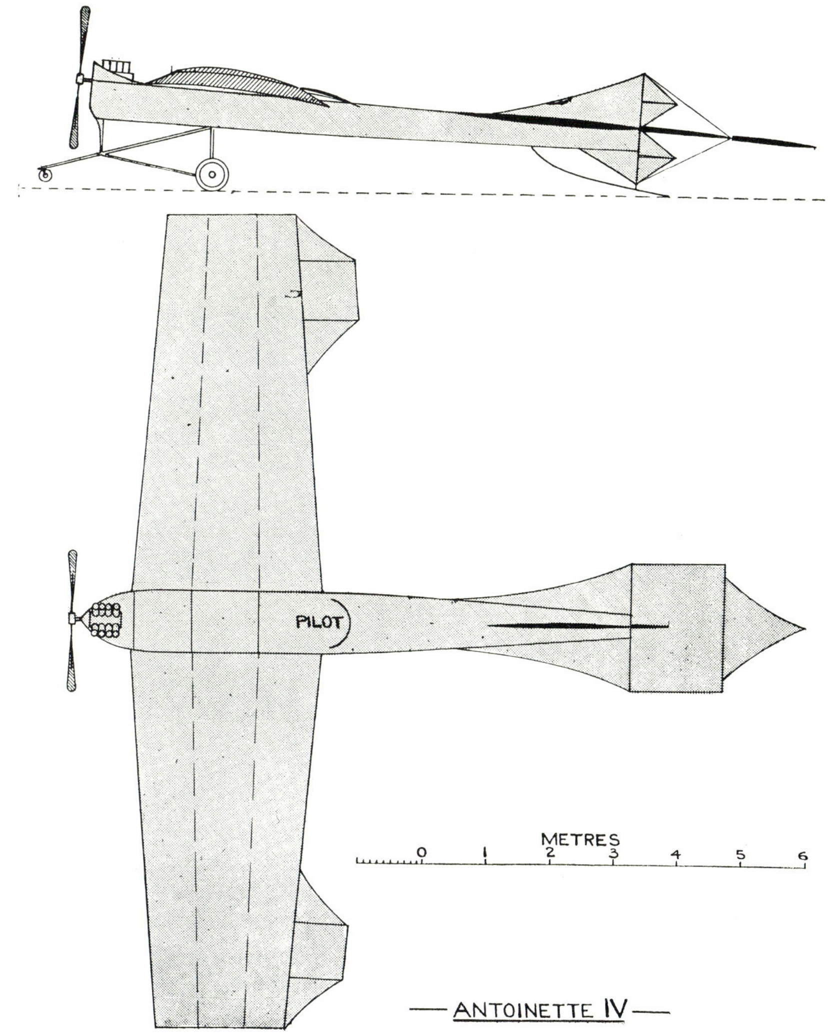

Maximum length, 37 feet / 11·30 m. **maximum breadth,** 40 feet / 12 m. **supporting surface,** 300 sq. feet / 27 m².

Total weight.—1,058 lbs. (480 kgs.) *including* aeranaut.

Body.—As in *Antoinette III.*, except that a pair of steadying wheels are carried on an outrigger below forward planes. Also a front wheel to save propeller in alighting suddenly.

Wings.—As in *Antoinette III.* Ailerons in rear of wing tips.

Motor.—50 h.p. Antoinette.

Speed.—46½ m.p.h (75 k.m.)

Propeller.—1 forward. Diameter, 7 feet (2·10 m.)

Steering.—As in *III.*

Remarks.—Has carried 2. In this, Hubert Latham made the first attempt to cross the Channel. This is a special racing model, and differs from *V* and *VI* only in the position of the triangular fins and being slightly shorter in consequence.

ANTOINETTE V & VI. Monoplanes.

Length, extreme, 37¾ ft. / 11·5 m. **extreme width,** 42 ft. / 12·8 m. **supporting surface,** 370 sq. ft. / 34 m². each. **total weight,** 1,144 lbs. / 520 kgs.

Motor.—50 h.p. Antoinette, 8 cylinder, 110/105 mm., 1,100 r.p.m. Placed at the fore end of the body.

Propeller.—2-bladed, direct driven, immediately before the motor. Diameter, 7·21 feet (2·20 m.)

Speed.—45 m.p.h. / 72 kilometres per hour.

Wings.—Hand polished cloth on a framework of aluminium. Span, 42 feet (12·80 m.) Depth at body, 9¾ feet (3 m.); at wing tips, 6½ feet (2 m.)

Body.—An aluminium framework ending in a stem piece at the front and a point at the rear. The vertical and horizontal planes and rudders are carried at this rear end. The body is covered with polished cloth; it contains the motor at the front end, and the aeronaut's seat abreast of the after edge of the wings. Wheels as in *Antoinette IV*.

Steering.—Effected by means of a vertical and a horizontal rudder at the rear end, and moving wing tips on the outer rear ends of the main wings.

Remarks.—A very successful machine. *V* belongs to Réné Demanest, *VI* to Capitaine Burgeat. This is the model on the market.

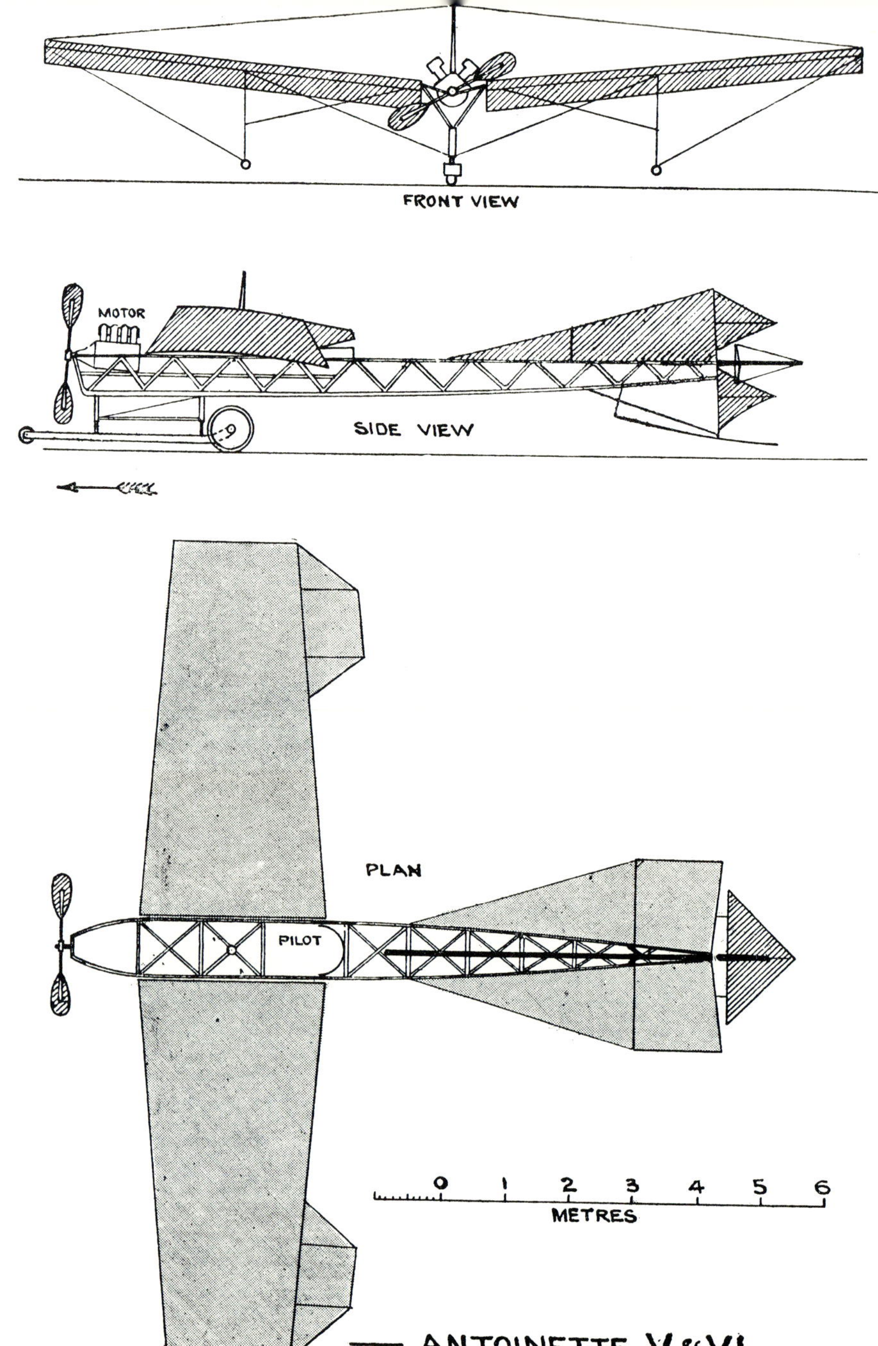

ANTOINETTE V & VI

ANTOINETTE VII.
("LATHAM.")

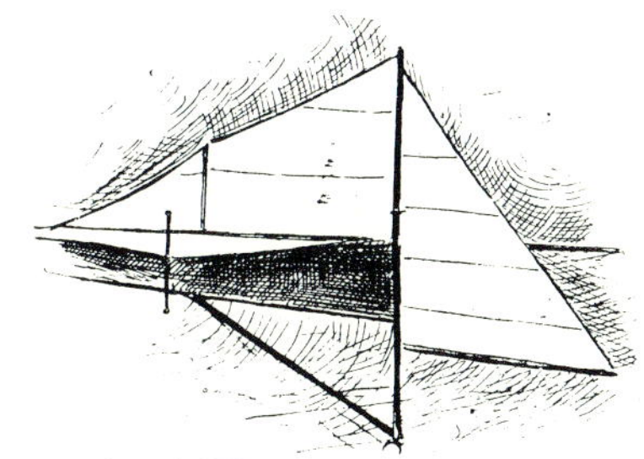

Tail as used on an *Antoinette* at Reims.

Maximum length, 37 feet / 11·30 m. **maximum breadth,** 42 feet / 12·80 m. **supporting surface,** 370 sq. feet / 34 m^2.

Total weight.—

Body.—*Usual.*

Wings.—Originally flexible, but now altered to usual Antoinette pattern.

Motor.—50 h.p. Antoinette.

Speed.—45 m.p.h. (72 km.)

Propellers.—One 2-bladed.

Steering.—*Usual.*

Remarks.—Latham made his second cross channel attempt on this machine. It is now simply an enlarged *IV*.

AUFFM-ORDT. Monoplane.

Photo, C. Malcuit.

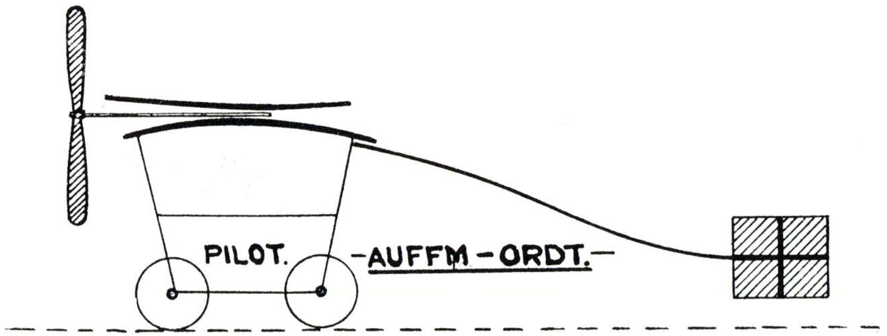

Maximum length, feet / m. **maximum breadth,** feet / m. **supporting surface,** sq. feet / m².

Total weight.—

Body.—

Wings.—Main surface consists of 3 planes slightly curved. Tail is a box-like structure with rudders.

Motor.—

Speed.—

Propellers.—1 forward. Diameter *about* 5 feet (1·5 m.)

Steering.—For horizontal steering rudder abaft tail. For elevating, etc., plane inside tail.

Remarks.—

BLERIOT Monoplanes.

An early Bleriot (*IV*). *Photo, C. Malcuit.*

Special features of Bleriots are fixed wings with rounded edges, and ailerons in rear of wings.

The Bleriots were first developed on Langley lines. The first successful one was a *V*, but subsequently Tatin features were embodied and a *Bleriot VIII* produced. This flew well till it was smashed up. All subsequent Bleriot monoplanes are developments of this machine.

Model 1908-9.
1 Bleriot IX
2 Bleriot XI (cross channel)
3 Bleriot XII
4 Leblanc, XI type
5 Delagrange, XI type
6 Cie Aerienne type (on sale)
7 ,, ,, ,,
8 ,, ,, ,,
9 ,, ,, ,,

Model 1908-9
10 Cie Aerienne, XI type (on sale)
11 Experimental (40 h.p. Anzani) (XII)
12 ,, (60 h.p. E.N.V.) (XII)
13 Parkinson
14 Balin Hinde
15 G. H. Cox (XII)
16 Grahame White (*for sale*)
17 Captain Windham (*for sale*)

In addition: 34 Bleriots are building in France for delivery this year; 50 building at the Humber Works, Coventry, England; 21 sold to owners who cannot be traced.

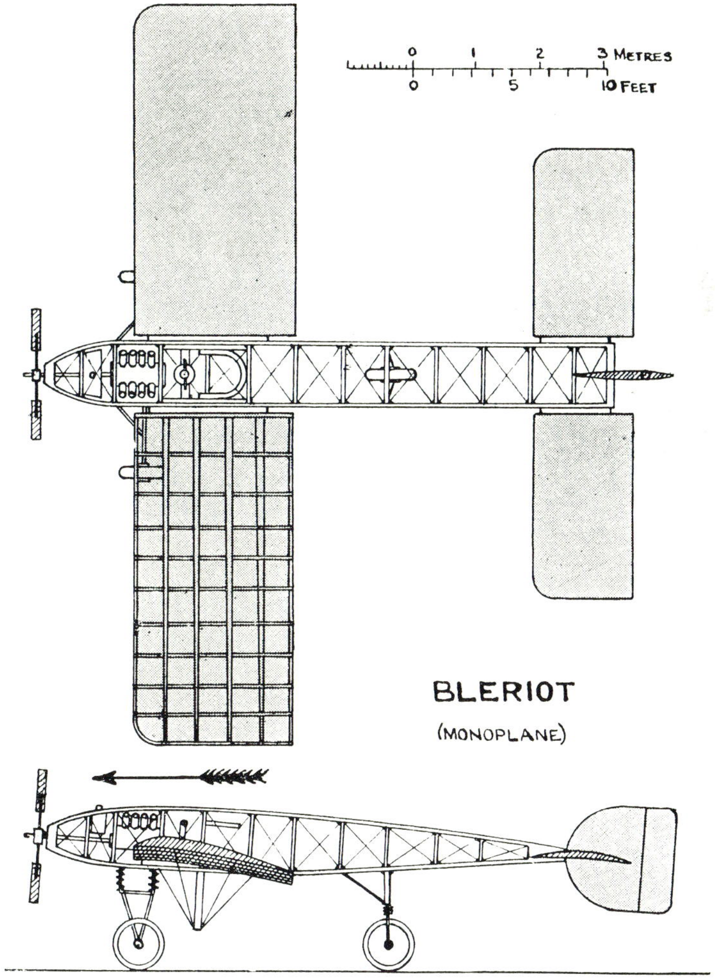

BLERIOT IV.

BLERIOT IX. Monoplane.

Maximum length, 39 ft. 4½ in. / 12 m. **maximum breadth,** 32 ft. 9½ in. / 10 m. **supporting surface,** 269 sq. ft. / 22·9 m².

Total Weight.—1,234 lbs. (558·7 kgs.)

Body.—A wooden girder frame, square section in front, growing to a triangular section in the rear.

Wings.—Built of a poplar wood framework, covered with parchment paper. At the end of each wing is a small stability plane pivoting on a horizontal axis. Angle of elevation of the wings, 8°.

Motor.—50 h.p. 16 cylinder Antoinette, with direct injection in lieu of carburettor.

Speed.—About 46½ m.p.h. (75 km. p.h.)

Propellers.—1 direct driven, 4-bladed. 7 feet (2 m., 10) diameter, 4¾ feet (1 m., 40) pitch. Metal blades, very flexible.

Steering.—Vertical rudder in rear of and above the tail. Small horizontal rudders, one on each side of the tail fin.

Remarks.—One of the features of all Bleriot machines is the "single lever control." The aeronaut holds a single lever attached to a universal cardan joint. Should the aeroplane heel over or tend to rise or dive, the fact of the aeronaut keeping his lever vertical will pull the necessary wires to counteract this motion. Another feature is the position of the radiator, which is built into the side of the body.

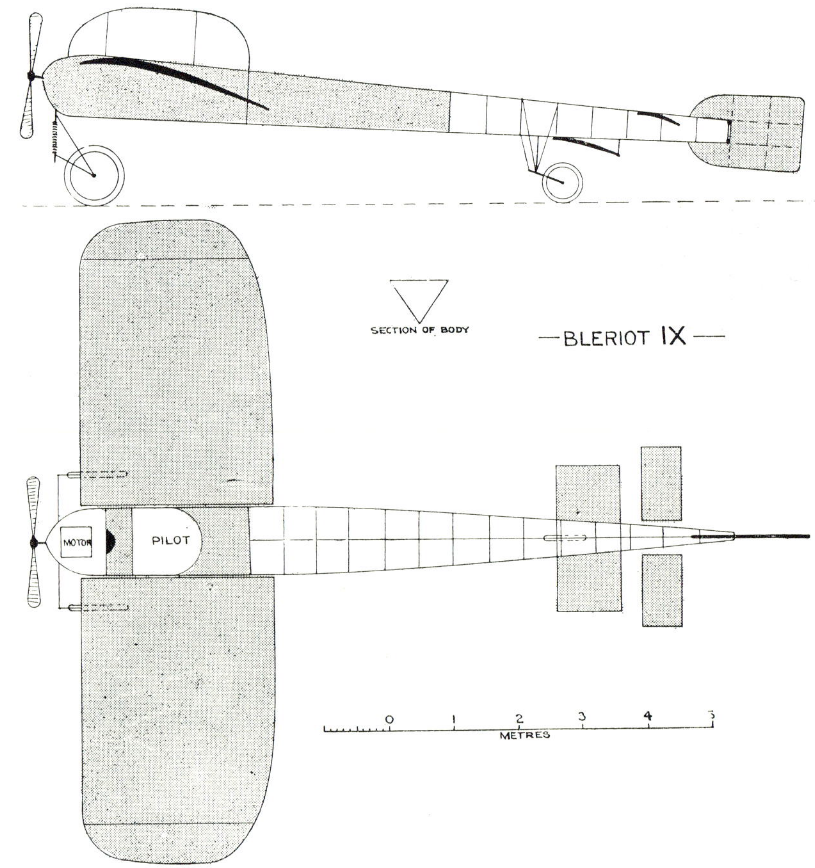

BLERIOT XI. Monoplane.

(Cross-Channel Bleriot.)

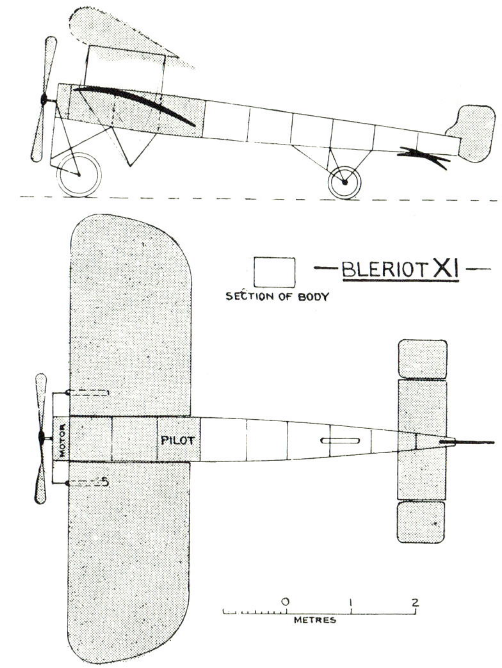

Maximum length, 23 feet. / 7 m. **maximum breadth,** 25⅔ feet. / 7·80 m. **supporting surface,** 150 sq. feet. / 14 m².

Total Weight.—660 lbs. (300 kgs.) *including* aviator.

Body.—As in *Bleriot IX*.

Wings.—Wings are pivoted for altering angle of inclination. No ailerons such as *IX* and *XII* have. Covered with light "Continental" rubbered fabric.

Motor. Anzani 3 cylinder 22-25 h.p. Bore, 103 m/m. Stroke, 120 m/m. Weight, 132 lbs. (60 kgs.)

Speed.—45 m.p.h. (72·4 km.) Has exceeded this. Lifts at 37 m.p.h. (60 km.)

Propellers.—One 4-bladed, direct driven. Chauviére. Diameter, 7 feet (2·10 m.) Made of French walnut. 1,600 r.p.m. Slip is 15% only.

Steering.—Tips of rear surface as elevators. Vertical rudder in tail.

Remarks.—This machine differs from *Bleriot IX* in being built much smaller and lighter throughout. Though it was found necessary slightly to enlarge its lifting surface, it has proved a very great success. A small triangular plane above the aeronaut ensures stability. This is the Cross Channel machine. Price of this model is £400.

BLERIOT XII. Monoplane.

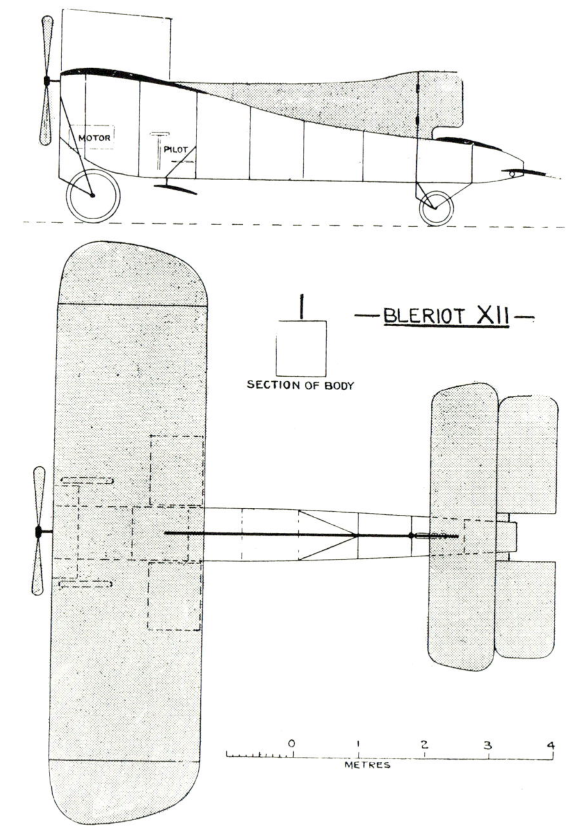

Maximum length, 26 feet / 7·92 m. **maximum breadth,** feet / m. **supporting surface,** 258 sq. feet / 24 m².

Total weight.—

Body.—Square in section, rising considerably forward and tapering aft. Aviator sits inside it entirely.

Wings.—Area, 92 sq. feet (9 m²) each. Large vertical fin on top of frame in rear of wings. Wings as *IX*.

Motor.—35 h.p. water-cooled. Weight of engine, 176 lbs. (80 kgs.) = 5 lbs. (2·26 kgs.) per h.p. Aluminium radiator.

Speed.—

Propellers.—1 in front, geared down considerably, chain driven. Diameter, 8 feet (2·43 m.) 700 r.p.m.

Steering.—Vertical rudder on top of frame in rear of vertical fin, and before rear surface. Elevator tail. Small elevators forward on lower frame about parallel with rear of wings.

Remarks.—This is generally known as "the large *Bleriot*." It has carried three. It is an improved *IX*. It has undergone very numerous alterations of tail and rudder, which accounts for the extraordinary variety of different photos of it extant. The above photo is its latest form.

CLEMENT-BAYARD. Monoplane.

(*Alias* **Bayard-Clement.**)
(*Alias* **Clement-Tatin.**)
(*Alias* **Tatin.**)

Building.

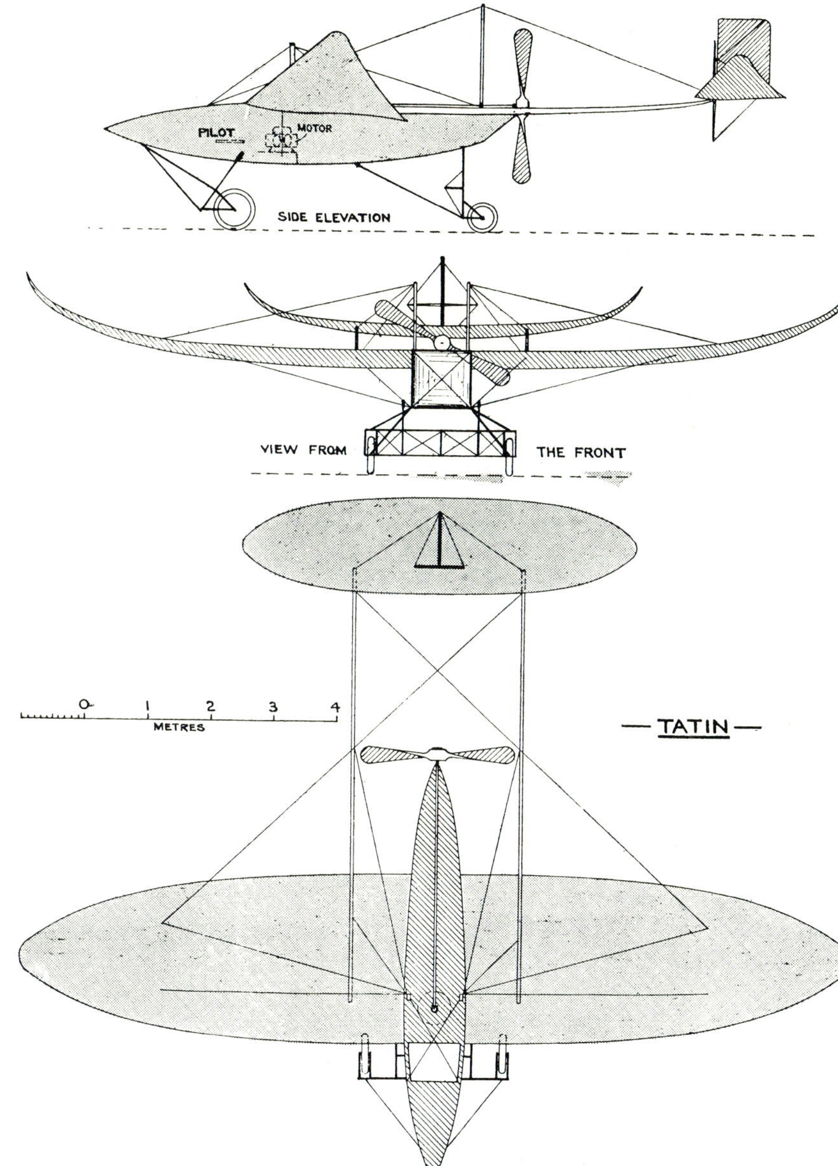

Maximum length, 35 feet (10·70 m.) **maximum breadth,** 43 feet (13·20 m.) **supporting surface,** 247 sq. feet (23 m².)

Total weight.—771½ (350 kgs.) *not including* driver.

Body.—20¼ feet (6·50 m.) long, square section; 3 feet (0·90 m.) sides. Built of wood covered with varnished Japanese silk.

Wings.—Span, 41 feet (12·50 m.) Width, 8¼ feet (2·50 m.) Attached to body by 8 vertical struts.

Tail.—Joined to body by two long wooden bars, between which is the propeller.

Motor.—(1) 50-60 h.p. radial. 7 cylinders. 115 × 100 m/m. 1200 r.p.m. Total weight, 154¼ lbs. (70 kgs.) = about 3·1 lbs. per h.p. *or* (2) a 4-cylinder vertical 43 h.p. at 1600 r.p.m. Dimensions, 100 × 120 m/m. Both are Clerget motors, and either can be fitted.

Speed.—

Propeller.—1 (see Tail). Diameter, 8 feet (2·40 m.) Pitch, 8¼ feet (2·50 m.) The propeller is gear driven at 1 rev. for 2 engine ones.

Steering.—Horizontal rudder and vertical rudder from the tail, 14½ feet (4·40 m.), abaft the main planes. Area of horizontal rudder is about 75¼ sq. feet (7 m².)

Remarks.—

BULOT. Monoplane.

Altering.

Maximum length, feet / m. **maximum breadth,** feet / m. **supporting surface,** sq. feet / m².

Total weight.—

Body.—

Wings.—

Motor.—

Speed.—

Propellers.—

Steering.—

Remarks.—Built at Tours. Has flown. No other details.

BLANC. Monoplane.

Photo, C. Malcuit.

Maximum length, feet / m. **maximum breadth,** feet / m. **supporting surface,** 323 sq. feet / 30 m².

Total weight.—

Body.—On 4 wheels. Long frame to which tail is attached. Planes are overhead.

Wings.—

Motor.—7 cylinder 35 h.p. R.E.P.

Speed.—

Propellers.—1 forward, 2-bladed. Diameter, 6½ feet (2 m.)

Steering.—By wheel. There is a considerable tail. Planes and tail all worked by the wheel.

Remarks.—

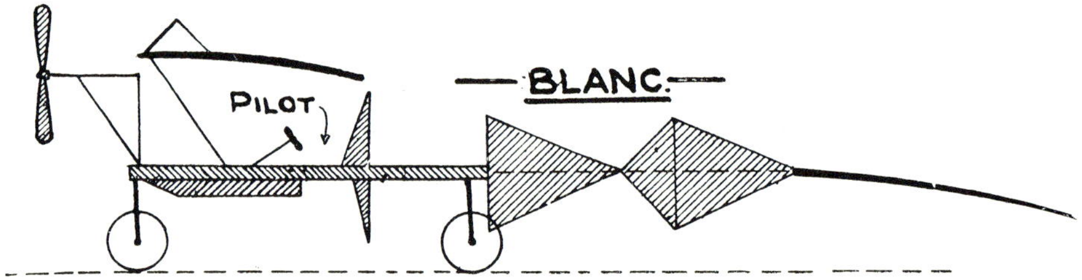

WEHRLÉ.

Building.

Maximum length, feet / m. **maximum breadth,** feet / m. **supporting surface,** sq. feet / m^2.

Total weight.—

Body.—On both skates and floats.

Wings.—

Motor.—

Speed.—

Propellers.—One.

Steering.—By a single wheel.

Remarks.—Laid down January, 1909.

BOURDARIAT. Monoplane (in triplicate).

Building.

Maximum length, $34\frac{1}{2}$ feet (10·50 m.) **maximum breadth,** 31 feet (9·50 m.) **supporting surface,** sq. feet / m^2.

Total weight.—

Body.—

Wings.—Forward planes, 31×5 feet ($9·50\times1·50$ m.) Middle planes span, $24\frac{1}{2}\times4\frac{1}{2}$ feet ($7·50\times1·44$ m.) Supporting surface, 241 sq. feet ($22\frac{1}{2}$ m^2.) Tail planes, $21\frac{1}{4}\times5$ feet ($6·50\times1·50$ m.)

Motor.—Antoinette.

Speed.—

Propellers.—One.

Steering.—

Remarks.—Building at Levallois-Perret. It is a combination of the *Langley* and *Chanute* types.

BORGNIS-DESBORDES.

Maximum length, feet / m. **maximum breadth,** feet / m. **supporting surface,** sq. feet / m².

Total weight.—

Body.—

Wings.—

Motor.—30 h.p. 6-cylinder.

Speed.—

Propellers.—

Steering.—

Remarks.—This is the *M. Borgnis* of the *Bousson-Borgnis.* The pair split, and Mr. Borgnis built this aeroplane in conjunction with M. Desbordes. He also built another by himself of which no data is available.

HENRIOT. Monoplane.

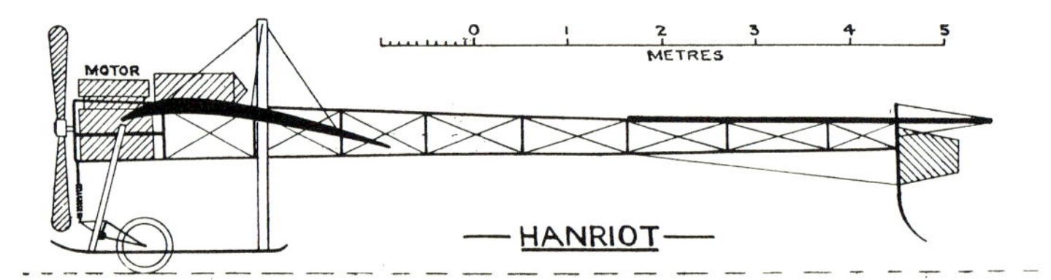

Maximum length, 33 feet / 10 m. **maximum span,** 31 feet / 9·50 m. **supporting surface,** sq. feet / m².

Total weight.—

Body.—On 2 wheels forward and a couple of skids astern and ahead of them.

Wings.—Warpable.

Motor.—50 h.p. Buchet, 6-cylinder, carried right forward. 1,800 r.p.m.

Speed.—

Propeller.—1 forward, 7 feet (2·20 m.) diameter. Direct driven.

Steering.—Elevator in tail.

Remarks.—Cross between a Bleriot and Antoinette. On the market at £800. Usual French flying guarantee.

"LA DEMOISELLE." Monoplane. XIX and XX.
(Santos Dumont.)

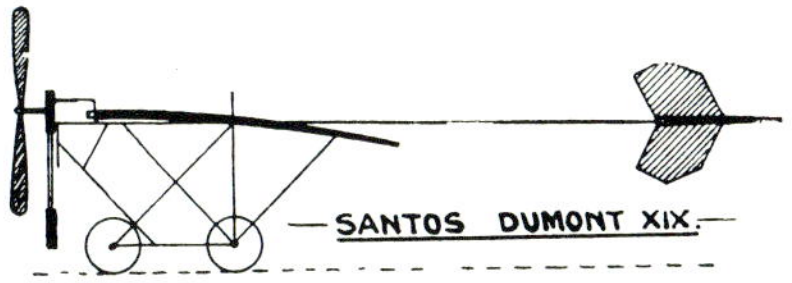

The smallest man carrying aeroplane in existence.

Total length, 19·7 feet. / 6 m. **total breadth,** 16·4 feet. / 5 m. **supporting surface,** 97 sq. feet. / 9 m^2. **total weight,** 330 lbs. / 150 kg.

Body.—Practically none. Mounted on 3 wheels.

Wings.—Aeroplane material sewn to wooden slats.

Motor.—17-25 h.p., Dutheil-Chalmers. 2 cyls., 125 × 127 m/m opposed horizontal.

Speed.—Up to 65 m.p.h. / 100 km. p.h.

Remarks.—This machine can easily be transported on a motor car. Has flown at 56 m.p.h. *XX* is identical with *XIX*.

DESCHAMPS AND BLONDEAU I.

Building.

Maximum length, feet / m. **maximum breadth,** feet / m. **supporting surface,** sq. feet / m^2.

Total weight.—
Body.—
Wings.—
Motor.—
Speed.—
Propellers.—
Steering.—
Remarks.—Building.

DESCHAMPS AND BLONDEAU II.

Building.

Maximum length, feet / m. **maximum breadth,** feet / m. **supporting surface,** sq. feet / m^2.

Total weight.—
Body.—
Wings.—
Motor.—
Speed.—
Propellers.—
Steering.—
Remarks.—

MAURICE FARMAN. Biplane.

(Voisin Type, without vertical surfaces.)

Photo, C. Malcuit.

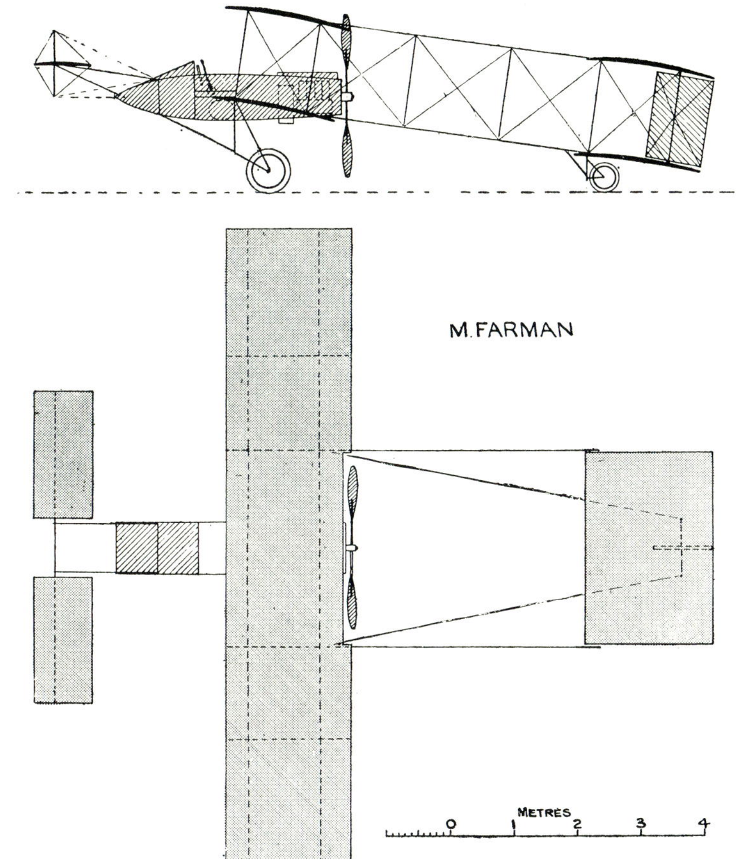

Total length, 35·1 feet. / 10·7 m. **total width,** 32·8 feet. / 10 m. **total surface,** 43·2 sq. feet. / 40 m². **total weight,** 990 lbs. / 450 kg.

Planes.—Upper and lower planes are similar, 32·8 feet. / 10 m. **wide,** 6·56 feet. / 2 m. **deep,** and 4·92 ft. / 1·5 m. apart.

Held rigid by 8 pairs of struts.

Tail.—A double tail of 2 fixed planes, with the vertical rudder between them.

Rudders.—Vertical rudder in the tail. Horizontal rudder right in front. The tips of the main planes can be warped for steering purposes.

Motor.—50-60 h.p., 8-cylinder Renault, weighing 392 lbs. / 178 kg. 1,600 r.p.m.

Propellers.—One 2-bladed. Gear driven at 800 r.p.m.

Lifts of the ground at 25 m.p.h. / 40 km. per h. Full speed about 50 m.p.h. / 80 km. per h.

GASTAMBIDE AND MENGIN. Monoplane.

Photo, C. Malcuit.

Maximum length, feet / m. **maximum breadth,** feet / m. **supporting surface,** sq. feet / m^2.

Total weight.—

Body.—Mounted on 4 wheels, the forward pair close together.

Wings.—

Motor.—

Speed.—

Propellers.—1, in front.

Steering.—

Remarks.—Built in 1907. First flight, January, 1908. First monoplane to carry a passenger. In July, 1908, flew 1,600 metres. The *Antoinettes* are developments of this machine.

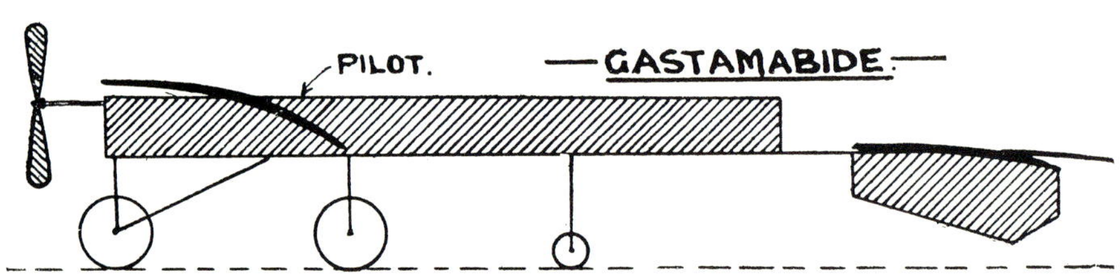

GREGOIRE-GYP. Monoplane.

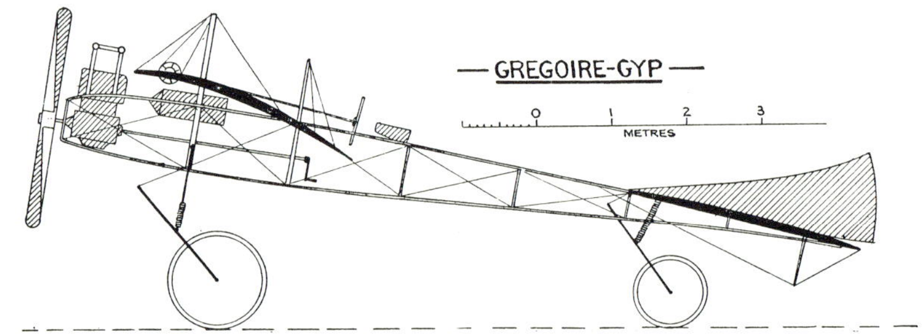

Maximum length, 36 feet / 11 m. **maximum breadth,** 36 feet / 11 m. **supporting surface,** 237 sq. feet / 22 m^2.

Total weight.—661 lbs. (300 kgs.) *not including* aviator.

Body.—On skids.

Wings.—Actuated on the "differential" system of motor cars. Also adjustable to suit the weight carried. Tail.

Motor.—40 h.p. 4-cylinder Gyp. Weight, 78 kgs. (172 lbs.) Special radiator.

Speed.—50 m.p.h.

Propeller.—1, forward.

Steering.—Vertical rudder in tail. Horizontal rudder, main planes.

Remarks.—The tail can be varied, but is not usėd for steering. On the market at £1,000.

GUILLEBAUD. Tandem Monoplane.

Photo, Théodoresco.

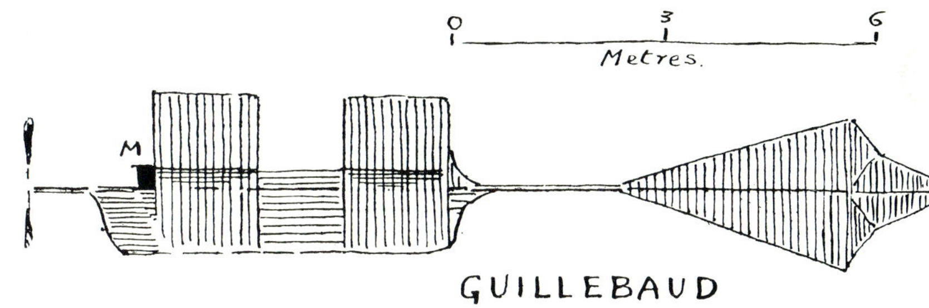

Maximum length, $42\frac{3}{4}$ feet / 13 m. **maximum breadth,** feet / m. **supporting surface,** sq. feet / m².

Total weight.—485 lbs. (220 kgs.)

Body.—Weighs $134\frac{1}{2}$ lbs. (61 kgs.)

Wings.—2 pairs, one behind the other. These are nearly semi-circular.

Motor.—11 h.p. Placed in front of body.

Speed.—

Propeller.—1, wooden. Diameter, $7\frac{1}{2}$ feet (2·30 m.)

Steering.—

Remarks.—There is room for a passenger in front of the motor. Designed by M. Guillebaud, of Rouen.

GUYOT. Biplane.

Building.

Maximum length, feet / m. **maximum breadth,** feet / m. **supporting surface,** 533 sq. feet / 50 m^2.

Total weight.—1000 lbs. (454 kgs.)

Body.—

Wings.—Planes, 36 × $5\frac{1}{2}$ feet (11 × 1·65 m.) They are $5\frac{1}{2}$ feet (1·65 m.) apart.

Motor.—40 h.p. 2-cylinder.

Speed.—

Propellers.—

Steering.—Horizontal rudders in front. Vertical rudder in tail.

Remarks.—

GRON.

Maximum length, feet / m. **maximum breadth,** feet / m. **supporting surface,** sq. feet / m^2.

Total weight.—

Body.—

Wings.—

Motor.—

Speed.—

Propellers.—

Steering.—

Remarks.—

D'EQUEVILLEY.

(Belongs to Le Marquis d'Equevilley.)

Photo, C. Malenit.

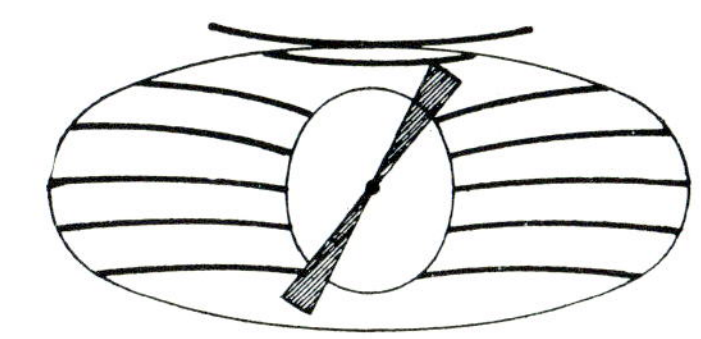

— DEQUEVILLEY (SECTIONAL VIEW)

Maximum length, 6·56 feet (2 m.) **maximum width,** 16·40 feet (5 m.) **supporting surface,** 268½ sq. feet (25 m².)

Total weight.—308½ lbs. (140 kgs.)

Body.—Length, 5·24 feet (1·60 m.)

Planes.—In two ovals.

Motor.—7 to 8 h.p. 1,500 revs.

Speed.—

Propellers.—500 revs.

Steering.—

Remarks.—

KAPFERER. Monoplane.

Photo, C. Malcuit.

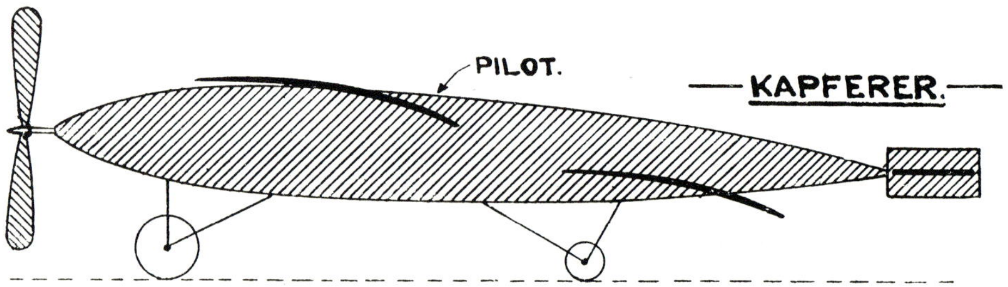

Maximum length, feet / m. **maximum width,** 9·84 feet / 3 m. **supporting surface,** 366 sq. feet / 34 m².

Total weight.—882 lbs. (400 kgs.)

Body.—On 2 wheels forward, 1 aft.

Wings.—Maximum span, 36 feet (11 m.)

Motor.—35 h.p. R.E.P., mounted right forward.

Speed.—

Propellers.—One 4-bladed, forward.

Steering.—Double rudder aft.

Remarks.—

LEVY-CAILLAT. Monoplane.

Maximum length, feet } m. **maximum breadth,** 24 feet } 7·30 m. **supporting surface,** sq. feet } m².

Total weight.—

Body.—Steel tube chassis on 4 wheels.

Wings.—One main plane, 24 × 6½ feet (7·30 × 2 m.)

Motor.—20 h.p. De Dion.

Speed.—

Propellers.—One 2-bladed in front.

Steering.—Short plane placed above main plane, acting as a combined lifting surface and horizontal rudder. This is mounted on a central axle. Vertical rudder aft.

Remarks.—

MILITARY Aeroplane.

Photo, Harlingue.

Maximum length, feet } m. **maximum breadth,** 38 feet } 11·5 m. **supporting surface,** 968½ sq. feet } 90 m².

Total weight.—660 lbs. (300 kgs.)

Body.—Consists of a 2-wheeled carriage in front carrying the motor, connected by a peculiar triangular girder to a light rear carriage of wide span. Right above the top member of the girder is fixed the aeroplane proper, a biplane rather like a Cody kite, with a central cell of triangular shape.

Wings.—

Motor.—43 h.p. Anzani. Weight,

Speed.—

Propellers.—One 2-bladed, built of steel tubing covered on both sides. The pitch is adjustable. Gear driven.

Steering.—A vertical rudder below the triangular tail, and a horizontal rudder just before this.

Remarks.—This machine, designed by Captain Dorand, is unique in having its supporting surfaces clear above the body. The idea seems to be to obtain stability by placing the centre of gravity as far as possible below the centre of lift. Numerous alterations have been made to this machine, and are still being tried.

MANGIN. Biplane.

(*Alias* **FERBER.**)

Photo, Branger.

Maximum length, 30 feet (9 m.) **maximum breadth,** $34\frac{1}{2}$ feet (10·50 m.) **supporting surface,** 430 sq. feet (40 m².)

Total weight.—880 lbs. (399 kgs.)

Body.—

Wings.—Span, $34\frac{1}{2}$ feet (10·50 m.) Dip slightly in centre.

Motor.—50 h.p 8-cylinder Antoinette, air-cooled. 110 × 105 m/m. 1100 r.p.m. Weight, 209 lbs. (95 kgs.) = 4·1 lbs. per h.p. (1·9 kgs.)

Speed.—

Propellers.—1. Diameter, $7\frac{1}{4}$ feet (2·20 m.)

Steering.—Triangular rudders.

Remarks.—Flown by Captain Ferber. Made several short flights. Very fair stability.

PASQUIER. Biplane.

Building.

Maximum length, feet (m.) **maximum breadth,** feet (m.) **supporting surface,** 432 sq. feet (40 m².)

Total weight.—882 lbs. (400 kgs.), *including* aviator.

Body.—Very light construction.

Wings.—Monoplane tail.

Motor.—Pasquier special, 60 h.p. Weight only 88 lbs. (40 kgs.) = 1·3 lbs. per h.p. (0·66 kgs. per h.p.)

Speed.—

Propellers.—2. Diameter, $6\frac{1}{2}$ feet (2 m.)

Steering.—

Remarks.—Designed by M. Pasquier, of Reims. Was to have been ready for trials early in 1909, but is apparently delayed in final completion.

MONIN. Helicopter.

(*Alias* MIEUSSET.)

Not yet completed.

Maximum length, feet / m. **maximum breadth,** feet / m. **supporting surface,** 194 sq. feet / 18 m².

Total weight.—794 lbs. (360 kgs.)

Body.—6-sided wooden frame.

Wings.—2 sets of lifting planes mounted on a vertical axis. Each set consists of 6 planes each 16 sq. feet (1·50 m².) and of variable aeliration, the planes being mounted at the 6 corners of the frame. These 2 sets are 4·92 feet (1·50 m.) apart and revolve in opposite directions at a low rate of speed.

Motor.—35 h.p.

Speed.—

Propellers.—

Steering.—

Remarks.—Built by Mieusset, of Lyons.

FOQUET. Biplane.

Building.

Maximum length, 29½ feet / 9·50 m **maximum breadth,** 33 feet / 10 m. **supporting surface,** 376 sq. feet / 35 m².

Total weight.—397 lbs. (180 kgs.)

Body.—Fuselage entirely boat shaped.

Wings.—Span 33 feet (10 m.)

Motor.—40 h.p.

Speed.—

Propellers.—2, chain driven.

Steering.—Tail combining both vertical and horizontal rudders.

Remarks.—Building by Pischoff & Koechlin to desigrer's plans.

ODIER. Biplane.

(*Alias* ODIER-VENDOME.)
(*Alias* LEPETIL.)
(*Alias* REGY FRÉRES.)
(*Alias* OGIER.)

Maximum length, feet / m. **maximum breadth,** 33 feet / 10 m. **supporting surface,** 523 sq. feet / 49 m^2.

Total weight.—990 lbs. (450 kgs.)

Body.—

Wings.—Span, 33 feet (10 m.) Triangular stabilising fins from tail towards main planes.

Motor.—18 h.p. Turcot-Mery.

Speed.—

Propellers.—One 4-bladed.

Steering.—Biplane tail as elevator. Vertical rudder inside it.

Remarks.—Completed July, 1909. Built by Regy Fréres. There are probably duplicates in existence, which accounts for the numerous names.

SOIRON (L.E.)

Maximum length, feet / m. **maximum breadth,** feet / m. **supporting surface,** sq. feet / m^2.

Total weight.—770 lbs. (kgs.)

Body.—Entirely aluminium.

Wings.—

Motor.—20 h.p.

Speed.—

Propellers.—

Steering.—

Remarks.—

PISCHOFF-KOECHLIN. Monoplane.

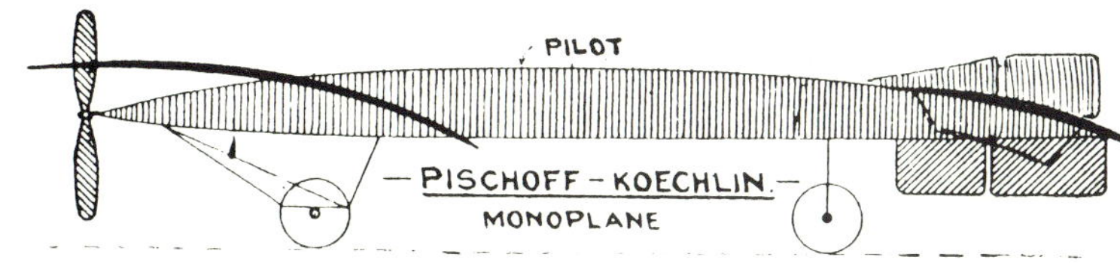

Photo. C. Malenit.

(An early *Pischoff-Koechlin* machine). *Photo, C. Malenit.*

Maximum length, feet / m. **maximum width,** $28\frac{3}{4}$ feet / 8·7 m. **supporting surface,** 301 sq. feet / 28 m².

Total weight.—771 lbs. (350 kgs.), *including* aeronaut.

Body.—

Wings.—Span, 26·2 feet (8 m.). Depth, $11\frac{3}{4}$ feet (3·5 m.)

Motor.—17-25 Dutheil-Chalmers. 2 cylinder. 125 × 120 m m.

Speed.—

Propellers.—1, in front.

Steering.—2 vertical rudders aft, 2 horizontal forward, 2 right aft.

Remarks.—

GANGLER. Monoplane.

Building.

Maximum length, feet / m. **maximum breadth,** 28 feet / 8·50 m. **supporting surface,** 301 sq. feet / 28 m².

Total weight.— lbs. (kgs.)

Body.—

Wings.—

Motor.—10 h.p.

Speed.—

Propellers.—One.

Steering.—

Remarks.—Building by Pischoff & Koechlin to the inventor's plans. In general design it follows the *P. K.* monoplane, but differs in some minor details, and is reported to have a special tail.

L. BREGUET. Biplane.

(Has also been known as BREGUET-RICHET III.)

Maximum length, 32¾ feet / 10 m. **maximum breadth,** 42¾ feet / 13 m. **supporting surface,** sq. feet / m².

Total weight.—1100 lbs. (500 kgs.), *including* aeronaut.

Body.—Oval. Frame is of tubular steel on runners. In addition, 1 large wheel (steerable) in centre line under engine, 2 small wheels aft, wheels at the extremities of the lower plane.

Wings.—The upper plane consists of 1 small fixed central plane. The large outer surfaces are fixed by steel tubes to the corresponding lower planes and tilt in unison with them. The span of upper planes is 42¾ feet (13 m.) and the lower planes 32¾ feet (10 m.) The lower planes have no central surface.

Motor.—50 h.p. 8-cylinder Renault, V shape, placed high on body.

Speed.—

Tractor.—One, 3-bladed, very large. Direct driven.

Steering.—Biplane elevator behind. Vertical rudders at its extremities. Control is by a single wheel.

Remarks.—A singularly neat and simply constructed machine.

R.E.P. (Robert Esnault-Pelterie.) Monoplane.

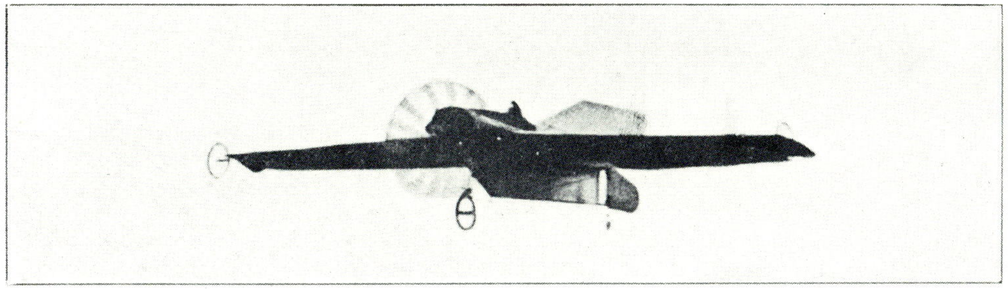

Photo, J. Hauser.

Maximum length, feet / m. **maximum breadth,** 31 feet / 9·5 m. **supporting surface,** 129 sq. feet / 12 m².

Total weight.—771½ lbs. (350 kgs.)

Body.—R.E.P. pattern.

Wings.—Span, 31 feet (9·5 m.), width of planes, 6½ feet (2 m.)

Motor.—30 h.p. R.E.P., 7-cylinder. Bore, ; stroke, 95 m/m.

Speed.—49½ r.p.h. (80 km.)

Propellers.—One 4-bladed (R.E.P. pattern.)

Steering.—R.E.P. pattern.

Remarks.—

REP II, bis. Monoplane.

Maximum length, 26 ft. 3 in. / 8 m. **maximum breadth,** 31 ft. 5 in. / 9 m. 60. **supporting surface,** 11½ sq. feet. / 15·175 m².

Total Weight.—926 lbs. (420 kgs.) with aeronaut.

Body.—Cigar shaped cage of welded steel tubes. The motor is in the extreme front, the aeronaut abaft it.

Wings.—The wings, which have a slight downward inclination, are built of wooden framing strengthened with steel. They are made to warp. They are strengthened each by 2 stays on the underneath side. Covered with rubbered silk

Motor.—Rep. 7 cylinders, 30 to 35 h.p. Weight, 150 lbs. (68 kgs.). 1,500 revs.

Speed.—About 37 m.p.h. (59·6 km. p.h.)

Propellers.—One 4-bladed, entirely of steel. Keyed directly on to the motor.

Steering.—Vertical and horizontal rudders in the tail. Warping the wings is also used for steering.

Remarks.—Control is effected by 2 handles, one for horizontal steering and one for vertical steering, and pedal which connects to the motor throttle. The machine starts by running on the ground on its front and rear central line wheels with the wheel at the end of one wing also touching. As it gathers speed the machine rights itself and runs on two wheels only till it leaves the ground.

PISCHOFF-KOECHLIN. Monoplane.

(Latest model.)

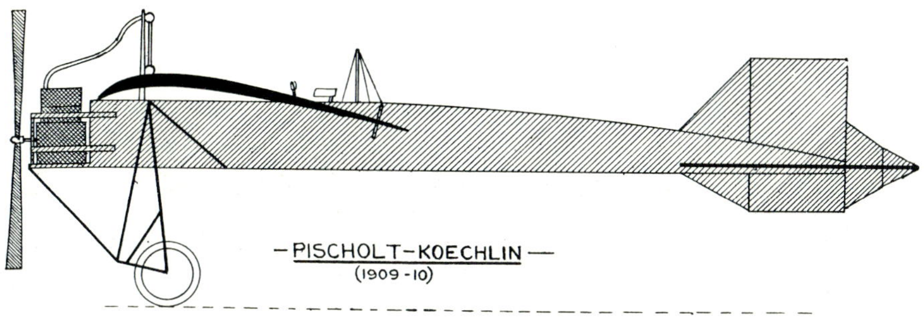

Maximum length, feet / m. } **maximum span,** feet / 8 m. } **supporting surface,** sq. feet / m^2. }

Total weight.—

Body.—Blunt nose, fine run aft.

Wings.—Flexed by the aviator moving laterally.

Motor.—28 h.p. Gyp. Bosch H.T. magneto.

Speed.—

Propeller.—1, forward.

Steering.—Both rudders aft, in tail. Control of both by a single wheel moved in accordance.

Remarks.—Flies well. On the market at £500.

TARISS-BUSSERON (Lieut.) Monoplane (or Biplane).

Completing on trial.

Maximum length, feet / m. } **maximum breadth,** feet / m. } **supporting surface,** sq. feet / m^2. }

Total weight.—

Body.—

Wings.—

Motor.—

Speed.—

Propellers.—

Steering.—

Remarks.—Reported ready for trials, August, 1909. *No data.*

CHARPENTIER. Monoplane.

Building.

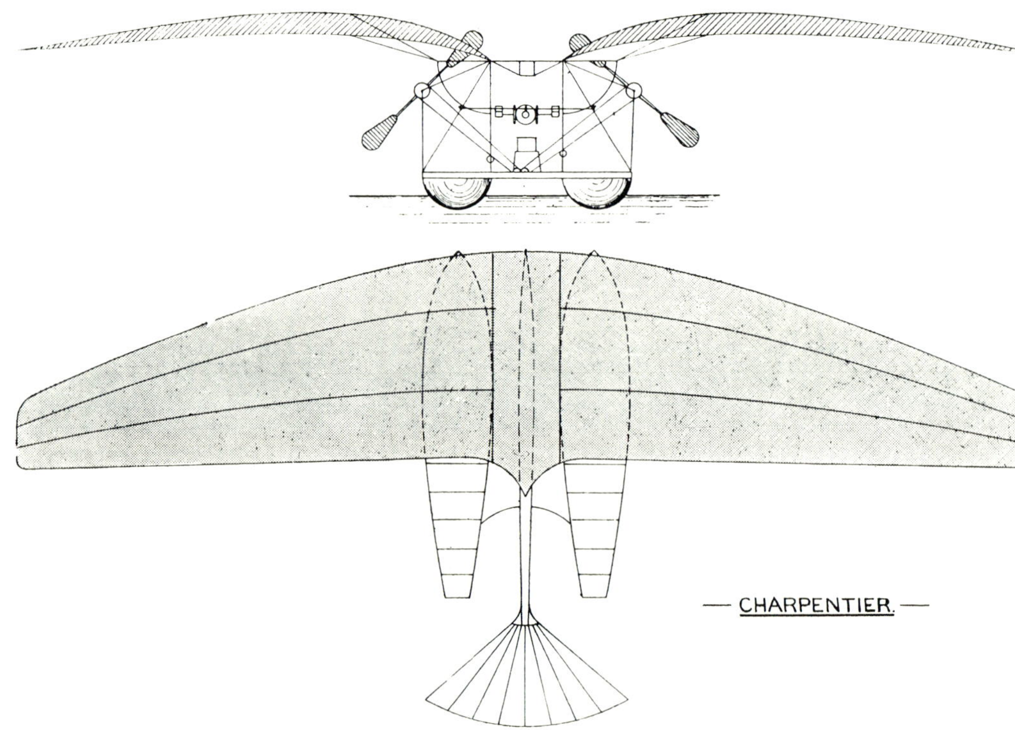

— CHARPENTIER. —

Maximum length, feet / m. **maximum breadth,** $49\frac{1}{4}$ feet / 15 m. **supporting surface,** 430 sq. feet / 40 m².

Total weight.— (450 kgs.)

Body.—Designed to rise off water ; mounted on floats (*see* plan).

Wings.—Span, $49\frac{1}{2}$ feet (15 m.) Flexible wings. Fan tail.

Motor.—50 h.p.

Speed.—50 m.p.h. (80 km.) *est.*

Propellers.—2. Diameter, 9 feet (2·80 m.) Pitch, 9·84 feet (3 m.) Run at 600 revs.

Steering.—The wings are used as elevators.

Remarks.—Built at St. Malo. Intended for naval use.

CHEDVILLE. Biplane.

Maximum length, feet / m. **maximum breadth,** feet / m. **supporting surface,** 554 sq. feet / 52 m².

Total weight.—1110½ lbs. (504 kgs.)

Body.—On 3 wheels.

Wings.—39⅓ feet (12 m.) span.

Motor.—60 h.p.

Speed.—

Propellers.—

Steering.—Horizontal rudder in front. Vertical rudder behind.

Remarks.—Flew at Flers de l'Orme, 380 yards (350 m.) at a height of 13 feet (4 m.)

VENDOME II. Biplane.

Maximum length, feet / m. **maximum width,** 26¼ feet / 8 m. **supporting surface,** 376 sq. feet / 35 m²

Total weight.—

Body.—

Wings.—

Motor.—

Speed.—43½ m.p.h. (70 k.m.)

Propellers.—

Steering.—

Remarks.—

VENDÔME III. Monoplane.

Max. length, $39\frac{1}{3}$ ft. / 12 m. **max. breadth,** 41 ft. / 12·50 m. **supporting surface,** 280 sq. ft. / 26 m². of which $21\frac{1}{2}$ sq. feet (2 m²) is tail.

Total weight.—$617\frac{1}{2}$ lbs. (280 kgs.) including petrol for four hours flight.

Body.—1 wheel in front, sledge runners behind and under each wing tip.

Wings.—These fold completely for housing on transport, and can be increased or reduced during flight, which confers the possibility of attaining very high speeds and at the same time rising or falling at moderate speed.

Motor.—50 h.p. 3 cylinder Anzani.

Speed.—Rises at 30 m.p.h. (48 k.m.)

Propellers.—One 2-bladed.

Steering.—

Remarks.—See note about wings.

RAOUL-VENDÔME II. Monoplane.

Maximum length, feet / m. **maximum breadth,** $39\frac{1}{3}$ feet / 12 m. **supporting surface,** 215 sq. feet / 20 m².

Total weight.—683 lbs. (310 kgs.)

Body.—

Wings.—$39\frac{1}{3}$ feet (12m.) span. Depth, $29\frac{1}{2}$ feet (9 m)

Motor.—50 h.p. Anzani. 3-cylinder. Stroke, 145 m/m. Bore, 150 m/m. 1200 r.p.m. Weight of engine, 237 lbs. (107 kgs.) = 4·7 lbs. (2·14 kgs.) p.h.p.

Speed.—

Propellers.—

Steering.—

Remarks.—

WITZIG-LIORE-DUTILLEUL. Monoplane.

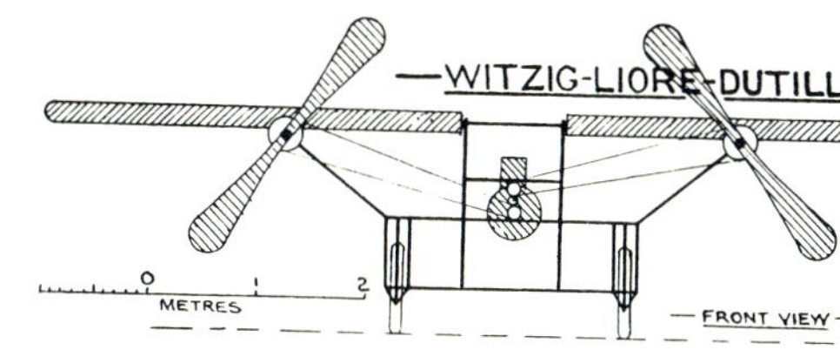

Maximum length, feet / m. **maximum breadth,** $28\frac{1}{2}$ feet / 8·70 m. **supporting surface,** 258 sq. feet / 24 m².

Total Weight.—

Body.—

Wings.—Forward half of each fixed; after half hinged for use as elevator or reduction of surface.

Motor.—35 h.p. Buchet.

Speed.—

Propellers.—2 in front. Special rubber fittings.

Steering.—Elevator tail, wings also. Vertical rudder universally jointed with tail elevator.

Remarks.—On the market at £519.

VUIA. Monoplane.

Photo, C. Malcuit.

Photo, J. Hauser.

Maximum length, feet / m. **maximum breadth,** feet / m. **supporting surface,** sq. ft. / m².

Total weight.—

Body.—On 4 wheels.

Wings.—These can be folded.

Motor.—4-cylinder, just abaft propeller. Direct drive.

Speed.—

Propellers.—One 2-bladed in front.

Steering.—Big rudder aft for horizontal steering.

Remarks.—

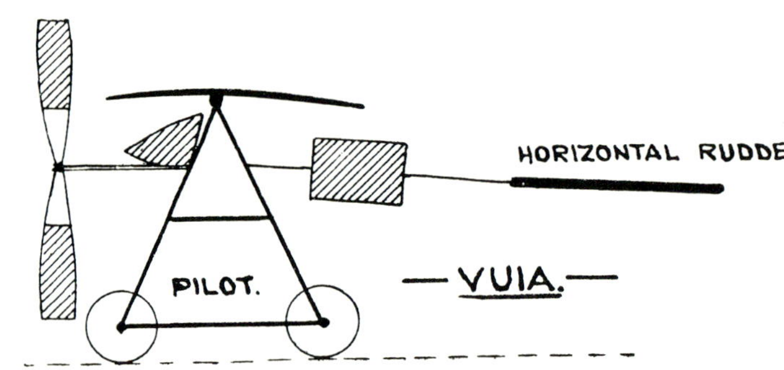

VERMOREL. Monoplane.

Building.

Maximum length, feet / m. **maximum breadth,** feet / m. **supporting surface,** 161½ sq. feet / 15 m^2.

Total weight.—172 (78 kgs.), *not including* aviator.

Body.—

Wings.—

Motor.—48 h.p. Vermorel.

Speed.—

Propellers.—

Steering.—

Remarks.—Building at Villefranche sur Saone.

VOSGIEN. Monoplane.

Maximum length, feet / m. **maximum breadth,** feet / m. **supporting surface,** sq. feet / m^2.

Total weight.—

Body.—On 4 wheels and skids.

Wings.—Warpable. Fixed tail, etc., like *Bleriot XI.*

Motor.—20-25 Dutheil-Chalmers. Placed above the wings.

Speed.—

Propeller.—1, right forward (Vosgiens).

Steering.—Twin vertical rudders. Otherwise much as *Bleriot XI.*

Remarks.—

ABRIS CALAS (Marseilles).

Building.

Maximum length, feet / m. **maximum breadth,** feet / m. **supporting surface,** sq. feet / m^2.

Total weight.—

Body.—

Wings.—

Motor.—4 cylinder Gregoire.

Speed.—

Propellers.—

Steering.—

Remarks.—Practically a Wright Bros. machine.

DE DION-BOUTON. Multiplane.

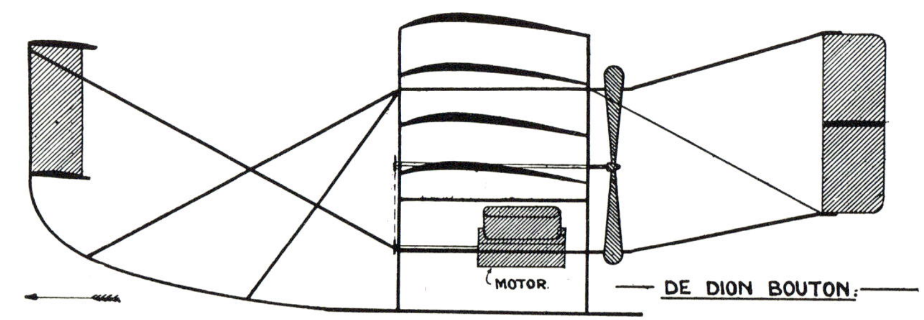

Maximum length, feet / m. **maximum breadth,** feet / m. **supporting surface,** sq. feet / m^2.

Total weight.—

Body.—A cube on long skids.

Wings.—Planes in series at a dihedral angle. Tail.

Motor.—De Dion.

Speed.—

Propellers.—4, in rear of planes.

Steering.—Elevator forward. Two vertical rudders.

Remarks.—

BRISSAND (Nice). Biplane.

Building.

Maximum length, feet, m. **maximum breadth,** feet, m. **supporting surface,** sq. feet, m^2.

Total weight.—

Body.—

Wings.—

Motor.—

Speed.—

Propellers.—

Steering.—

Remarks.—Nothing known.

FERNANDEZ. Biplane I & II.

Maximum length, feet, m. **maximum breadth,** 28 feet, 9·50 m. **supporting surface,** $268\frac{1}{2}$ sq. feet, 25 m^2.

Total weight.—496 lbs. (225 kgs.) *not including* aviator.

Body.—On 3 wheels, two behind. Front wheel braked.

Wings.—Ailerons in rear.

Motor.—42 h.p. E.N.V. 8-cylinder. Placed behind the driver.

Speed.—

Propeller.—1, behind planes.

Steering.—Single plane elevator forward. Vertical rudder aft.

Remarks.—Combination of a Wright Bros. and Curtis.

BONNET-LABRANCHE II. Biplane.

Photo, J. Hauser.

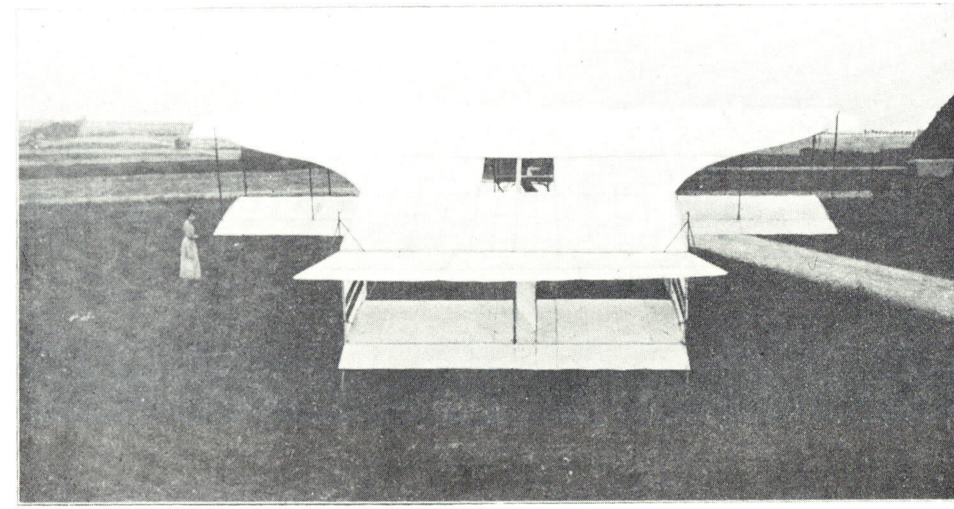

View from behind.

Maximum length, feet / m. **maximum breadth,** feet / m. **supporting surface,** 533 sq. feet / 50 m^2.

Total weight.—

Body.—On 4 wheels, the rear two carrying the tail.

Wings.—Considerable curve to upper plane, which extends all the way back to the tail.

Motor.—80 h.p.

Speed.—

Propellers.—1 in rear.

Steering.—Double single plane elevator forward. Vertical rudder in tail.

Remarks.—

CRINI AND BERTRAND. Biplane.

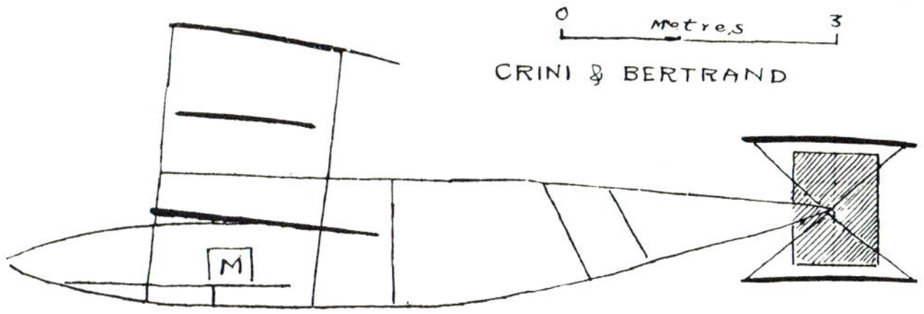

Maximum length, 31 feet / 9·4 m. **Maximum breadth,** $35\frac{1}{2}$ feet / 10·8 m. **Supporting surface,** 533 sq. feet / 50 m².

Total weight.— lbs. (550 kgs.)

Body.—Usual biplane construction; built of wood. Mounted on runners (may subsequently be mounted on wheels).

Wings.—Superposed; $35\frac{1}{2}$ feet (10·8 m.) wide, and $7\frac{1}{4}$ feet (2·2 m.) from front to rear.

Motor.—50 h.p. 3-cylinder Anzani. 9 litres of petrol carried.

Speed.—

Propellers.—2. Steel paddles mounted on steel tubes. Diameter, $8\frac{1}{4}$ feet (2·5 m.) Belt driven 550 r.p.m.

Steering.—The horizontal tail surfaces can be inclined for vertical steering. Horizontal steering is affected by means of the small moving planes placed one each side between the tip of the main planes.

Remarks.—

SANCHIS-BESA. Biplane.

Maximum length, feet / m. **maximum breadth,** $39\frac{1}{2}$ feet / 12 m. **supporting surface,** 527 sq. feet / 49 m².

Total weight.—600 lbs. (272 kgs.)

Body.—

Wings.—Span, $39\frac{1}{2}$ feet (12 m.) Depth of planes, $6\frac{1}{2}$ feet (2 m.)

Motor.—Anzani.

Speed.—

Propellers.—2 in rear.

Steering. Cellular rudder.

Remarks.—First trials at Issy, France, August, 1909. Machines of this make were at Reims:—1. Kluytmans. 2. G. Bailly. 3: Fernandez. 4. Clemenceau.

BLERIOT X. Biplane.

Maximum span, 39⅓ feet / 12 m. **maximum width of planes,** 8¼ feet / 2·5 m. **supporting surface,** 646 sq. feet / 60 m^2.

Total weight.—2,268 lbs. (500 kgs.) *including* aeronaut.

Body.—

Planes.—These are 26¾ feet (8 m.) in width and 6½ feet (2 m.) apart. Between the 2 main planes are 4 fixed vertical planes. The two inner ones are the special Bleriot Radiators; the two outer are fixed triangular planes, with their bases in front.

Motors.—50 h.p. Antoinette, 8-cylinder, 119 × 105 m/m., 1,100 revs. Weight, 209 lbs. (95 kgs.) = 4·1 lbs. (1·9 kgs.) per h.p.

Speed.

Propellers.—One 4-bladed, of wood. Diameter, 9·84 feet (3 m.) Pitch, 9·84 feet (3 m.)

Steering.—Front vertical rudder consists of 3 vertical planes, side by side. These move together. In rear a tail of 2 fixed triangular planes, above and outside of which are 2 horizontal rudders of 32¼ sq. feet. (3 m^2.) each.

Remarks.—Main features resemble the Wright Bros.' machine, except that vertical rudder is in front and the horizontal governor (2 auxiliary planes) in rear.

FERNANDEZ. Biplane.

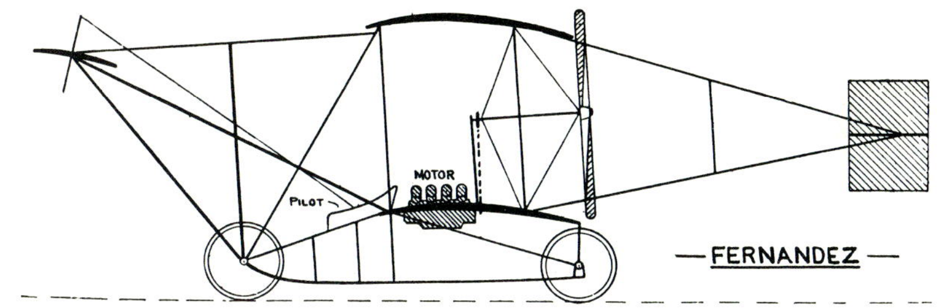

Maximum length, feet / m. **maximum breadth,** feet / m. **supporting surface,** 533 sq. feet / 50 m^2.

Total weight.—1,058 lbs. (480 kgs.) *including* aviator.

Body.—On 4 wheels.

Wings.—Warping.

Motor.—50 h.p. 8-cylinder. Antoinette.

Speed.—

Propeller.—1 Chauvière. Diameter, 6½ feet (2 m.)

Steering.—

Remarks.—

DURAY MATTHYS.

Building.

Maximum length, feet / m. **maximum breadth,** feet / m. **supporting surface,** sq. feet / m^2.

Total weight.—

Body.—

Wings.—

Motor.—

Speed.—

Propellers.—

Steering.—

Remarks.—

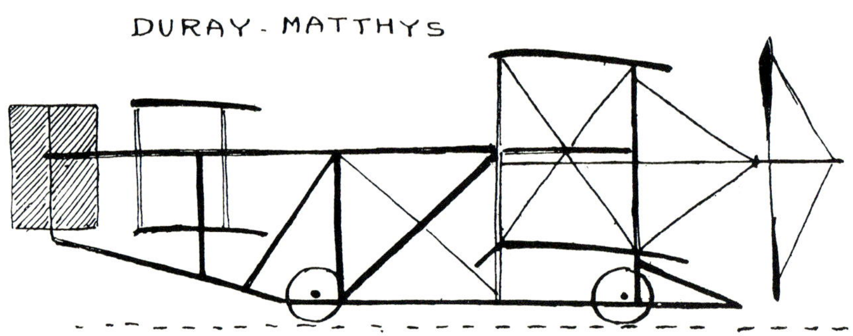

DUTHEIL AND CHALMERS. Biplane.

Maximum length, 33 feet / 10 m. **maximum breadth,** 33 feet / 10 m. **supporting surface,** sq. feet / m^2.

Total weight.—

Body.—

Wings.—Planes are $33 \times 4\frac{1}{2}$ feet ($10 \times 1{\cdot}36$ m.)

Motor.—4-cylinder 42 h.p. Dutheil and Chalmers, horizontal, water-cooled, 1,000 r.p.m. Bore and stroke, 125×120 m/m. 2 ignitions. *Features*: Overhead exhaust ports, fly-wheel with fan, cast iron pistons, copper water jackets fastened by a patent process.

Speed.—

Propellers.—1 wooden, direct driven.

Steering.—

Remarks.—Price complete, £600.

HENRI FARMAN II. Biplane.

Maximum length, 39½ feet / 12 m. **maximum breadth,** 26¼ feet / 8 m. **supporting surface,** sq. feet / m².

Total weight.—800 lbs. (363 kgs.) *not including* aviator.

Body.—On 6 wheels; two pairs in front, two small wheels behind. Long runners to front wheels. Body entirely unenclosed.

Wings.—2 main planes, 6½ feet (2 m.) apart. 26¼ × 6½ feet (8 × 2 m.) Biplane tail without vertical surfaces. Stabilisers in rear of planes.

Motor.—50 h.p. Gnome or Vivinius.

Speed.—

Propellers.—One 2-bladed, chain driven, mounted in rear of lower plane.

Steering.—Single plane, horizontal rudder in front. 2 vertical rudders at back of tail.

Remarks.—This is the type flown at Reims (1909) by Sommer and Cockburn. H. Farman's first biplane was subsequently converted into a triplane and sold to an Austrian Syndicate (see Austria).

Of this type there are in existence:—

1. Sommer.
2. Cockburn.
3. G. Grahame White (*order*).

HENRI FARMAN III. Biplane.

Maximum length, $42\frac{1}{2}$ feet / 3 m. **maximum breadth,** $34\frac{1}{2}$ feet / 10·50 m. **supporting surface,** 409 sq. feet / 38 m^2.

Total weight.—About 800 lbs. (363 kgs.)

Body.—As in *F II.*

Wings.—Main planes are $34\frac{1}{2} \times 6\frac{1}{2}$ feet (10·50 × 2 m.) Stabilisers in rear of planes.

Motor.—45 h.p. Gnome, rotary.

Speed.—

Propellers.—1, wooden, 2-bladed in rear of planes. Diameter, $7\frac{1}{2}$ feet (2·3 m.) Direct driven.

Steering.—Horizontal rudder in front. Vertical rudder behind the tail.

Remarks.—Differs from *II* mainly in the side surfaces to tail and in a tiller being used instead of a wheel.

Several of this type built or building.

One for M. Farman's own use has a 160 h.p. motor.

GOUPY II. Biplane. (Also a Triplane.)

(Built by Bleriot.)

GOUPY, Biplane.

GOUPY, Triplane. *Photo, Malcuit.*

GOUPY, Triplane.

Maximum span, 19·7 feet (6 m.) **supporting surface,** 281 sq. feet (26 m².)

Total weight.—461 lbs. (209 kgs.), *not including* aeronaut; with him, *about* 639 lbs. (290 kgs.)

Body.—Mounted on three wheels.

Planes.—Upper plane is a little in advance of the lower one. 2 vertical planes between the main planes. Width of planes, $5\frac{1}{4}$ feet (1·60 metres).

Tail.—2 horizontal planes, top one a little in advance of lower one.

Motor.—R.E.P. 25 h.p.

Speed.—

Propellers.—One 4-bladed tractor, right forward.

Steering.—Vertical rudder in rear; just forward of this, 2 vertical side planes.

Remarks.—Follows type of *Bleriot* monoplane, but biplane instead. There is also a *Goupy I* triplane, which never flew, built to specifications by Voison Fréres. Carries 35 litres petrol + 10 litres oil (enough for 2 hour's flight).

JERME. Biplane.

Maximum length, feet m. **maximum breadth,** $39\frac{1}{3}$ feet 12 m. **supporting surface,** 129 sq. feet 12 m^2.

Total weight.—882 lbs. (400 kgs.)

Body.—On 4 wheels.

Wings.—Span, $39\frac{1}{3}$ feet (12 m.) Rigid flat planes of equal size. Distance between planes, *about* 7 feet (2·13 m.) Lower deck is cut to admit of aviator's body.

Motor.—3-cylinder 50 h.p. Anzani. Mounted directly over back wheels.

Speed.—37 m.p.h.

Propellers.—1 very large 4-bladed in rear of main planes, chain driven.

Steering.—Elevator in front, with triangular vertical stabilising planes. 2 oblong vertical rudders in rear.

Remarks.—Completed July, 1909. Of very simple and rigid construction.

DOUTRE. Biplane.

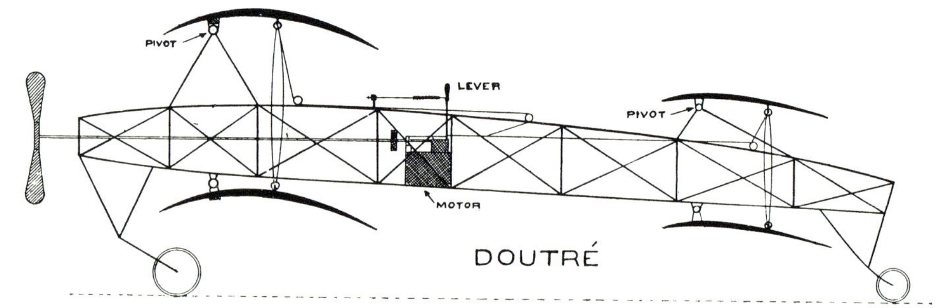

Maximum length, feet m. **maximum breadth,** feet m. **supporting surface,** sq. feet m^2.

Total weight.—

Body.—

Wings.—

Motor.—

Speed.—

Propellers.—

Steering.—

Remarks.—

LASTENAS.

Maximum length, feet / m. **maximum breadth,** feet / m. **supporting surface,** sq. feet / m^2.

Total weight.—

Body.—

Wings.—

Motor.—

Speed.—

Propellers.—

Steering.—

Remarks.—

PENTEADO.

(*Alias* **CHAUVIÈRE.**)

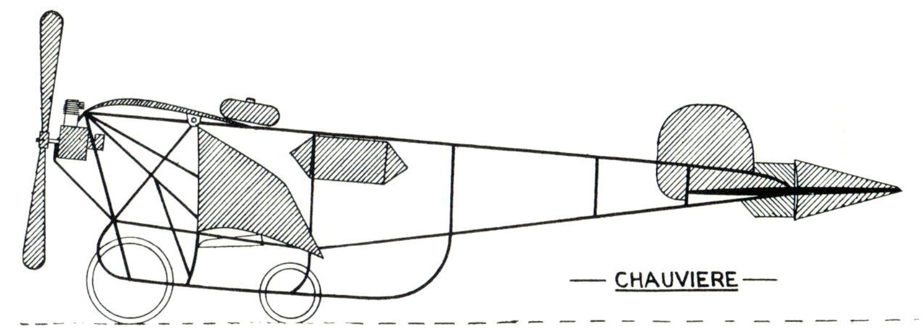

Maximum length, feet / m. **maximum breadth,** feet / m. **supporting surface,** sq. feet / m^2.

Total weight.—

Body.—

Wings.—Lower wings (which are flexible?) set at a dihedral angle. The ends rise up to and behind the upper deck.

Motor.—R.E.P.

Speed.—

Propeller.—1, forward. Chauvière.

Steering.—Elevator in rear.

Remarks.—Built by Chauvière to M. Penteado's design. The machine is on monoplane lines, and in many respects suggest a *Bleriot XII* with the lower forward stabilisers converted into wings.

LEJEUNE I. Biplane.

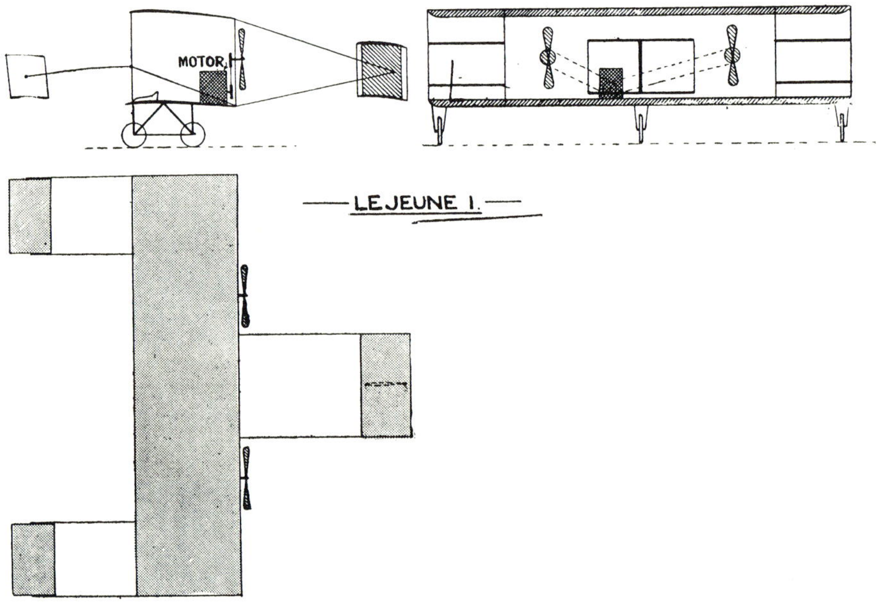

Maximum length, 20½ feet / 6·20 m. **maximum breadth,** 19¾ feet / 6·50 m. **supporting surface,** 269 sq. feet / 25 m^2.

Total weight.—386 lbs. (175 kgs.)

Body.—Practically none. The lower plane is mounted on a wheel at either forward extremity, a third wheel in the centre line and a fourth just behind it. Driver sits on lower plane, a little to port of centre line.

Wings.—Planes on bamboo frame, covered with unvarnished linen. They both curve downwards in rear.

Motor.—3-cylinder 10-12 h.p. Buchet, mounted directly on the lower plane, a little to starboard of centre line.

Speed.—

Propellers.—Two in rear of planes, chain driven.

Steering.—Two elevators in front. These are placed ahead of the ends of the planes. Vertical rudder inside a biplane tail.

Remarks.—Built for M. Lejune by M. M. Pischoff and Koechlin at Billancourt. It is of extremely light construction throughout.

PISCHOFF-KOECHLIN. Biplane.

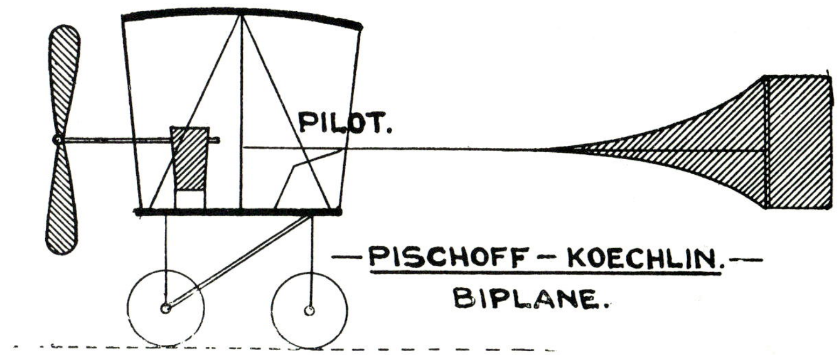

Photo, C. Malcuit.

Latest type.

Maximum length, feet m. **maximum width,** $21\frac{3}{4}$ feet 6·5 m. **supporting surface,** $268\frac{1}{2}$ sq. feet 25 m^2.

Total weight.—120 lbs, (54·5 kgs.)

Body.—

Wings.—Upper plane, $21\frac{3}{4} \times 5\frac{1}{4}$ feet ($6·5 \times 1·6$ m.) Lower deck much less span than upper.

Motor.—12 h.p. Buchet.

Speed.—

Propellers.—1, forward.

Steering.—2 vertical rudders aft. 2 horizontal.

Remarks.—

Privately owned in France, built, building, or on order:—

Model 1908-9.

1 Pischoff-Koechlin experimental.

2 M. Piquerez (*special, q.v.*)

3 M. Lejune (*special, q.v.*)

4 M. Fouquet (*special, q.v.*)

Note.—The Pischoff-Koechlin works are at Billancourt. About 12 men employed at present. Foreign orders, British and Russian in hand (*q.v.*)

HENRI ROBART (Amiens). Biplane.

Maximum span, $39\frac{1}{3}$ feet / 12 m. **supporting surface,** 533 sq. feet / 50 m². (may be slightly more).

Total weight.—882 lbs. (400 kgs.)

Body.—On 3 wheels.

Wings.—The upper small plane is 75 sq. feet (7 m².) Lower plane curved considerably.

Motor.—40-50 Antoinette, 8 cylinder.

Speed.—

Propellers.—2 wooden, $7\frac{1}{4}$ feet (2·20 m.) in diameter. Chain driven: in front, 2 chains, one crossed.

Steering.—Horizontal rudder abaft the horizontal bird-like tail. Vertical rudder above this.

Remarks.—The features of this aeroplane are that the main wings have a unique upward curvature, and that a plane, with an area of 75·35 sq. feet (7 m²) is superposed above the centre of the main plane. The machine runs on 3 wheels, 2 abreast in front, and a single one in rear.

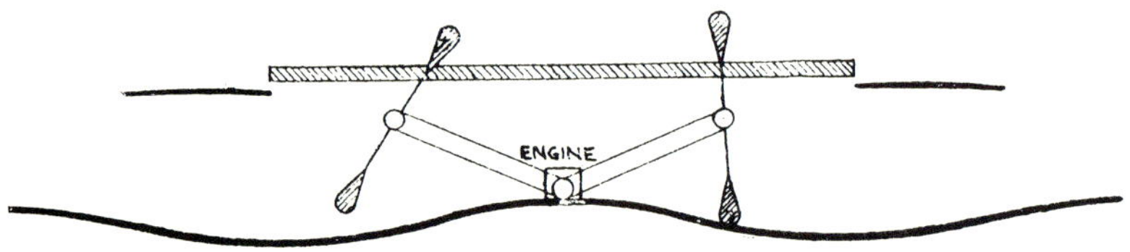

AIMÉ-SALMSON. Biplane.

Maximum length, feet. / m. **maximum breadth,** feet. / m. **supporting surface,** sq. feet / m².

Total Weight.—

Body.—

Wings.—Upper plane slightly curved up; lower plane curved up in a semi-circle to meet it.

Motor.—

Speed.—

Propellers.—2 athwartships (to drive air against the planes). 2 aft.

Steering.—

Remarks.—

BREGUET RICHET II, bis.

Maximum length, feet / m. **maximum breadth,** feet / m. **supporting surface,** sq. feet / m².

Total weight.—

Body.—The car carrying the mechanism is independent of the rest of the frame; attached by cables with springs at the extremities to absorb shock on coming to earth. Chassis built of steel tubing.

Planes.—Spread about 40½ feet (12 m.) Lifting power, 12 cwt. (600 kgs.)

Motor.—Gobron-Brille, 80 h.p. 8 cylinders in pairs, 2 opposed pistons in each cylinder. Cylinders form an "X" round a central crank chamber, which contains a two-throw crankshaft. Camshaft consists of only one large eccentric cam, which revolves with the crankshaft. Rocker levers operate valves. No gears. No camshaft. 2 carburettors. 2 H.T. magnetos. Automatic inlet valves. Forced lubrication. Pump for water circulation. Weight, 330 lbs. (149·6 kgs.), 1,600 r.p.m.

Speed.—43½ m.p.h. (70 k.m.) mominal, but 56 m.p.h. can be reached.

Propellers.—One 2-bladed, 8 feet 2 inches (2·50 m.) diameter. Weight of this 13 lbs. (6 kgs.)

Steering.—

Remarks.—

GASNIER.
(René.)

Maximum length, 29·68 feet / 9·5 m. **maximum breadth,** 32·8 feet / 10 m. **supporting surface,** 376 sq. feet / 35 m².

Total weight,—992 lbs. (450 kgs.)

Body.—Voisin type.

Wings.—

Motor.—50 h.p. Antoinette.

Speed.—

Propellers.—

Steering.—Elevator in front. Fixed horizontal tail in rear.

Remarks.—Flew in November, 1908. Smashed up, and since reconstructed.

JACQUELIN. Miscellaneous.

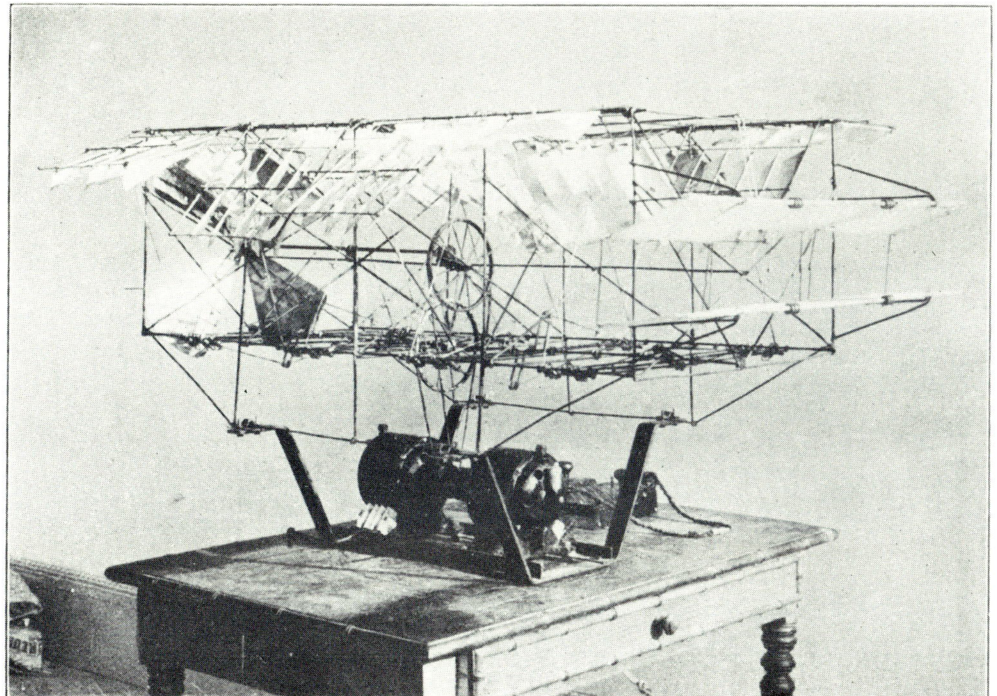

Model.

Maximum length, feet / m. **maximum breadth,** feet / m. **supporting surface,** sq. feet / m².

Total weight.—

Body.—

Wings.—

Motor.—

Speed.—

Propellers.—

Steering.—

Remarks.—

BEAUFEIST.

Maximum length, feet / m. **maximum breadth,** feet / m. **supporting surface,** sq. feet / m².

Total weight.—

Body.—

Wings.—

Motor.—

Speed.—

Propellers.—

Steering.—

Remarks.—Designed by a private soldier of the 3rd Regiment of Engineers, at Verdun.

VERMOREL. Triplane.

(Givaudin Type.)

(*Alias* **GIVAUDAN II.**)

Maximum length, $19\frac{3}{4}$ feet / 6 m. **maximum breadth,** 13 feet / 4 m. **supporting surface,** 194 sq. feet / 18 m^2.

Total weight.—882 lbs. (400 kgs.) *including* pilot.

Body.—Tubular steel. Mounted on 4 wheels, the first pair steerable.

Wings.—Maximum span, 13 feet (4 m.) Made of thin wood.

Motor.—40 h.p 8-cylinder Vermorel. 1,200 r.p.m. Air-cooled. 2 carburettors. H.T. magneto ignition.

Speed.—$40\frac{1}{2}$ m.p.h. (65 km.)

Propellers.—1, forward. Diameter, $7\frac{10}{12}$ feet (2·40 m.)

Steering.—The forward cell is universally jointed and is the sole rudder.

Remarks.—This is an improvement on the *Givaudan* (page 158).

LANDÉ.

Building.

Maximum length, feet / m. **maximum breadth,** feet / m. **supporting surface,** sq. feet / m^2.

Total weight.—

Body.—

Wings.—

Motor.—Moto bloc.

Speed.—

Propellers.—

Steering.—

Remarks.—Designed by M. Landé, at Libourne.

VOISIN Type. Biplane.

Later type. *Photo, C. Malcuit.*

Max. length, 37·8 ft. / 11·4 m. **max. span,** 32·8 ft. / 10 m **max. breadth, of** *planes,* 6·5 ft. / 2 m. **supporting surface,** In *earlier type.* 537 sq. ft. / 49 m². In *later type.* 597 sq. ft. / 55 m².

Total weight.—*About* 1200 lbs. (544 kgs.), *including* aeronaut.

Body.—Mounted on 4 wheels, two in front, two behind. These are sprung to give about 2 feet (60 c/m) on landing. Front wheels are steerable. Angle of incidence, 8°.

Planes.—Space between main planes, 5 feet (1·50 m.) In later models, 6¾ feet (2·10 m.)

Tail.—Two small planes, span 8 feet (2·43 m.) Width, 6½ feet (2 m.) Distance between planes, 5 feet (1·50 m.)

Motor.—36-39 h.p. in some; up to 80 h.p. E.N.V. in others.

Speed.—Up to 42 m.p.h. (67·5 k.m.), 1100 revs. Control by throttle and cutting out cylinders. The machine rises at about 30 m.p.h. (48 k.m.)

Propellers.—Renard type. One of steel behind front planes in centre of gravity. Weight, 30 lbs. (13·60 kgs.) Diameter, 7½ feet (2·28 m.) Pitch, 5 feet (1·52 m.) Blades are curved. Size, 2·62 feet (0·75 m.) by 9 inches (23 c/m) wide. Direct drive from engine.

Steering.—Horizontal rudder in front, 54 sq. feet (5 m²), controlled by a slide ending in a motor car type steering wheel. Vertical rudder in rear, *about* 10½ sq. feet (1 m²). This also is connected to steering wheel, the control wires going to drum, so that both horizontal and vertical steering is effected by the same wheel.

Remarks.—This type is well tried and safe. It follows the general design of M. O. Chanute of 1898-1900.

Weights: Fuselage, 250 lbs. (114 kgs.); main planes, 180 lbs. (81·5 kgs.); tail, 118 lbs. (53·5 kgs.); elevator, 32 lbs. (14·5 kgs.)

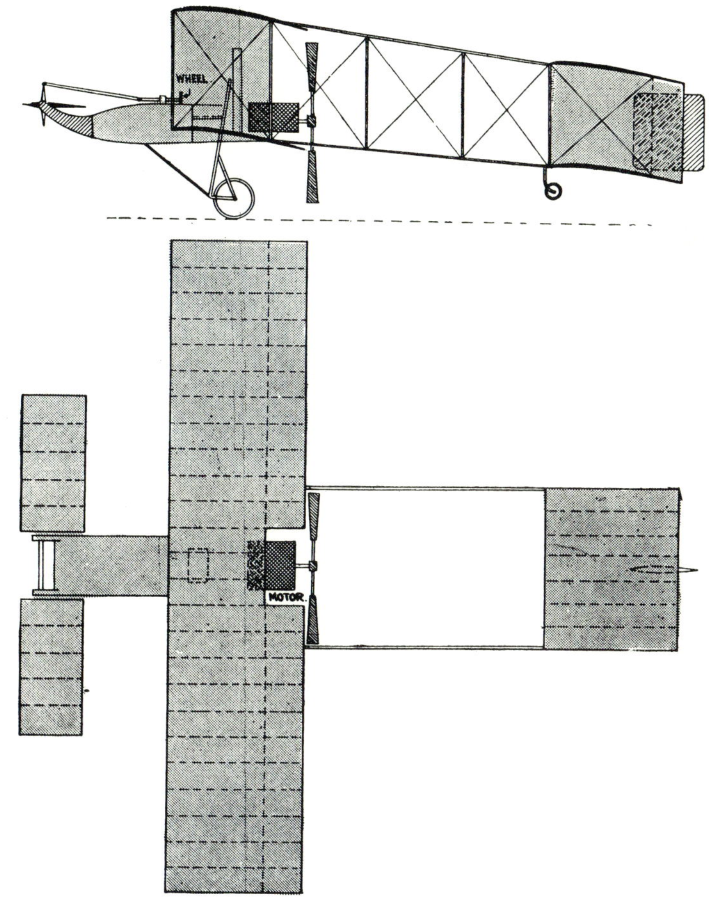

The 1910 type will have no elevator and a special new kind of tail.

VOISIN Biplanes. Mostly fitted with E. N. V. engines.

Full list of machines of this type in existence, privately owned (September, 1909):—

1 Farman
2 Count Von Hedburg (*ex* Delagrange)
3 Baron de Caters
4 *late* De Rue (Captain Ferber)
5 Rougier
6 L. Paulhan
7 Zipfel
8 Mortimer Singer
9 Anzani
10 Bunau-Varilla
11 Salvert
12 Louis Gaudart
13 Jean Gobron
14 H. Fournier
15 M. Colliex
16 La Ligue Aerienne
17 Alberto Braniff
18 Legagneux
19 Sanchis Besa
20 Edward Bello
21 Legrand
22 A. George *ex* Moore-Brabazon
23 Florio
24 Prince Bibesco
25 Aero Club, Odessa
26 Prince Bolotoff
27 Dadé
28 August Euler
29 de Frettes
30, 31 } Compagnie Aérienne (*on sale*)
32 Goupy
33 Dr. Hansen
34 Prince Stourdza
35 Seymour
36 F. Simms
37 Koch
38 G. Grahame-White
39 C. Moering
(Many others on order).
See also *Avis* Co., Italy.

Photo, J Hauser.

Note.—2.—Wrecked June '09. 6.—Fitted with 7-cylinder Gnome rotary engine. 22.—*Bird of Passage.*

WRIGHT BROS. Biplanes. (*see U.S.A.*)

French Concessionaires:—Astra Cie. (Compagnie Générale de Navigation Aériénne.)

Of this make there are in France, built, building, or to order:—

Model 1908-9.	Model 1908-9.	Model 1908-9.
1 Comte de Lambert	21	41
2 P. Tissandier	22	42
3 Juvissy Aerodrome	23	43
4 Paul Tissandier (II)	24	44
5 René Gasnier	25	45
6 Alfred Leblanc	26	46
7 Comte de Lambert (II)	27	47
8 Henri Kapferer	28	48
9 *late* Lefebvre	29	49
10 Juvissy II	30	50
11 } Soc. Aeriel (*sale*)	31	51
12 }	32	52
13 Capt. Girardville	33	53
14 — Schrek	34	54
15 — Barratoux	35	55
16	36	56
17	37	57
18	38	58
19	39	59
20	40	

Note.—The machine on which Wilbur Wright made his original French records is now in a museum.

WRIGHT type.

ZENS. Biplane.

Maximum length, feet m. **maximum breadth,** feet m. **supporting surface,** 301 sq. feet m². including tail.

Total weight.—

Body.—

Wings.—The tail is one plane only.

Motor.—50 h.p. Antoinette, with F.I.A.T. carburettor and special thin copper radiator.

Speed.—

Propellers.—One.

Steering.—There is only 1 rudder. This is placed in front, and used for both vertical and horizontal steering.

Remarks.—Extremely simple in construction.

COLLOMB. Flapper.

Maximum length, feet m. **maximum breadth,** feet 12 m. **supporting surface,** sq. feet 24 m².

Total weight.—

Body.—

Wings.—2 rocking planes, formed of latices which open on up stroke and close on down stroke.

Motor.—40 h.p.

Speed.—

Propellers.—

Steering.—

Remarks.—Designed by M. Collomb, of Lyons.

BORGNIS-DE SAVIGNON I. Triplane.

Maximum length, feet / m. **maximum breadth,** $47\frac{3}{4}$ feet / 14·50 m. **supporting surface,** 861 sq. feet / 80 m^2.

Total weight.—1,257 lbs. (570 kgs.)

Body.—Tubular steel. Mounted on 4 wheels in pairs.

Wings.—Curved planes, reduced in depth at the wing tips. Ailerons to middle planes. No tail.

Motor.—28 h.p. 6-cylinder.

Speed.—

Propellers.—1. Diameter, 8 feet (2·50 m.) Chain driven and geared down. Placed behind the main planes.

Steering.—Single plane elevator forward, on level with lower plane. Steering in horizontal plane by ailerons only.

Remarks.—This is M. Borgnis, of the Bousson-Borgnis and Borgnis-Devordes partnership. Built by M. Borgnis and the Comte de Savignon. A *No. II* is building, embodying slight variations.

DUFAUX. Triplane.

Maximum length, feet / m. **maximum breadth,** feet / m. **supporting surface,** sq. feet / m^2.

Total weight.—

Body.—Complete nacelle like a dirigible.

Wings.—Two series of triplane surfaces at dihedral angle. Corresponding biplane tail.

Motor.—

Speed.—

Propellers.—2 very large, behind front planes.

Steering.—

Remarks.—

THEROUL. Biplane.

Maximum length, feet / m. **maximum breadth,** feet / m. **supporting surface,** 512 sq. feet / 48 m².

Total weight.—

Body.—

Wings.—Span, 46 feet (14 m.)

Motor.—50 h.p.

Speed.—

Propellers.—

Steering.—

Remarks.—Reached Issy, June, 1909, for trials.

GILLY. Triplane.

Maximum length, feet / m. **maximum breadth,** feet / m. **supporting surface,** 1076 sq. feet / 100 m².

Total weight.—

Body.—

Wings. Span, feet (16 m.)

Motor.—100 h.p.

Speed.—

Propellers.—Two.

Steering.—

Remarks.—Commenced first trials Aug., 1909. Pilot for trials: M. Geurcin.

BOUSSON-BORGNIS. Triplane.

Maximum length, feet / m. **maximum breadth,** feet / m. **supporting surface,** sq. feet / m^2.

Total weight.—

Body.—

Wings.—

Motor.—34 h.p. Buchet.

Speed.—

Propellers.—

Steering.—

Remarks.—

VANNIMAN (Melville). Triplane.

Photo, J. Hausier.

Maximum length, feet / m. **maximum breadth,** feet / m. **supporting surface,** sq. feet / m^2.

Total weight.—

Body.—On 4 wheels.

Wings.—

Motor.—

Speed.—

Propellers.—

Steering.—Rudder for horizontal steering is placed forward of top and middle planes. Elevator also forward. Side tips to middle planes.

Remarks.—Several other models reported building. This one has flown successfully.

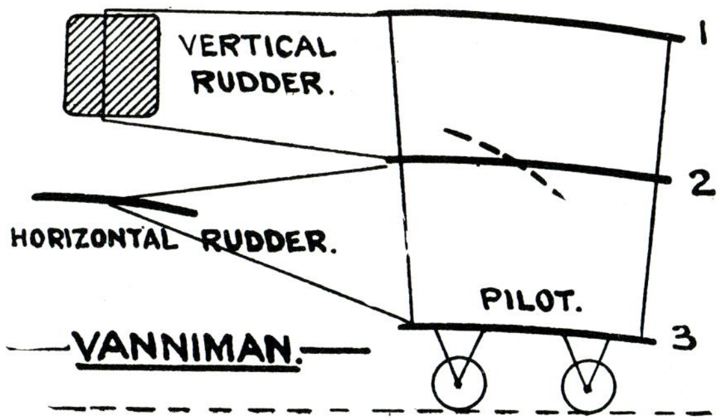

WITZIG-LIORE-DUTILLEUL.

(Nondescript.)

Photo, C. Malcuit.

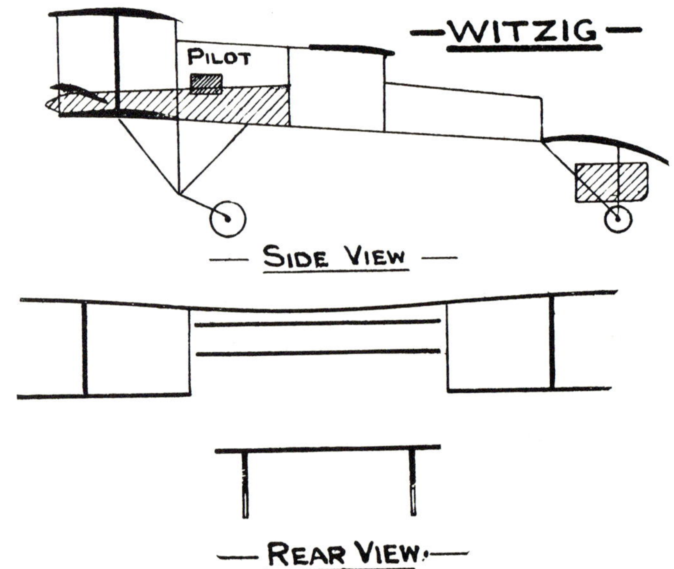

Maximum length, 37 feet / 11·27 m. **maximum breadth,** feet / m **supporting surface,** sq. feet / m^2.

Total weight.—

Body.—Chassis forward is very high. Mounted on 4 wheels.

Wings.—There are 4 *en escalier* on the upper surface. 2 forward on the lower surface for rising, etc. 2 side steadying planes have been added to these (see end-on plan). Maximum span *about* 33 feet (10 m.) Height of forward planes above chassis, 5 feet (1·5 m.)

Motor.—50 h.p. Resault.

Speed.—37 m.p.h. (60 km.)

Propellers.—One.

Steering.—2 aft, $4\frac{1}{2} \times 2$ feet ($1{\cdot}36 \times 0{\cdot}61$ m.), for horizontal steering. 2 elevator planes forward.

Remarks.—This machine, which it is impossible to classify, has been altered several times. The plans give its latest form with the third top plane removed.

PIQUEREZ. Biplane.
(*Alias* **PISCHOFF-KOECHLIN, 1909 Model.**)

Maximum length, 35½ feet / 10·80 m. **maximum span,** 35 feet / 10·60 m. **supporting surface,** 840 sq. feet / 78 m^2.

Total weight.—1,323 lbs. (600 kgs.) *including* driver and passenger.

Body.—An entirely enclosed monoplane body mounted on 4 wheels.

Wings.—Span, 35 feet (10·60 m.) Ailerons to upper plane tips.

Motor.—40 h.p. 4-cylinder Dutheil-Chalmers.

Speed.—

Propellers.—2, carried well in rear of main planes and mounted one on either side of the body.

Steering.—Biplane elevator forward. Box kite tail is a combined elevator and vertical rudder.

Remarks.—Built by Pischoff-Koechlin. As will be seen in the photograph, the passenger sits behind the pilot.

OUVIERE. Helicopter.

Building.

Maximum length, feet / m. **maximum breadth,** feet / m. **supporting surface,** sq. feet / m^2.

Total weight.—849 lbs. (385 kgs.)

Body.—

Wings.—2 lifting wings. **Lift,** 992 lbs. (450 kgs.)

Motor.—80 h.p.

Speed.—

Propellers.—

Steering.—By varying the inclination of the wings.

Remarks.—

BREGUET-RICHET II bis. Biplane-Helicopter.

Maximum length, 29½ feet / 9 m. **maximum breadth,** 46 feet / 14 m. **supporting surface,** 646 sq. feet / 60 m².

Total weight.—1,212½ lbs. (550 kgs.) *not including* aviator.

Body.—Enclosed. There are 7 wheels.

Wings.—Upper plane: span, 33 feet (10 m.); lower: span, 46 feet (14 m.) Depth of planes, 5 feet (1·50 m.)
Tail: upper plane, 33 × 5 feet (10 × 1·50 m.); lower, 19 × 5 feet (5·80 × 1·50 m.)

Motor.—

Speed.—

Propellers.—Two 4-bladed. Diameter, 14 feet (4·25 m.), inclined at 40° for rising.

Steering.—Large vertical rudder aft.

Remarks.—

VUITTON HUBER. Helicopter.

(M. G. Vuitton).

Maximum length, 14 feet / 4·25 m. **maximum breadth,** 13 feet / 4 m. **supporting surface,** sq. feet / m².

Total weight.—

Body.—Slight frame on 4 small wheels.

Wings.—2 large fans *about* 13 feet (4 m.) in diameter. Superposed direct drive.

Motor.—50 h.p. Farcot radial engine.

Speed.—

Propellers.—1 behind *about* 5 feet (1·50 m.) diameter. Gear driven.

Steering.—By inclination of driver's body.

Remarks.—

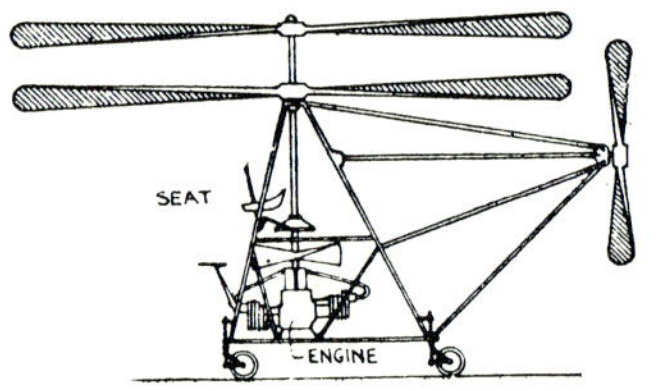

BERTIN. Helicopter.

Maximum length, feet / m. **maximum breadth,** feet / m. **supporting surface,** sq. feet / m².

Total weight.—463 lbs. (210 kgs.)

Body.—

Helicopters.—The upper is of 9·8 feet (3 m.) diameter driven direct. The lower is $25\frac{1}{2}$ feet (7·80 m.) diameter gear driven at 120 revs. p.m.

Motor.—55 h.p., 3 cylinder Bertin, 1,000 revs.

Speed.—

Propellers.—2, superposed on the same vertical axis; revolving in opposite directions.

Steering.—

Remarks.—

BAYLAC. Helicopter.

Maximum length, feet / m. **maximum breadth,** feet / m. **supporting surface,** sq. feet / m².

Total weight.—

Body.—

Wings.—2 lifting screws.

Motor.—

Speed.—

Propellers.—

Steering.—2 tubing screws, also planes.

Remarks.—

CORNU. Helicopter.

Photo, J. Hausier.

Maximum length, feet / m. **maximum breadth,** feet / m. **supporting surface,** sq. feet / m^2.

Total weight.—

Body.—On 4 small bicycle wheels.

Wings.—2 large 2-bladed fans, one forward, one aft.

Motors.—

Speed.—

Propellers.—

Steering.—

Remarks.—Has flown successfully once or twice.

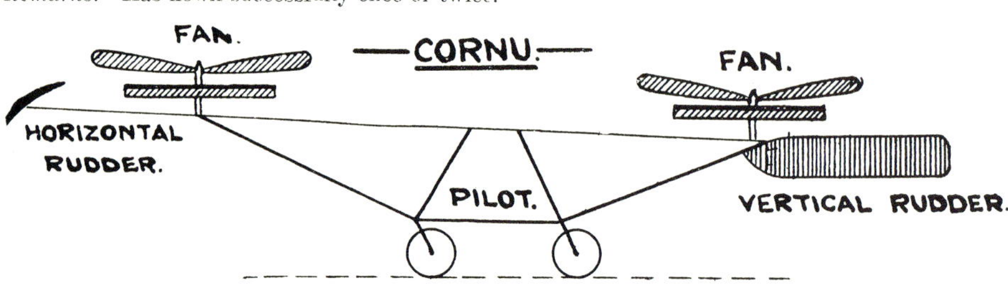

DENNISELL AND GODVELLE. Helicoplane.

Building.

Maximum length, feet / m. **maximum breadth,** feet / m. **supporting surface,** sq. feet / m^2.

Total weight.—

Body.—

Wings.—

Motor.—

Speed.—

Propellers.—

Steering.—

Remarks.—Building. Nothing known.

JUGE AND ROLLAND. Helicopter.

Building.

Maximum length, feet m. **maximum breadth,** feet m. **supporting surface,** sq. feet m^2.

Total weight.—

Body.—

Wings.—

Motor.—

Speed.—

Propellers.—

Steering.—

Remarks.—

OBRÉ. Biplane.

Photo, Rol.

Maximum length, $29\frac{1}{2}$ feet 9 m. **maximum breadth,** feet m. **supporting surface,** sq. feet m^2.

Total weight.—

Body.—Very light girder frame, mounted on two rather large, heavily tyred wheels in front, and a very small wheel amidships.

Wings.—Lower plane slightly curved upwards at edges. Has a span of 33 feet (10 m.) Small ailerons at the wing tips. The upper plane is only $11\frac{1}{2}$ feet (3·50 m.) span.

Motor.—50 h.p. 3-cylinder Anzani

Speed.—

Propellers.—One 2-bladed. Diameter *about* 7 feet (2·10 m.)

Steering.—Vertical rudder of special shape amidships. Single horizontal rudder right aft.

Remarks.—The small upper deck is the special feature of this machine. Met with an accident in its first trials.

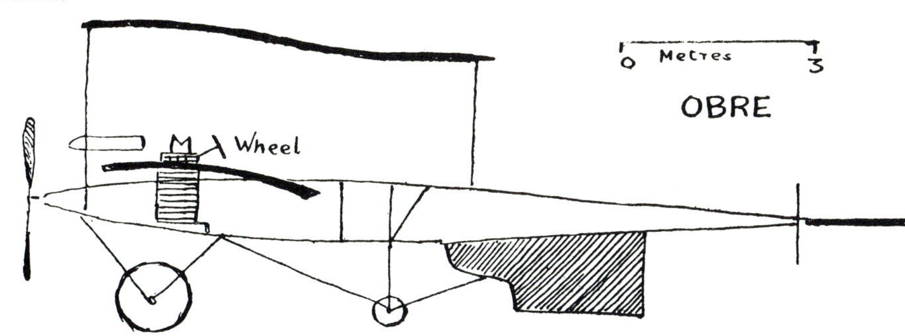

GIVAUDAN.

Maximum length, feet / m. **Maximum breadth,** feet / m. **Supporting surface,** sq. feet / m^2.

Total Weight.—

Body.—Girder. Motor just abaft front drums. Aeronaut abaft the motor *(see photo).* Mounted on 4 wheels: the 2 front ones secured to body, the 2 rear ones to rear outer drum.

Wings.—2 concentric drums at either end joined by a triangular girder.

Motor.—8 cylinder 40 h.p. Vermorel-V shape. 90 m/m bore, 120 m/m stroke. Weight of engine, 176$\frac{1}{3}$ lbs. (80 kgs.) (*about* 4$\frac{1}{2}$ lbs. per h.p.) Air-cooled. This engine has a single cam shaft: valves in a hemispherical combustion chamber. This engine has run several hours without overheating.

Speed.—

Propellers.—1 in front. Diameter, 7 ft. 10 in. (2·40 m.) Gear driven.

Steering.—By slewing the front drums.

Remarks.—The advantage of this system is that side gusts of wind have no effect whatever, and that heeling over when turning is impossible.

Property of the Vermorel Cie, Villefranche, Rhone.

FRENCH DIRIGIBLES.

LEBAUDY CLASS.

DISTINCTIVE CHARACTERISTICS:

The cars are short and suspended from a long keel which is suspended close up to the gas bag, and is mostly covered in with fireproof canvas.

The rear end of the keel is expanded into fixed vertical and horizontal fins, and carries a vertical and a horizontal rudder.

The rear end of the gas bag is fitted with thin fixed planes (compare with the pear shaped or tubular fins of the "Astra" Class).

The cars are provided underneath with an extraordinarily strong conical structure, which takes the shock of striking the ground and distributes it over the whole car.

Aeroplanes are now fitted, one each side of the keel, well forward.

Ships of this class:—

LEBAUDY I	***French Military Air-ship.***
LEBAUDY II	***,, ,,***
PATRIE	***Lost in a storm.***
REPUBLIQUE	***French Military Air-ship.***
LA RUSSIE	***Sold to Russian Government.***
LIBERTÉ	***French Military Air-ship.***
ONE "PROJECTED"	***,, ,,***

LEBAUDY II *ex* I.
(LEBAUDY I rebuilt.)

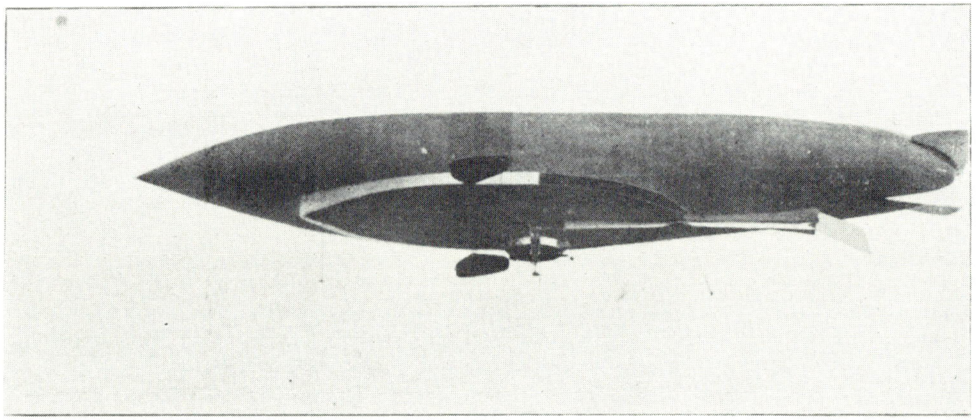

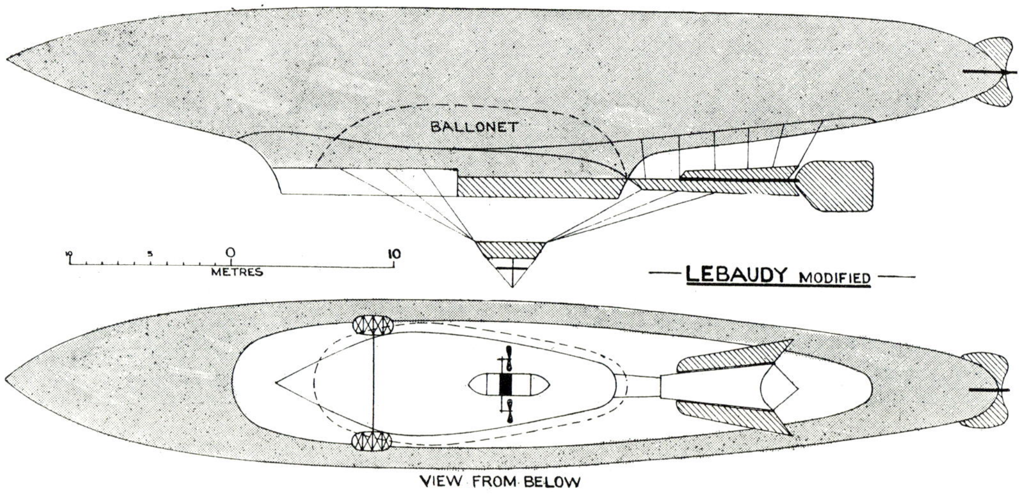

Maximum length, 188·65 feet. / 57·5 m. **Maximum diameter,** 34¼ feet. / 10·38 m. **volume,** 116,545 c. feet / 3,300 m^3.

Total Lift.—9325 lbs. (3,630 kgs.) **Useful lift,** lbs. (kgs.)

Gas Bags.—Lebaudy material as in *Republique.* Ballonet, 233 c. feet (6·60 m^3.)

Motors.—70 h.p. 4 cylinder Panhard.

Speed.—28 m.p.h. (46·45 km. p.h.)

Propellers.—2, one each side of the car; each is fitted with a clutch and also with a brake so as to be able to stop it instantly on touching the ground if a man should get dangerously near.

Steering.—Horizontal rudder and vertical rudder in rear of the keel. An aeroplane each side of the keel, well forward.

Remarks.—This ship has been altered repeatedly, particularly the dimensions of the gas bag. She has made many long voyages, and can go at full speed for 10 to 12 hours.

REPUBLIQUE (*Wrecked 1909*).

Photo, R. Tuck.

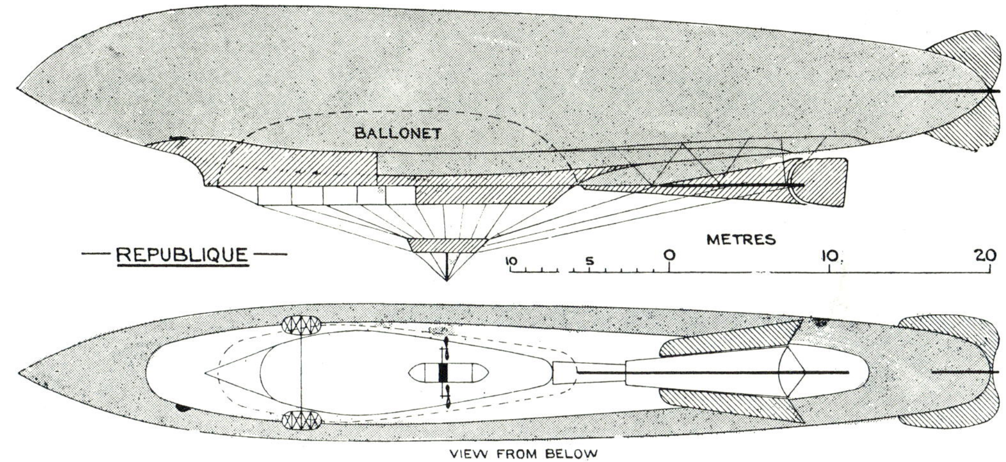

Belongs to the French Government.

Length, 200 feet. / 61 m. **diameter,** 35½ feet. / 10·8 m. **volume,** 131,000 cubic feet. / 3,700 m³. **total lift,** 9,000 lbs. / 4,080 kg. **useful lift** (fuel, passengers, etc.), 2,700 lbs. / 1,225 kg.

Ballonet.—23,000 cubic feet. / 650 m³. Divided into three compartments by two transverse perforated bulkheads of balloon material.

Gas Bag Material.—Continental rubber cloth.

(*a*) Outer layer of cotton cloth covered with lead chromate	3·25	oz. per sq. yard.	108·3	grammes per sq. metre.
(*b*) Layer of vulcanised rubber	3·25	,, ,,	108·3	,, ,,
(*c*) Layer of cotton cloth	3·25	,, ,,	108·3	,, ,,
(*d*) Inner layer of vulcanised rubber ...	0·73	,, ,,	24·3	,, ,,
Total weight of material	10·48	,,	349·3	,, ,,

Motor.—60-70 h.p., 4-cylinder Panhard & Levassor. 1,000 r.p.m. Fuel carried, about 60 gallons.

Propellers.—2, one each side of the engine, turning in opposite directions at 1,200 r.p.m.

Speed.—30 miles per hour = km. per hour.

Car.—Boat shaped, of nickel steel tubes, cloth covered, 16 feet (4·88 m.) long, 5 feet (1·5 m.) wide, 2½ feet (·76 m.) high, 11 feet (·3 m.) below the gas bag.

Suspension.—The car is hung by 30 steel cables to an elliptical frame of nickel steel tubes. This frame is secured by toggles to a hempen net, which is hung by toggles from a canvas band sewn to the gas bag.

LIBERTÉ (French Army).

(Lebaudy Type.)

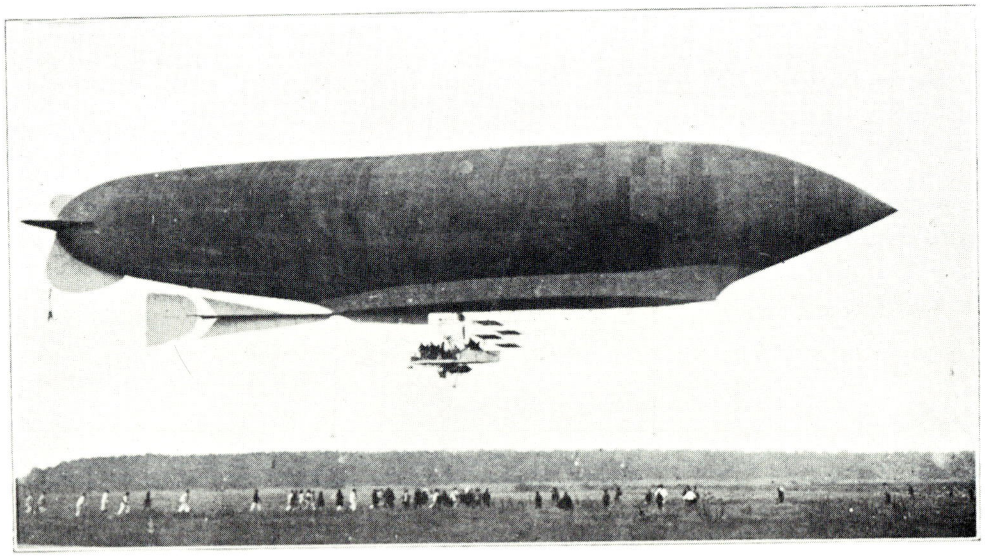

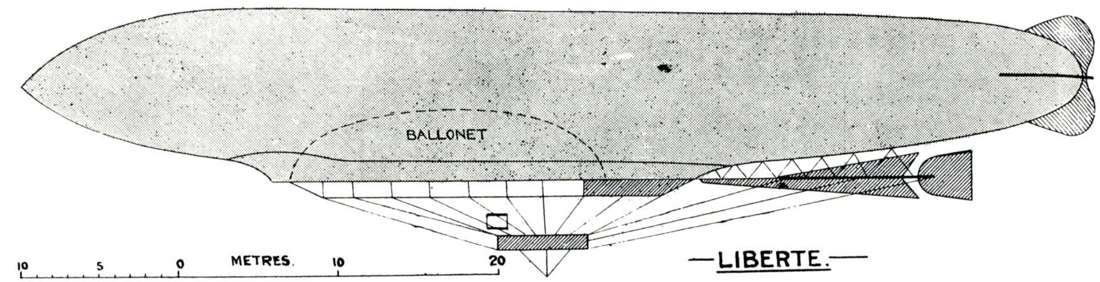

Maximum length, 219·8 feet. (67 m.) **maximum diameter,** $35\frac{3}{4}$ feet. (10 m., 8.) **volume,** 148,333 c. feet. (4,200 m³.)

Total lift.—10,185 lbs. (4,620 kgs.) **Useful lift,** lbs. (kgs.)

Gas bags.—Made of Lebaudy stuff, 2 layers of cotton with a layer of rubber in between, a layer of rubber on the inside, and a coating of yellow lead chromate on the outside. Ballonet about one-fifth of the total capacity, of slightly lighter material.

Motor.—4-cylinder Panhard, 135 h.p.

Speed.—31·3 m.p.h. (50·3 km.)

Propellers.—Two 2-bladed wooden propellers, one on either side of the car, and somewhat above it. Driven by shaft gearing. Each propeller is fitted with a clutch and brake.

Steering.—Large single balanced vertical rudder in rear of the keel and fins. Horizontal aeroplanes above the fore part of the car.

Remarks.—The car is of steel, thinly plated in front, and covered with wire netting in rear. The suspension is of the usual Lebaudy type, by many cords from a rigid keel, which is supported by further cords from suspension strips each side. The keel and its suspensions are covered with fireproof material.

FRENCH DIRIGIBLES.

ASTRA CLASS.

Dirigibles of the Astra class are of the Non-Rigid type, the weights being distributed by means of a long girder hung under the gas bag.

The only feature possessed by this type alone is the use of inflated stabilising shapes at the rear end of the balloon.

These shapes were originally cylindrical in the *Ville de Paris*. In the *Clement-Bayard I*, pear shaped bags were substituted for these cylinders, and in succeeding ships these shapes have been given finer and finer lines.

The use of a long girder enables unusually generous accommodation to be provided for the passengers and crew, but makes the ships less easy to transport by waggon than most other Semi or Non-Rigid types, although the girder is made to take to pieces.

VILLE DE PARIS.
(Astra Type.)

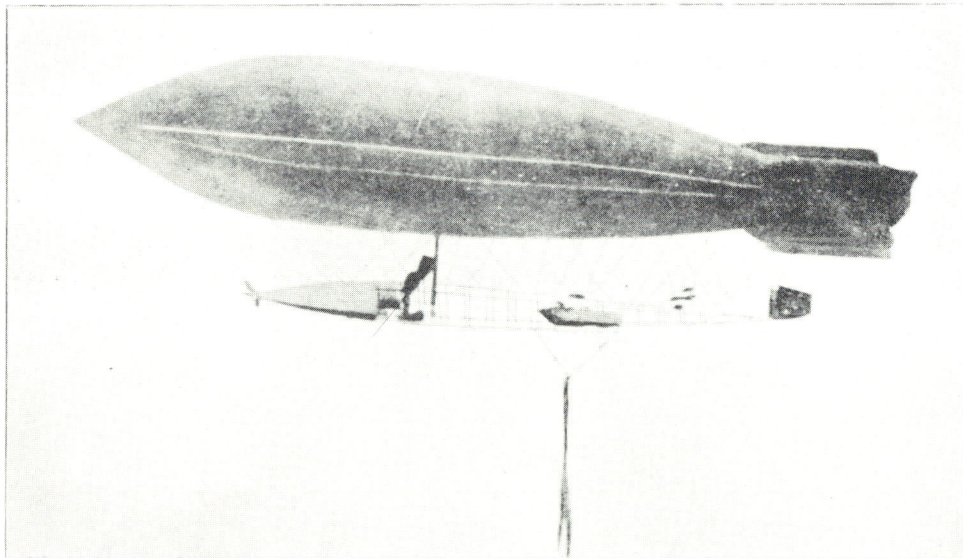

Photo, J. Hauser.

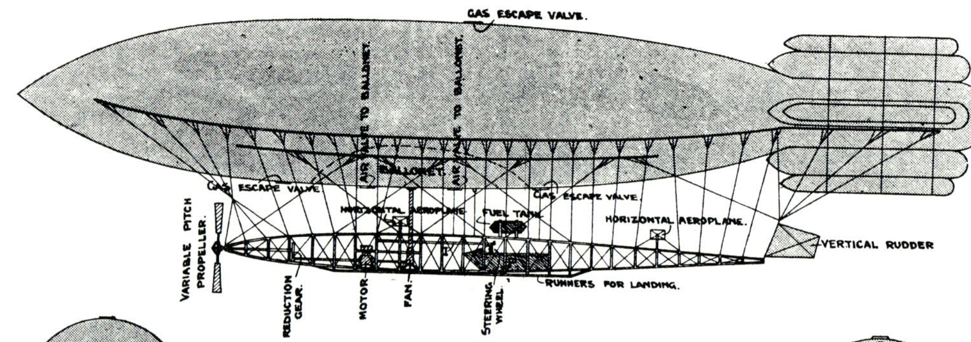

— SIDE ELEVATION —

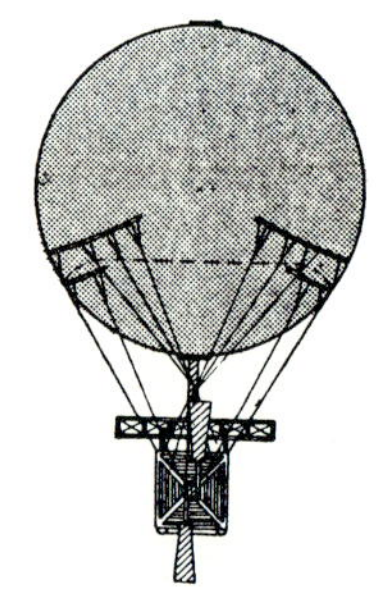

— FRONT VIEW —

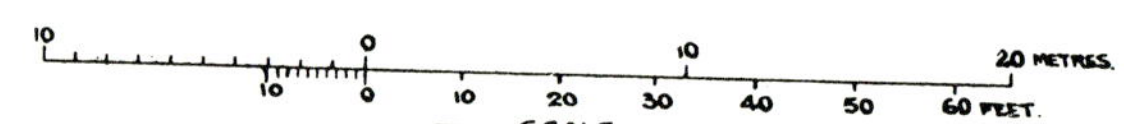

— SCALE —

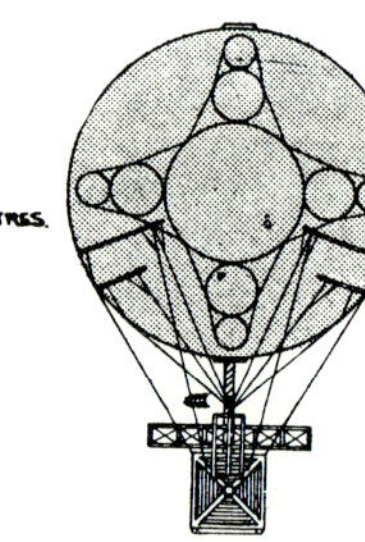

— BACK VIEW —

P. = Pilot's station.
I. = Speaking tube from pilot to engine.

Maximum length, 198½ feet. (60 m., 42.) **maximum diameter,** 34·3 feet. (10 m., 50) **volume,** 112,836 c. feet. (3,195 m³.)

Total lift.—7,760 lbs. (3,520 kgs.) **Useful lift,** lbs. (kgs.)

Gas bags.—Continental fabrics, 2 layers of cotton with a layer of rubber between them and another on the inside. Coated with yellow lead chromate outside. Surface, sq feet (1,828 m².) Weight, 1,827½ lbs. (820 kgs.) Stabilising fins consist of 4 sets of 2 cylinders each, the inner cylinders being 31 feet (9 m., 46) long by about 5 feet (1 m., 60) diameter, the outer cylinders slightly shorter and about 3 (0 m., 93) diam.

Motor.—1 Chenu motor, 70 h.p. (900 r.p.m.)

Speed.—25 m.p.h. (40·2 k.m.)

Propellers.—One 2-bladed propeller at the front of the car, 19·68 feet (6 m.) in diameter. 180 r.p.m.

Steering.—Double vertical rudder in rear of and just above the car. 2 double horizontal aeroplanes above the car, one-third from forward, the other one-quarter from aft.

Remarks.—Ballonet 17,650 (500 m³.) filled by a fan, working normally at 1,530 r.p.m., designed to deliver 254,283 c. feet (7,200 m³.) of air per hour at 30 mm of water pressure. This historic airship was built for the private use of M. Deutsch, and was presented by him to the French Military Authorities after the Military Airship *Patrie* had been lost.

VILLE DE BORDEAUX. *(Astra Type.)* *VILLE DE NANCY* the same.

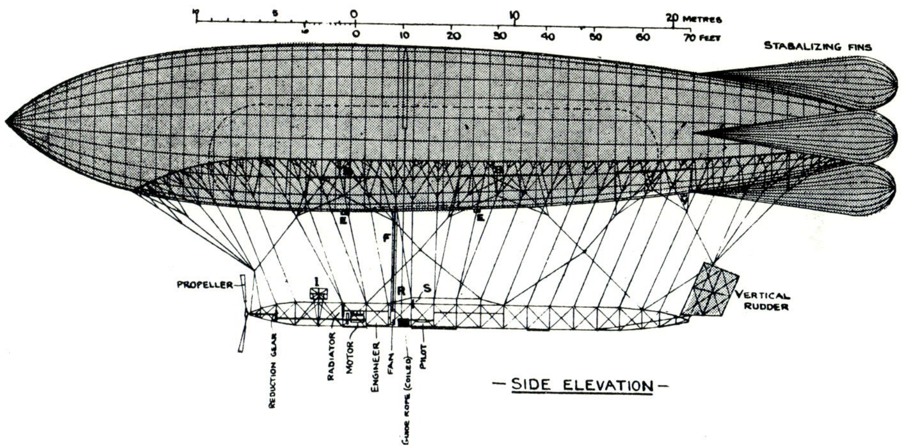

— SIDE ELEVATION —

Maximum length, 173·8 feet. (53 m.) **maximum diameter,** 32·8 feet. (10 m.) **volume,** 105,950 c. feet. (3,000 m³.)

Total lift, 7,275 lbs. (3,300 kgs.) **Useful lift,** lbs. (kgs.)

Gas bags.—Continental balloon material, chrome yellow. Weight, 1,653½ lbs. (750 kgs.) Pear-shaped stabilising fins at the rear end, connecting with the interior of the gas bag by small holes. Ballonet, 22,955 c. feet (650 m³.)

Motor.—1 Renault, 80 h.p., 4-cylinders. Mounted on 4 quarter-elliptic leaf springs placed transversely and shackled to car.

Speed.—33 m.p.h. (53 km. p.h.)

Propellers.—1, in front of the car; 16·4 feet (5 m.) diameter; 12 feet (3·6 m.) pitch. Driven at 360 r.p.m. through reduction gearing.

Steering.—Double vertical rudder of 193½ sq. feet (18 m².) in rear of and just above the car. Treble horizontal aeroplane 172 sq. feet (16 m²) just above the car and before the motor.

Remarks.—Steel girder car, similar to that of *Bayard-Clément.* Suspension by means of wires adjustable in the car, which secure to a wood-cored suspension strip which is sewn to the gas bag. Ballonet pressure 25 mm. Valves open at 30 mm. This ship is very similar in all details to the *Bayard-Clément.* She differs conspicuously from the *Ville de Paris* in the shape of her stabilising fins. *Ville de Nancy* has been bought by the Societeé de Transport Aerienne.

COLONEL RENARD.

(Astra class.)

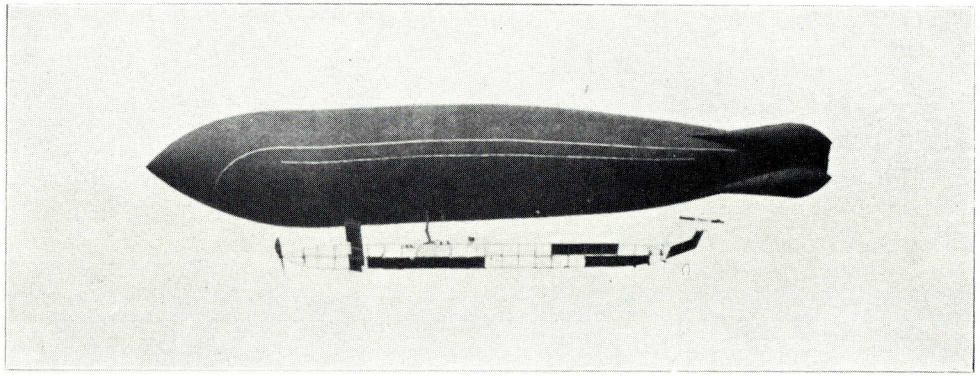

Maximum length, 197 feet / 60 m. **maximum diameter,** $35\frac{1}{3}$ feet / 10·75 m. **volume,** 43,057 c. feet / 4,000 m^3.

Total lift.—9,700 lbs. (4,400 kgs.) **Useful lift,** lbs. (kgs.)

Gas bags.—Yellow coloured rubber proofed Continental fabric.

Motor.—One 100 h.p. 4-cylinder Panhard.

Speed.—29 m.p.h.

Propellers.—1, at the front end of the car, of wood, "Integrale" type.

Steering.—As in *Clement Bayard I* and *Ville de Nancy.*

Remarks.—The two side stabilising shapes are duplicated, as they are in the *Ville de Paris.* A webbing stretched on steel tubes is introduced between the inner edges of the 4 main stabilising shapes to provide extra stabilising surface.

CAPPAZZA Type.

(To be built by Astra, works to the design of M. Cappazza.)

It appears doubtful whether this design will be proceeded with, owing to the danger from rain.

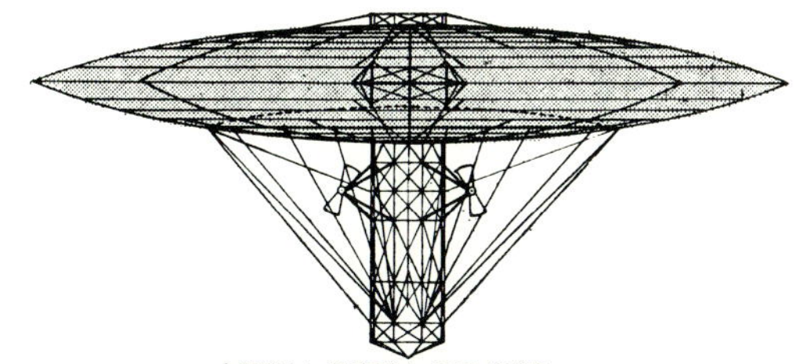

— VIEW FROM BEHIND —

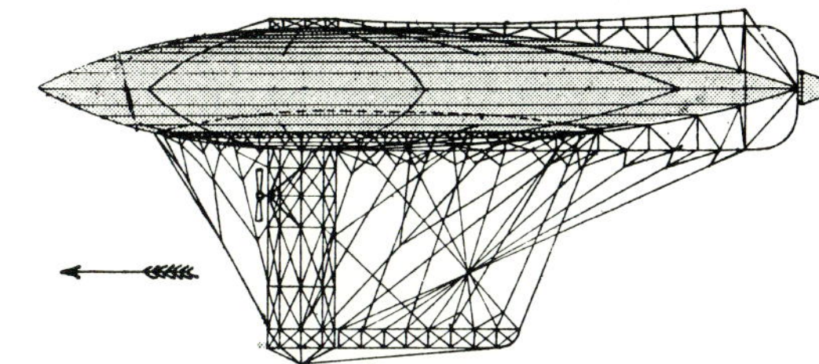

— SIDE ELEVATION —

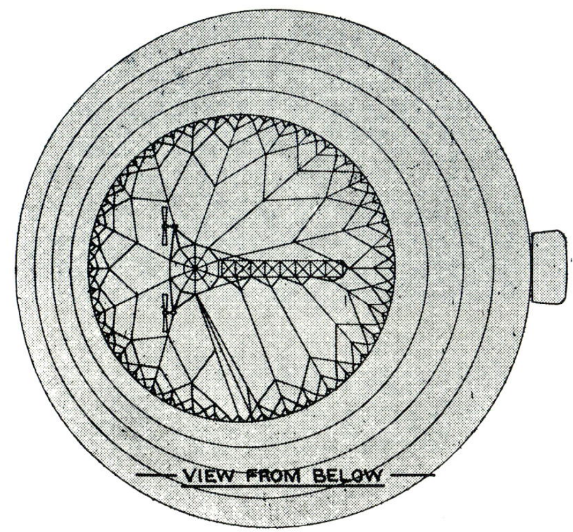

— VIEW FROM BELOW —

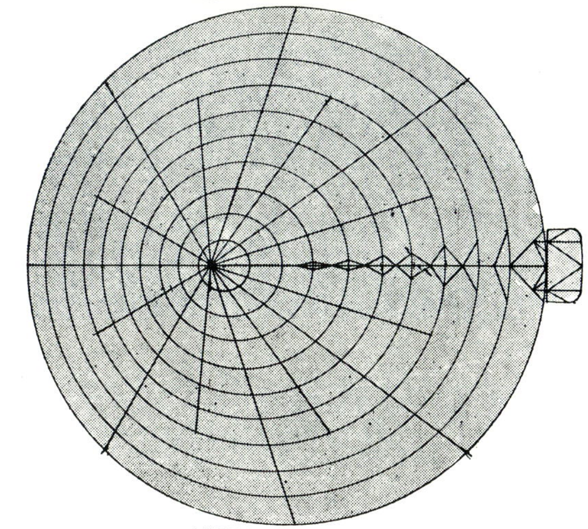

— VIEW FROM ABOVE —

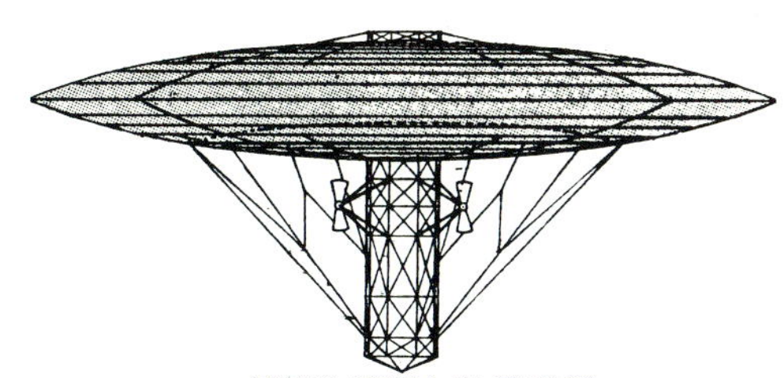

— VIEW FROM IN FRONT —

Max. length, 141½ feet / 43 m. **max. width,** 137½ feet / 41 m. **max. thickness,** 21·32 feet / 6·5 m. **max. height,** 58.23 feet / 17·75 m. **volume,** 332,612 c. ft. / 9,418 m3.

Total lift.—22,843 lbs. (10,360 kgs.) **Useful lift,** lbs. (kgs.)

Gas bags.—Of Continental fabric, but shaped like a circular plaice, or like two saucers placed with their edges together. Ballonet of 67,102 c. feet (1,900 m3.) Gas bag and ballonet both divided into compartments.

Motors.—Two 123 h.p. Clement, placed in the centre line of the keel; they are coupled together and drive a single bow propeller when both motors are in use. When only one is in use it is coupled to two propellers carried on brackets on each side of the keel tower, close under the gas bag.

Speed.—Problematical. 31 m.p.h. (50 km.p.h.) hoped for.

Propellers.—3. The bow one direct driven by both motors; the other two being close under the gas bag, on either side of the keel tower, and gear driven from either motor. All three can be used simultaneously. Efficiency of large propeller, 77%. Efficiency of upper propellers, 66%. Diameter, 6·88 feet (2·1 m.)

Steering.—Horizontal rudder at the rear of the circumference of the gas bag. Vertical rudders on either side of the horizontal rudder. No horizontal fins. Vertical fin in the midship line, above and below the gas bag, in the rear part.

Keel and car.—Below the gas bag, and slightly before the centre, is a 10-sided steel girder tower, 32·80 feet (10 m.) high. From its rear lower part a girder car extends aft for 29·52 feet (9 m.) The top of the tower is suspended from the gas bag by short wires. Further wires support the tower at every point, and the base of the tower and the car are suspended from a large number of crow's feet sewn all round below the rim of the flat gas bag.

Remarks.—The object of adopting this entirely original shape is to obtain a very large aeroplane effect while presenting little resistance to forward motion. The resistance of the keel and suspension appears excessive, but otherwise the design seems promising. The crew consists of 6 men, and the total weight, including crew, is 12,125 lbs. (5,500 kgs.), leaving 10,229 lbs. (4,646 kgs.) for fuel, food, and ballast.

FRENCH DIRIGIBLES.

ZODIAC CLASS.

These dirigibles are intended specially for private pleasure purposes. Consequently their cost is kept down to a minimum—about £700—and they are designed to fly when filled with coal gas if necessary.

Every effort is made to render them easily transportable; the long girder frame by which the weight is distributed is made to take to pieces.

It is held in France that numbers of this class of vessel would form an invaluable asset in time of war, as each could be transported in a single cart, filled with a very few bottles of hydrogen, and when so filled could manœuvre for some 6 hours at a speed which compares favourably with that of the standard types.

When the service of reconnaissance was performed, the vessel could be packed up and sent out of harm's way in an hour, whereas this could scarcely be done with a larger vessel on account of the quantity of hydrogen that would be required if it had to be filled afresh for each service. The mooring of an airship in the open during war requires such an amount of preparation and attention as to be a serious drawback to the alternative plan of keeping such vessels filled, while the sending of an airship back to its distant shed on each occasion means doubling the work that the ship is called upon to perform.

PETITS-JOURNAL I (or ZODIAC I).
(Zodiac Type.)

Designed by "La Societé francaise des Ballons Dirigeables."

Maximum length, 98½ feet / 30 m. **Maximum diameter,** 23 feet / 7 m. **Volume,** 24,722 c. feet / 700 m³.

Total lift.—1,697½ lbs. (770 kgs.) if filled with hydrogen. **Useful lift,** lbs. (kgs.)

Gas bag.—Cotton, varnished outside and inside. 5,861 sq. feet (550 m².) of surface. Fitted with 3 manœuvring valves and a ripping strip.

Motor.—Clerget, 16 h.p., 4-cylinder. Weight, 220½ lbs. (100 kgs.)

Speed.—*Circa* 15½-17 m.p.h. (25-28 km. p.h.)

Propellers.— At the rear of the car. 8·2 feet (2 m. 30) in diameter. Pitch 2·95 feet (0·90 m.)

Steering.—Horizontal rudder in front of the car. Vertical rudder in rear, between the gas bag and the car.

Remarks.—Built by Maurice Mallet. The main feature is the ease with which the ship can be dismantled and transported by cart or train. Further, it is designed to fly when filled with coal gas, which reduces the cost of filling enormously. It will carry 1 man at full speed for 3 hours, or 2 men if a mixture of coal gas and hydrogen is used. Three-quarter hour required to deflate it and pack it up. The car divides into 3 parts. Fuel for 3 hours carried and 165 lbs. (75 kgs.) of ballast. Cost about £1,000.

DE LA VAULX I (or ZODIAC II).

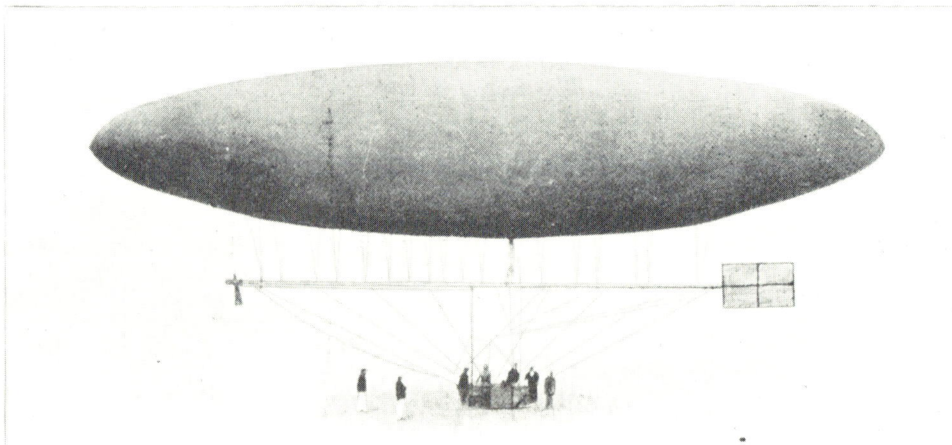

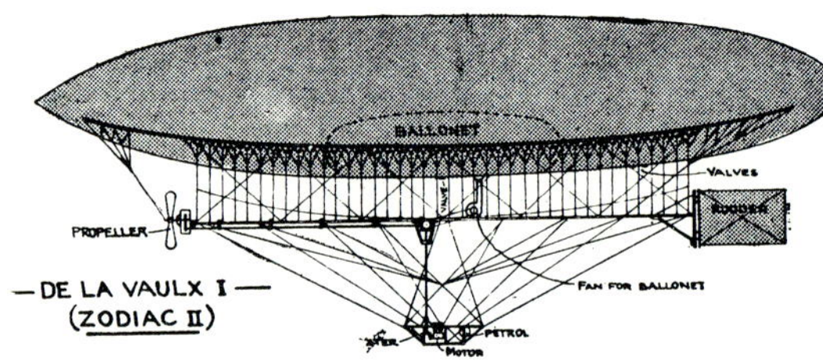

— DE LA VAULX I —
(ZODIAC II)

Maximum length, 106 feet / 32 m., 30 **maximum diameter,** 21 feet / 6 m., 50 **volume,** 31,785 c. feet / 900 m³.

Total lift.—2,182½ lbs. (990 kgs.) **Useful lift,** lbs. (kgs.)

Gas bags.—French silk, varnished.

Motor.—45 h.p. Ader.

Speed.—21¾ m.p.h. (35 km. p.h.)

Propellers.—1, in front of the keel front which the car is hung.

Steering.—Horizontal rudder consisting of 2 wings. Vertical rudder abaft the keel.

Remarks.—Built in 1906. Has since been reconstructed and enlarged. Is of the Zodiac type, and so can be rapidly deflated and taken to pieces. Has recently been further enlarged to 1,000 m³. The motor has also been changed for experiment, and yet another motor is about to be fitted.

ZODIAC III.

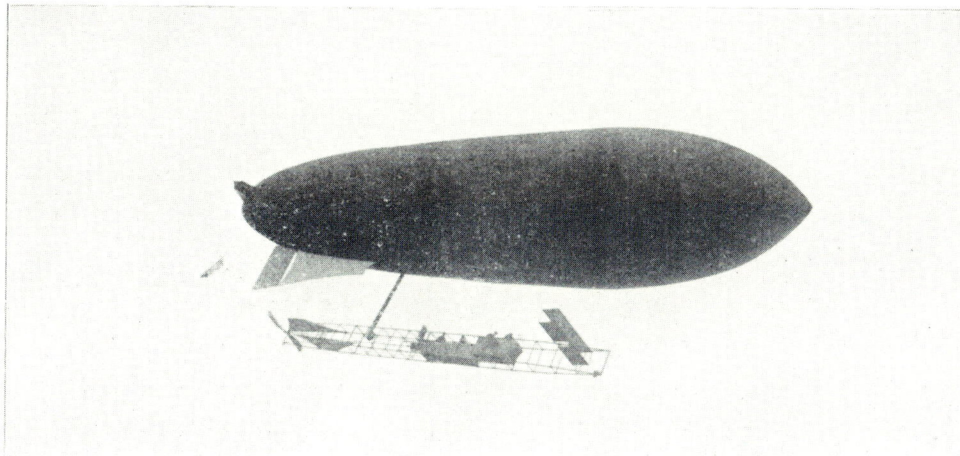

Maximum length, 134 feet / 40·8 m.; **maximum diameter,** 28 feet / 8·5 m.; **volume,** 49,443 c. feet / 1,400 m³.

Total lift.—3,395 lbs. (1,540 kgs.) **Useful lift,** lbs. (kgs.)

Gas bags.—Light continental rubbered fabric.

Motor.—Ballot, 4-cylinder, 40-45 h.p., 1,200 r.p.m.

Speed.— m.p.h. (45 km.p.h.)

Propellers.—Driven at 600 r.p.m. Integral type, 12¼ feet (3·75 m.) in diameter. Pitch, 6½ feet (2 m.) In rear of the car.

Steering.—Vertical balanced rudder in rear of the vertical fin, under the rear of the gas bag. Double horizontal rudder above the fore end of the car. Horizontal fins of material spread on iron frames on either side of the rear end of the car.

Remarks.—The car consists of a 130 feet (40 m.) long wooden girder, which can be divided into 4 separate parts of 13 feet (4 m.) each. The suspension is by steel wires fitted with adjusting screws at the lower ends and toggles at the upper ends, by which they connect to the crows' feet which are sewn to the suspension strips. This number *III* marks a great advance on number *II*, but the result is merely obtained by increase of volume.

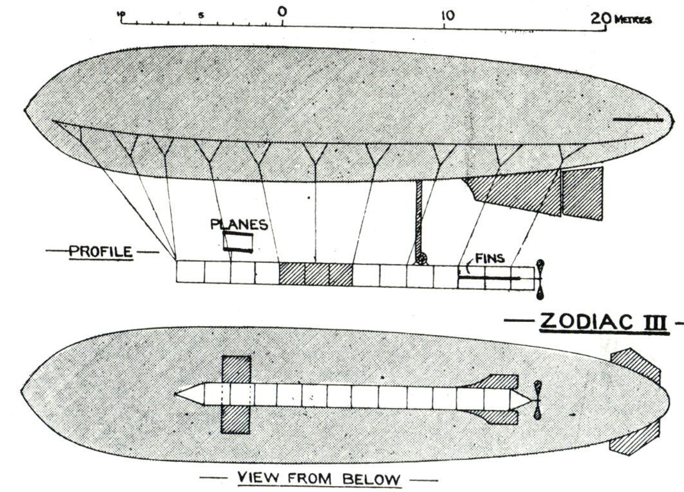

DETAILED WEIGHTS OF *ZODIAC III*.

	kgs.	lbs.
Gas bag (*including* ballonet)	330	727½
Valves	12	26½
Suspension wires and gear	15	33
Tail fins	24	53
Horizontal rudder	10	22
Vertical rudder	10	22
Girder car	168	370¼
Motor (*including* pump, magneto, lubricating gear, etc.)	275	606¼
Motor bearer and gear	22	48½
Petrol tank	10	22
Radiator	25	55
Reduction gearing	12	26½
Shafting	15	33
Fan	9	20
Steering gear	5	11
Water	8	17¾
Petrol	20	44
Miscellaneous : 4 men	300	661
Total	1,270	2,800 *about*
Ballast	270	595
Total weight	1,540	Total lift 3,395

MISCELLANEOUS FRENCH DIRIGIBLES.

FAURE.

(Zodiac pattern.)

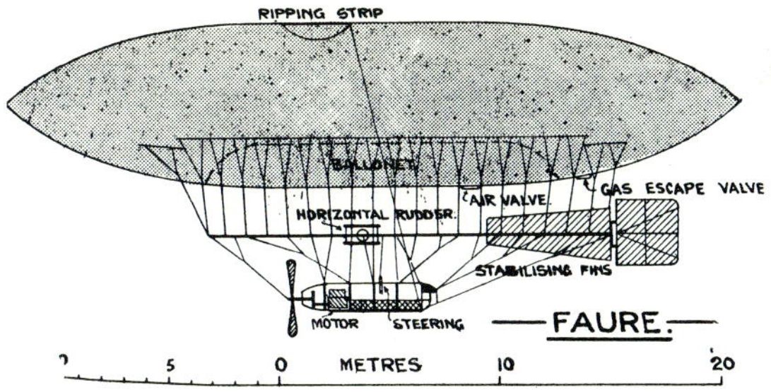

Maximum length, 108·3 feet. / 33 m. **maximum diameter,** 24 feet. / 7·35 m. **volume,** 36,553 c. feet. / 1,035 m³.

Total lift.—2,293 lbs. (1,140 kgs.) **Useful lift,** lbs. (kgs.)

Gas bags.—Silk, varnished. Ballonet, about 6,180 c. feet (175 m³.), fitted with automatic valves.

Motors.—Buchet, 28 h.p.

Speed.—About 20 m.p.h. (32 km.)

Propellers.—In front of the car $10\frac{3}{4}$ feet (3·2 m.) diameter. 13 feet (3·9 m.) pitch. 340 r.p.m.

Steering.—Vertical rudder in rear of keel, 86 sq. feet (8 m².) Horizontal double rudder mounted on the centre of the keel, 70 sq. feet (6·5 m².) Fins at the rear of the keel, 118·4 sq. feet (11 m².)

Remarks.—The private property of M. Jacques Faure. The car is of steel tubes, $61\frac{1}{2}$ feet (18·75 m.) long, square in section, the sides being 4 feet (1·20 m). It is suspended by wires from the keel, a beam $61\frac{1}{2}$ feet (18·75 m.) long, which is suspended from strips sewn to the gas bag. Total weight, 1,499 lbs. (680 kgs.). Made to take completely to pieces. It is designed to fly with coal gas. The lift here given is with hydrogen.

MALECOT.

Photo, J. Hauser.

Maximum length, $111\frac{1}{2}$ feet / 34 m. **maximum diameter,** $24\frac{1}{4}$ feet / 7·4 m. **volume,** 35,317 c. feet / 1,000 m³.

Total lift.—2425 lbs. (1,100 kgs.) **Useful lift,** lbs. (kgs.)

Gas bags.—Rubbered fabric.

Motor.—30 h.p. 4-cylinder. 1,200 revs. per min.

Speed.—20 m.p.h. (32 km.)

Propellers.—One 2-bladed, in rear of the car, driven at 400 r.p.m.

Steering.—Vertical steering by means of a small basket hung some 60 feet below the keel, and suspended from its two ends. Horizontal steering by means of a vertical rudder in rear of the heel.

Remarks.—The feature of this ship is the large horizontal surface which she exposes. Below the gas bag, close up, is slung a wooden girder, of triangular section, base upwards. It is on the base of this girder that the horizontal surface is spread. The car is built on below this girder.

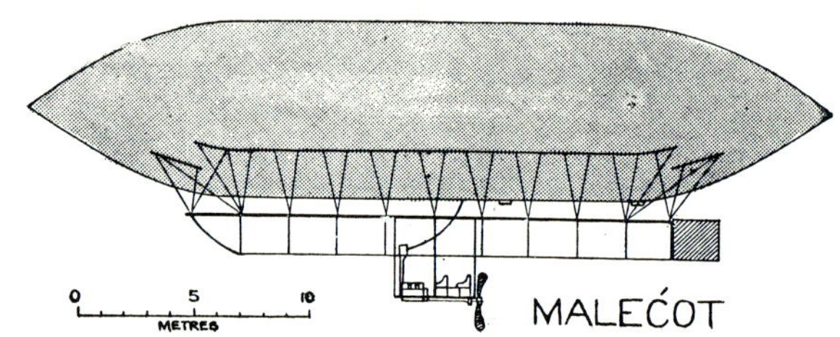

MARCAY.

Photo, J. Hauser.

Maximum length, 100 feet / 30·47 m. **Maximum diameter,** $12\frac{1}{2}$ feet / 3·80 m. **Volume,** 12,000 c. feet / 340 m^3

Total lift, 1852 lbs. (840 kgs.) **Useful lift,** lbs. (kgs.)

Gas bags.—Varnished silk.

Motor.—Only experimental motors have as yet been fitted. The motor is carried under the central division, very close up.

Speed.—

Propellers.—(See *Remarks*).

Steering.—Single vertical rudder in rear of the keel. Small horizontal fins fixed to the rear of the gas bag.

Remarks.—The gas bag is divided into 2 entirely separate halves, terminating in a large "bicycle" wheel with a rim of steel tubing. These two wheels are connected by 4 U-shaped steel connections, the bases of the U's being outwards. Between these wheels revolves another wheel carrying 4 propeller blades. By this means a very large blade area can conveniently be obtained. This model is purely experimental. The dimensions are too small to admit of really useful results.

Sister Ship to LA FLANDRE. (for La Compagnie Transaérienne).

(Astra class.) *Transaérien II.*

Maximum length, 246 feet / 75 m. **maximum diameter,** $48\frac{1}{4}$ feet / 14·7 m. **volume,** 69,914 c. feet / 6,500 m³.

Total lift.—15,763 lbs. (7,150 kgs.) **Useful lift,** lbs. (kgs.)

Gas bags.—Continental rubbered fabric, yellow. Ballonet, 16,146 c. feet (1,500 m³.)

Motors.—2 Pipe motors of 110 h.p. each, placed in line with each other in the fore and aft line, and with clutches and the necessary gearing in between them.

Speed.—35 m.p.h.

Propellers.—3, namely, one at the fore end, driven by the two motors when coupled together, and two placed above and on either side of the centre of the car, for use when only one motor is running. Made of wood, and of the "Integrale" type.

Steering.—Vertical steering by means of a large double aeroplane fixed above the car, about a third from the front. Horizontal steering by means of a double vertical rudder above the rear end of the car. Stability is secured by the usual Astra pear shaped stabilising gas bags, with fins of rubbered cloth spread between the inner edges of these shapes.

Remarks.—The distinctive feature of this ship is the arrangement of the propellers. Both motors can be coupled either on to the front propeller or on to the two rear propellers, or on to all three together, but they are actually intended only to drive the front one. On stopping either motor the other is connected to the two rear propellers, which are designed for a slower speed of translation than the front one, with the result that the running motor does not find itself overloaded as it would if the same propeller had to serve both for one and for two motors.

GERMAN.

General Note.—In Germany the greatest attention has been bestowed upon dirigibles, and heavier-than-air machines have been, comparatively speaking, neglected. Few of those in existence appear to have flown.

Aerial Societies:—

Deutscher Luftshiffer Verband (10,235 members).
Berliner Verein für Luftshiffahrt.
Münchener „ „
Hanover Aeronautical Club.
Oberrheinischer Verein für Luftshiffahrt.
Augsburger. „ „
Niederrheinischer „ „
Posener „ „
Ostdeutscher „ „
Frankischer „ „
Mittelsheinisher „ „
Kolner Klub für Luftshiffaht.
Physikalischer Verein im Frankfort.
Motorluftshiffstudiengesellschaft.
Rhine Aero Club.
German Aerial Navy League.

Aerial Journals:—

Illustrierte Aeronautische Mitteilungen.
Flug Motor Technik.

Dirigible Hangars at

Cologne.
Metz.
Frankfurt.
Strasburg.
Konigsberg.
Buhn.
Berlin.
Friedrichafen.

Flying Grounds:—

Near Berlin.
Heilbroun.
Crailshenn.

Dirigible Services to be installed:—

1. Dresden—Chemnitz—Wurzburg—Monaco.
2. Monaco—Ulm—Stuttgard—Mannhei n—Coblenz—Cassel.
3. Berlin—Bremen—Hamburg—Lubeck.
4. Berlin—Stetten—Danzig—Konigsberg.
5. Strasburg—Metz—Frankfurt—Magdburg—Berlin.

PARSEVAL (Major). Monoplane.

Completing.

Maximum length, 39⅓ feet / 12 m. **maximum breadth,** 46 feet / 14 m. **supporting surface,** 450 sq. feet / 40 m²:

Total weight.—2,205 lbs. (1,000 kgs.) *with* passengers.

Body.—Tubular steel, extending right back to the tail. Long float under centre and a couple each side.

Wings.—Bent up at front and back extremities, connected to frame by springs.

Motor.—100 h.p. Daimler.

Speed.—

Propellers.—Two.

Steering.—

Remarks.—Built for approval of the German War Office. Preliminary trials, September, 1909. Designed to carry three. Intended to rise off water.

FRITZCHE (Leutnant). Monoplane.

[Leut. Fritzsche has purchased either the Farman monoplane or a duplicate of it.]

(*For details see French Aeroplanes.*)

STAECKEL. Monoplane.

Photo, Consée.

Maximum length, 20 feet / 6 m. **maximum breadth,** feet / m. **supporting surface,** 172 sq. feet / 16 m^2.

Total weight.—

Body.—Steel U shaped frame, carrying engine and driver in its centre on a wooden frame. Mounted on 2 wheels amidships and a runner aft.

Wings.—The wings are mounted on top of U frame.

Motor.—

Speed.—

Propellers.—1, belt driven from engine.

Steering.—Vertical rudder aft of frame, *about* 3 × 2 feet (0·90 × 0·60 m.)

Remarks.—A low centre of gravity is the principal aim of this design.

BEILHARZ. Monoplane.

Building.

Maximum length, feet / m. **maximum breadth,** feet / m. **supporting surface,** sq. feet / m^2.

Total weight.—

Body.—Large body, porpoise shaped; aviator and all mechanism inside it.

Wings.—Curved and pivotted—and connected with front propeller (*q.v.*)

Motor.—

Speed.—

Propellers.—2, one at either end, each mounted on a universal joint. The forward one has a longitudinal and the after one a latitudinal movement.

Steering.—In the vertical plane, by tilting the front propeller. In the horizontal plane, by shifting the rear propeller. The big steering wheel controls the fore propeller; the small upper wheel, the after one.

Remarks.—Building.

SCHÜLKE.
(Julius & Willig.)

SCHÜLKE model.

Maximum length, feet / m. **maximum breadth,** feet / m. **supporting surface,** sq. feet / m^2.

Total weight.—

Body.—Cigar shaped.

Wings.—2 fixed planes.

Motors.—

Speed.—

Propellers.—2 large 4-bladed.

Steering.—

Remarks.—Not yet tried. Building. Models reported to fly well.

SCHOLZ. Ornithopter.

Photo, Maeckel.

Maximum length, feet / m. **maximum breadth,** feet / m. **supporting surface,** sq. feet / m^2.

Total weight.—

Body.—

Wings.—

Motors.—

Speed.—

Propellers.—

Steering.—

Remarks.—Not yet flown.

HUTH (Dr. Fritz). Biplane.

Building.

Maximum length, feet / m. **maximum breadth,** feet / m. **supporting surface,** 600 sq. feet / 56 m².

Total weight.—

Body.—

Wings.—2 pairs, one behind the other; all with variable angles of incidence.

Motor.—

Speed.—

Propellers.—

Steering.—No horizontal rudder; main planes being tilted instead. A single lever controls both horizontal and vertical steering.

Remarks.—Property of the Flying Machine and Motor Co., Berlin.

JOSPE. (?) (or KOPSCHE).

Nothing known.

Maximum length, feet / m. **maximum breadth,** feet / m. **supporting surface,** sq. feet / m².

Total weight.—

Body.—

Wings.—

Motor.—

Speed.—

Propellers.—

Steering.—

Remarks.—This machine has actually flown at Dresden, and a company formed to exploit it. Favourably reported on by the military authorities. No details known. It appears to be the same as the *Kopsche*, but this is not certain.

FLÜGEL.

Building. No data.

Maximum length, feet m. **maximum breadth,** feet m. **supporting surface,** sq. feet m^2.

Total weight.—

Body.—

Wings.—

Motor.—

Speed.—

Propellers.—

Steering.—

Remarks.—Bought by Cosmographical Institute of Breslau.

RUTHENBERG. Helicopter.

Building. No data.

Maximum length, feet m. **maximum breadth,** feet m. **supporting surface,** sq. feet m^2.

Total weight.—

Body.—

Wings.—

Motor.—

Speed.—

Propellers.—

Steering.—

Remarks.—

JORCH.

Building. No data.

Maximum length, feet / m. **maximum breadth,** feet / m. **supporting surface,** sq. feet / m^2.

Total weight.—

Body.—

Wings.—

Motor.—

Speed.—

Propellers.—

Steering.—

Remarks.—Designed to carry 4 people and a large supply of petrol.

ERHARD.

(Military).

Building (?)

Maximum length, feet / m. **maximum breadth,** feet / m. **supporting surface,** sq. feet / m^2.

Total weight.—

Body.—Bullet proof body.

Wings.—

Motor.—

Speed.—

Propellers.—

Steering.—

Remarks.—Designed for military purposes. No information.

GRADE. Triplane.

Maximum length, feet / m. **maximum breadth,** feet / m. **supporting surface,** sq. feet / m^2.

Total weight.—
Body.—On 3 wheels, two in front.
Wings.—Planes are about 2 feet (0·60 m.) apart. Top one straight, the lower ones slightly curved upwards at ends.
Motors.—
Speed.—
Propellers.—
Steering.—All rudders behind.
Remarks.—This machine has flown on several occasions.

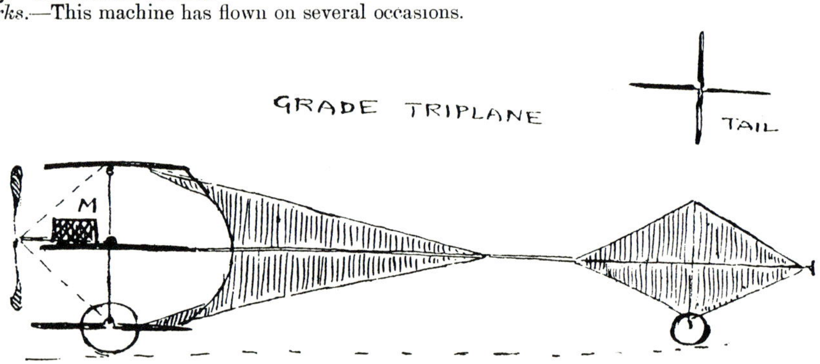

DEGN. Helicopter-Flapper.

Building.

Maximum length, feet / m. **maximum breadth,** feet / m. **supporting surface,** sq. feet / m^2.

Total weight.—1,000 lbs. (453 kgs.)
Body.—
Wings.—Flapping. Special method of connection.
Motor.—
Speed.—
Propellers.—Helices working in conjunction with the flapping wings for raising, and then for propulsion.
Steering.—By the helices only.
Remarks.—

JATHO.

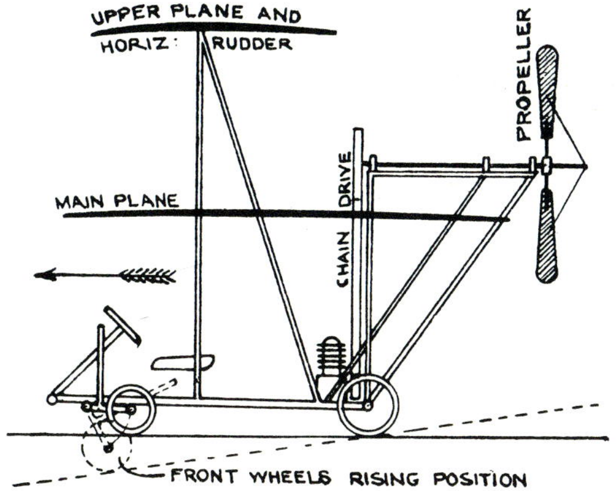

—JATHO BIPLANE—

Maximum length, 14·76 feet / 4·5 m. **maximum breadth,** 26·24 feet / 8 m. **supporting surface,** 484 sq. feet / 45 m².

Total weight.—

Body.—A skeleton wooden truck, with 4 small wheels. The axles of the front wheels are carried on 2 short arms which may be either horizontal or vertical. For starting they are horizontal; the planes are then parallel to the ground and the machine quickly gathers speed; the arms are then turned vertical, lifting the front of the machine, when it should leave the ground.

Wings.—2 curved planes, 26·24 feet (8 m.) wide, the upper one 8·20 feet (2·5 m.), and the lower one 13·12 feet (4 m.), from front to rear.

Motor.—35 h.p., 4-cylinder Koerting, water-cooled. Weight, 176 lbs. (80 kgs.) = 5 lbs. per h.p. (2·3 kgs. per h.p.) Originally it was fitted with a 12 h.p. Buchet.

Speed.—32 m.p.h. (51·5 km.)

Propellers.—One, 2-bladed. Chain driven at 500 r.p.m. Placed centrally behind the planes.

Steering.—Steering in the vertical plane by altering the inclination of the upper plane. No vertical rudder yet fitted.

Remarks.—Only short flights have been accomplished as yet, and longer flights seem improbable until the machine has been considerably altered.

SCHNELL. Monoplane.

THE MODEL.

Maximum length, feet / m. **maximum breadth,** feet / m. **supporting surface,** 161½ sq. feet / 15 m².

Total weight.—331 lbs. (150 kgs.) *including* aviator.

Body.—

Wings.—The main plane curves slightly backward, at the tips forward, thence cut off at an angle of 45 degrees to the rear of it.

Motor.—

Speed.—

Propellers.—

Steering.—

Remarks.—Building at Linden. The model flies very successfully.

LORENZEN. Helicopter.

Building. No data.

Maximum length, feet / m. **maximum breadth,** feet / m. **supporting surface,** sq. feet / m².

Total weight.—

Body.—

Wings.—

Motor.—

Speed.—

Propellers.—

Steering.—

Remarks.—

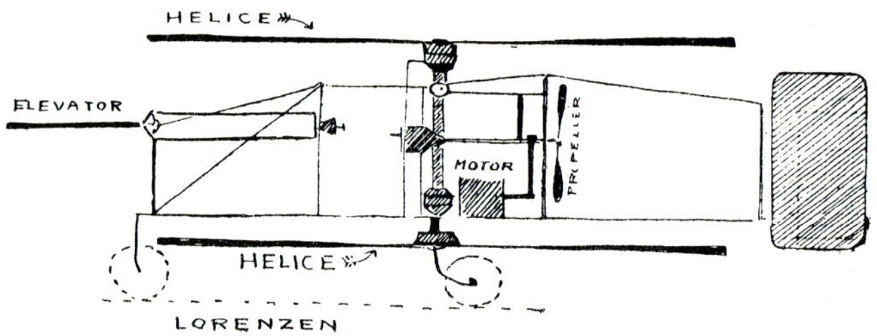

H. FARMAN. Monoplane.
(Bought by Lieut. FRITSCHE.)

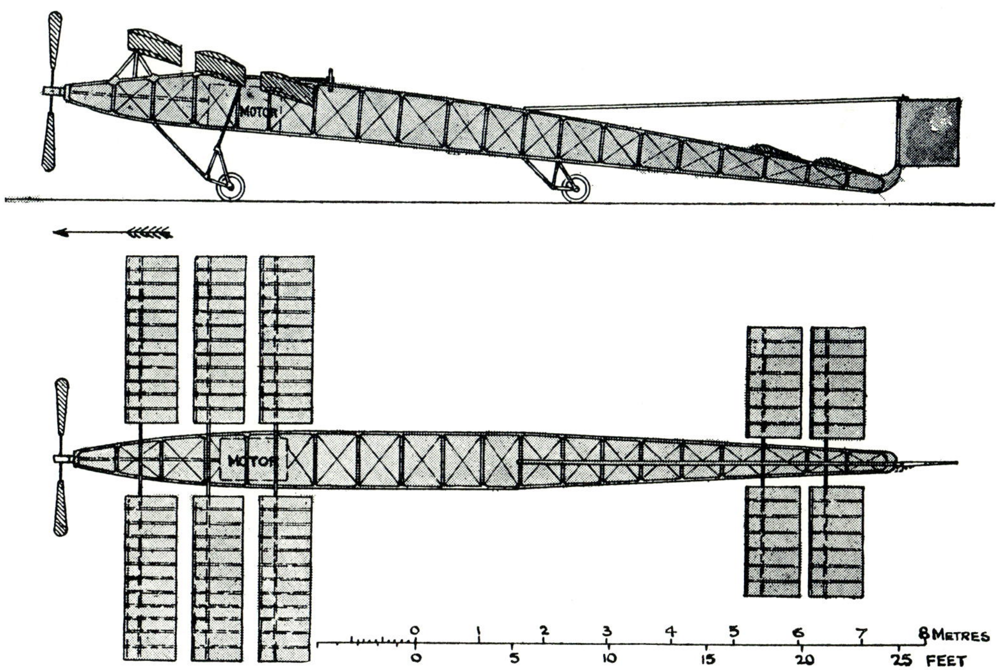

Maximum length, 46½ feet / 14·2 m. **maximum breadth,** 20·2 feet / 6·28 m. **supporting surface,** 258 sq. feet / 24 m².

Total weight.—1,323 lbs. (600 kgs.) *including* aeronaut.

Body.—Wooden girder structure 44 × 29 feet (13·5 m.) long, covered with fabric, and containing the motor and aeronaut, 2 small wheels in front, one half-way along the body.

Wings.—3 front sets and 2 tail sets, each set being slightly above the level of the one abaft it, and the extremities being lighter than the centre. Built of shaped wooden framework, covered both sides, with a tubular steel axis. The wings can be adjusted to any desired angle. Front wings 8·52 × 2·95 feet (2·6 m. by ·9 m.) Tail wings 5·24 × 2·95 feet (1·6 m. by ·9 m.)

Motor.—Renault. 8 air-cooled cylinders, arranged in a V, 90 m/m bore by 120 m/m stroke. 50 to 60 h.p.

Speed.—Lifts at 45 m.p.h. (72 km. p.h.) Speed in flight 46·7 to 50 m.p.h. (75 to 80 km. p.h.)

Propellers.—4-bladed. Diameter 8·2 feet (2·5 m.); pitch 4·1 feet (1·25 m.)

Steering.—Vertical rudder 2·6 × 2·6 feet (·8 m. by ·8 m.) abaft and above the rear end of the tail. Rear tail plane used as horizontal rudder.

Remarks.—Owing to Mr. Farman being so successfully occupied with the development of biplanes, this machine has not come very prominently before the public. Since it was designed, so many monoplanes have been constructed of lighter weight and simpler construction that it is probable that this will undergo considerable alteration before it can compete with them.

***VOISIN* Biplanes.** *(see France).*

German Concessionaire:—August Euler, Frankfort-on-Main.

Usual Voisin type, but fitted with Simms V engines, 50 h.p.

Of this make there are in Germany, built, building or on order:-

Model 1908-9.

1 August Euler

2

3

Voisin type.

***WRIGHT BROS.* Biplanes.** *(see U.S.A.)*

German Concessionaires:—Wright Flying Machine Co., Berlin.

Of this make there are in Germany, built, building, or on order:—

Model 1908-9.

1 Demonstration machine (built)

2 Ganz Fabrice

Building.
3
4
5
6
7
8
9
10
11
12

NOLTE.

Nothing known.

Maximum length, feet / m. **maximum breadth,** feet / m. **supporting surface,** sq. feet / m².

Total weight.—
Body.—
Wings.—
Motor.—
Speed.—
Propellers.—
Steering.—
Remarks.—

REISSNER (Professor). Biplane.

Altering.

Maximum length, feet / m. **maximum breadth,** feet / m. **supporting surface,** 360 sq. feet / 33.42 m².

Total weight.—
Body.—Steel tubes chiefly used.
Wings.—
Motor.—40 h.p.
Speed.—
Propellers.—
Steering.—No elevator; a special tail used instead.

Remarks.—Rose to 15 feet on April 21st, on the Branderheide, and flew 40 yards easily. The aviator then stopped his engine to prevent rising higher and fell rather suddenly, damaging the machine.

GANZ FABRICE. Monoplanes I and II.

Building.

Maximum length, feet / m. **maximum breadth,** feet / m. **supporting surface,** sq. feet / m^2.

Total weight.—
Body.—Bird like.
Wings.—Flexed by movement of driver's body.
Motor.—10 h.p. French make.
Speed.—
Propeller.—
Steering.—
Remarks.—Both identical.

SCHULER. Biplane.

Building.

Maximum length, feet / m. **maximum breadth,** 39½ feet / 12·50 m. **supporting surface,** 565 sq. feet / 53 m^2.

Total weight.—
Body.—Welded steel frame. Mounted on 4 wheels like a R.E.P.—one central in front, one central in rear, one under each wing tip.
Wings.—Maximum span, 39½ feet (12·50 m.)
Motor.—
Speed.—
Propellers.—2 in rear of the planes, chain driven in opposite directions.
Steering.—Horizontal rudder, like *Wright*, in front; vertical at rear, in tail, Voisin style.
Remarks.—Built by Max Schuler, of Chemnitz, Saxony. Is a species of compromise between a Wright and a Voisin.

MESCHNER. Biplanes.

Completing.

Maximum length, feet / m. **maximum breadth,** feet / m. **supporting surface,** sq. feet / m^2.

Total weight.—
Body.—Chassis of aluminium tubes. Welded steel ribs.
Wings.—Like *Wright*. Warping wing tips.
Motor.—70 h.p. Koerting, water-cooled.
Speed.—37 m.p.h. (60 km.)
Propellers.—One, 3-bladed.
Steering.—Elevator forward, vertical rudder aft.
Remarks.—Engine, etc., tried at Tempelhof, December, 1908.

GERMAN DIRIGIBLES.

German Dirigibles consist of four distinct classes and a number of miscellaneous types.

Best known is the Zeppelin type, with a rigid framework of aluminium, a thin outer cover, and a number of drum-shaped gas bags.

Three of this type have been built and destroyed; three have been built and are still in existence. These belong to the German War Office.

One is under construction for the Government, subject to satisfactory trials, and two are under construction for a private company which is about to undertake an Aerial Passenger service.

It is understood that the Government will not order any more vessels of this type for the present.

Also of the rigid type are the *Schütte* airships.

These represent such a vast improvement on the Zeppelin type that the abandonment by the Government of the former is probably not due to a dislike to rigid vessels so much as to a liking for the *Schütte* ships.

Two of these are under construction, both for the Government, and arrangements are being made for very speedy construction of a numerous class.

The features of the class are the use of wood, wonderful strength, and a novel arrangement of gas bags.

The military type, known as the *Gross* airships, are Semi-Rigid, and designed and built by the military authorities.

Three have been built and are still in existence; a fourth is nearing completion, and a fifth is under design.

The Parseval class of Non-Rigid dirigibles comprises four vessels built and available for service, and two under construction.

Among miscellaneous airships are the *Clouth*, the *Ruthenberg*, and one or two other vessels being constructed privately.

It seems probable that by the summer of 1910 Germany's Aerial Fleet will comprise:—

6 or 7 Rigids.
4 Military Semi-Rigids.
5 PARSEVAL Non-Rigids.
5 Miscellaneous Vessels. ***Total—21 or 20 Vessels.***

ZEPPELIN type. Rigid.

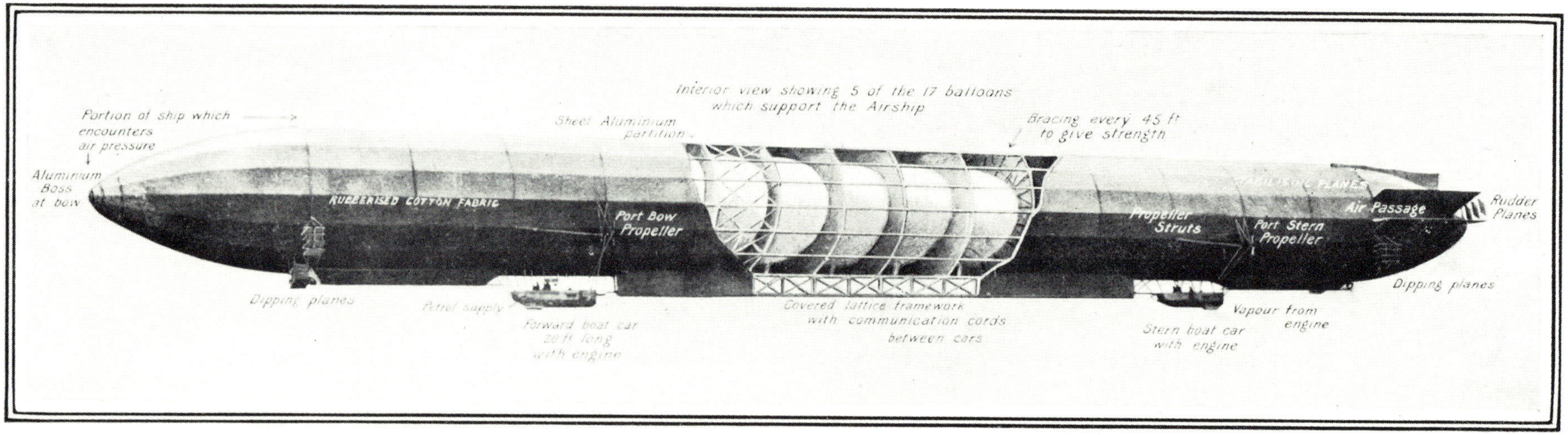

Popular Diagramatic Sketch of the Zeppelin System.

1st series. Old numbers.	Length.		Diameter.		Volume.		Total lift.		Motor.	Speed.		Completed.	Fate.
	ft.	m.	ft.	m.	c. feet.	m³	Tons.	Kg.		m.p.h.	km.p.h.		
Zeppelin I.	420	128	38	11·66	399,082	11,300	12·5	12,500	2 16 h.p. Daimler	20	27·4	1900	Broken up on account of small speed and radius of action.
Zeppelin II.	420	128	38	11·66	399,082	11,300	12·5	12,500	2 80 h.p. Daimler	28	45	1906	Destroyed by a storm after a very short life.
Zeppelin III.	420	128	38¾	11·7	403,707	11,431	12·575	12,575	2 85 h.p. Daimler	28	45	Oct., 1906	Reconstructed 1907. Laid aside for 18 months. Now afloat as *Z. I.*
Zeppelin IV.	440	134	44	13·4	460,000	13,000	14·5	14,500	2 110 h.p. Daimler	30		1908	Short and brilliant life. Destroyed by fire and storm.

General Notes on the type.—Where speeds of 40 m.p.h. are aimed at, a rigid nose and stern are essential.

The earlier models must be regarded as experiments necessary before the first useful vessel could be built. All that they did could have been better done by a semi or non-rigid vessel of the same cost, but the ratio of dead weight to useful weight decreases with the size, and the ill fated *Zeppelin IV* must be considered as the first really useful model.

Even *Zeppelin III*, now known as *Zeppelin I*, in spite of its success, represents a vast expense in comparison with the results attained.

In *Zeppelin II* (new numbering) however, the dividing line has been crossed; what she can do could scarcely be done by a non or semi-rigid vessel, however big and costly.

Other rigid types have entered the field, and at present the distinguishing features of the original type are: (1) aluminium frame, as opposed to wood; (2) parallel body, as opposed to other shapes.

2nd SERIES.

Zeppelin I. Late *Zeppelin III*, improved and slightly lengthened. Taken over by War Office. To be stationed at Berlin.

Zeppelin II. Similar to *Zeppelin IV* Taken over by War Office. To be stationed at Metz.

Zeppelin III. More lightly built and of increased horse power. Taken over by War Office. To be stationed at Cologne.

Zeppelin IV. Similar to *Zeppelin III.* Ordered by War Office, subject to satisfactory trials.

Zeppelin V. } Under construction. Designed for passenger
Zeppelin VI. } service.

ZEPPELIN I.
(New Series.)

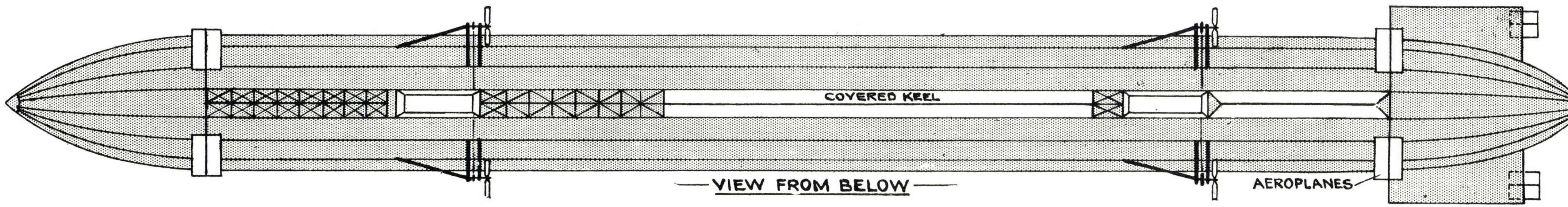

ZEPPELIN I

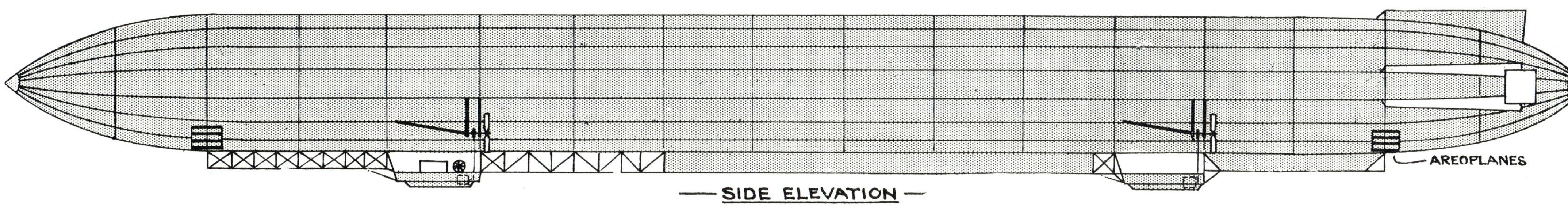

This is the original *Zeppelin III.*, twice re-constructed. The re-construction has considerably reduced her weight.

Length, 426 ft. / 128 m. **diameter,** 36·3 ft. / 11·7 m. **volume of gas bag,** 403,700 c. ft. / 11,431 m³. **total lift,** 12·575 tons / 12·575 kg. **useful lift** (fuel, ballast, crew, equipment), 3·5 tons / 3500 kg.

Gas bags.—17 of continental material. Filled to about 95% for an ascent. Probable leakage about 2% of volume per diem.

Frame.—Aluminium longitudinal and circumferential girders, with an outer and inner network of diagonal wires. A frame of wires right across the section between every two bags.

Outer skin.—Pegamoid linen (*i.e.* treated with celluloid) on top. Silk underneath.

Gondolas.—2 of aluminium. Rigidly connected to frame and to keel.

Motors.—1 in each gondola. 85 h.p. Mercedes Daimler. Each drives 2 propellers, one each side.

Fuel.—Variable, as ballast can largely be replaced by fuel. Maximum, 2 tons.

Speed.—28 miles per hour = 45 km. per hour.

Note.—Stationed at Metz. Has been very successful, but has the limitations of the Rigid type without the advantages which this type possesses when constructed on a larger scale.

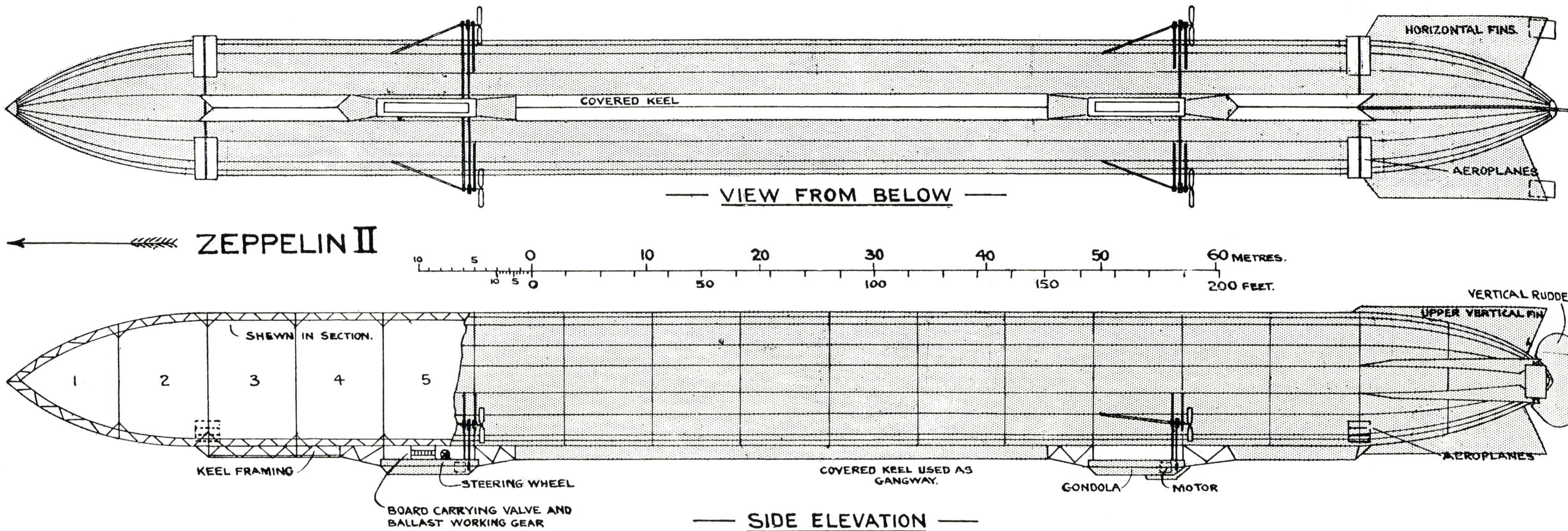

Built on the plans of the late *Zeppelin IV*.

A rigid double frame of aluminium, covered with waterproof material, containing 16 separate gas bags, and supporting 2 gondolas.

Length, 446 feet. / 136 m. **diameter,** $42\frac{1}{2}$ feet. / 13 m. **volume,** 460,000 c. feet. / 13,000 m. **total lift,** 32,000 lbs. / 14,500 kg. **useful lift,** 8,500 lbs. / 3,850 kg.

Gas bags.—16 in number, of Continental material, 2 layers of cotton and 2 of rubber, weighing ·039 lbs per sq. foot.

Motors.—1 in each gondola, each 110 h.p. Daimler-Mercedes. Weight, 700 lbs. each. / 318 kg. Speed, 30 m.p.h. / km. p.h.

Propellers.—Four 3-bladed metal propellers, 12 feet / 3·65 m. in diameter, one pair to each motor. These are driven by shafting and placed in line with the centre of resistance.

Steering.—Vertically, by 2 sets of horizontal rudders, 1 set each end. Each set consists of 4 aeroplanes on each side, amounting to 300 sq. feet. / 27·7 m².

The function of the foremost set is to alter the height of the ship, while the after set keeps the ship horizontal or otherwise.

Horizontally, by a large vertical rudder at the stern, and 2 pairs of small vertical rudders, one between each pair of horizontal tail planes.

Table of approximate weights.—

Frames	3 tons.	Petrol	1·00 tons.	Fixed planes ...	·3 tons.
Wire	1 ,,	Outer skin (upper half pegamoid, lower half silk)	·5 ,,	Aeroplanes ...	1·0 ,,
Keel	·50 ,,			Crew	·8 ,,
Propellers and frame	·50 ,,	Gas bags	2·1 ,,	Miscellaneous ...	·5 ,,
Motors	·70 ,,	Gondolas	·6 ,,	Ballast	1·5 ,,

Note.—1 ton = 1000 kilos. approximately.

Remarks.—The motors and gondolas are the same as were used in the ill-fated *Zeppelin IV*.

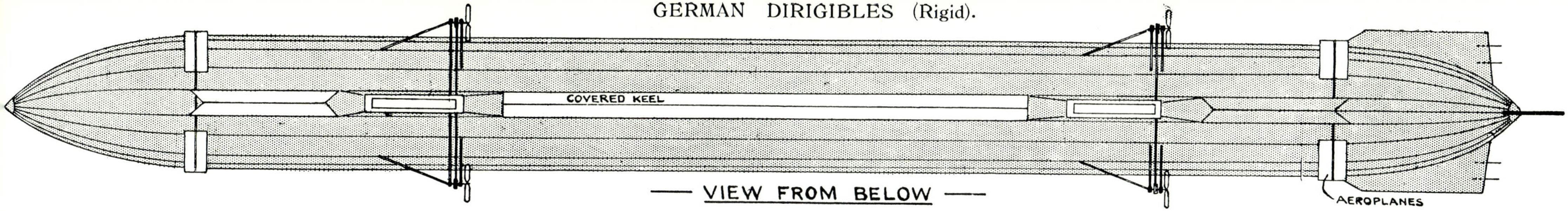

ZEPPELIN III

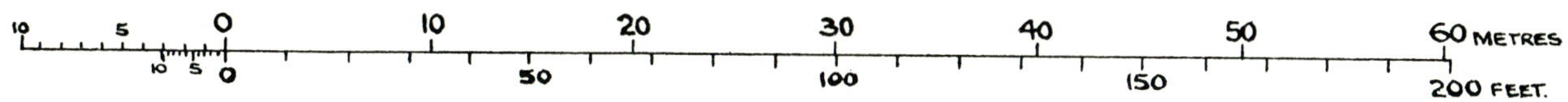

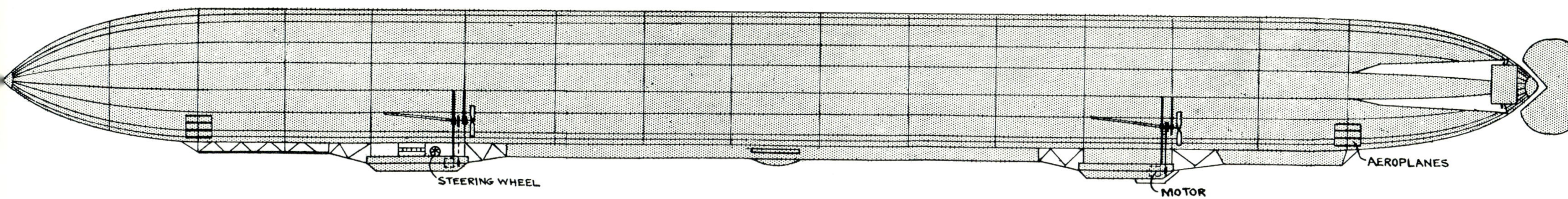

SIDE ELEVATION

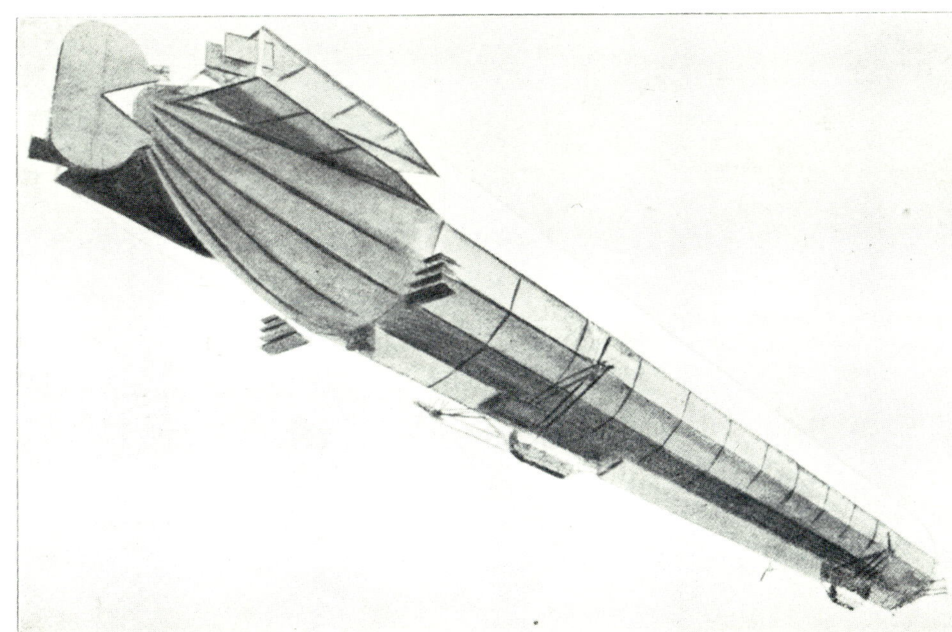

Maximum length, 446 feet / 136 m. **maximum diameter,** 42½ feet / 13 m. **volume,** 460,000 c. feet / 13,000 m³.

Total lift.—32,000 lbs. (14,500 kgs.) (14½ tons). **Useful lift,** 8,500 lbs. (3,850 kgs.) (*about* 3¾ tons).

Gas bags.—Continental material. 16 in compartments.

Motors.—Two 130 h.p. Daimler-Mercedes, one in each gondola.

Speed.—35 m.p.h. (56 km.)

Propellers.—Four 3-bladed, metal, a pair on either side to each motor. Shaft driven. Diameter, 12 feet (3·65 m.)

Steering.—As in *Zeppelin II*, but the small vertical rudders differ somewhat.

Remarks.—Differs from *Zeppelin II* in the absence of vertical tail fins, and the horizontal fins are slightly different, also other small details. She has also a semi-circular sponson on either side of the keel amidships. Presumably these are for guns. Reported that an upper vertical fin has been added.

SCHÜTTE I; *also* SCHÜTTE II.

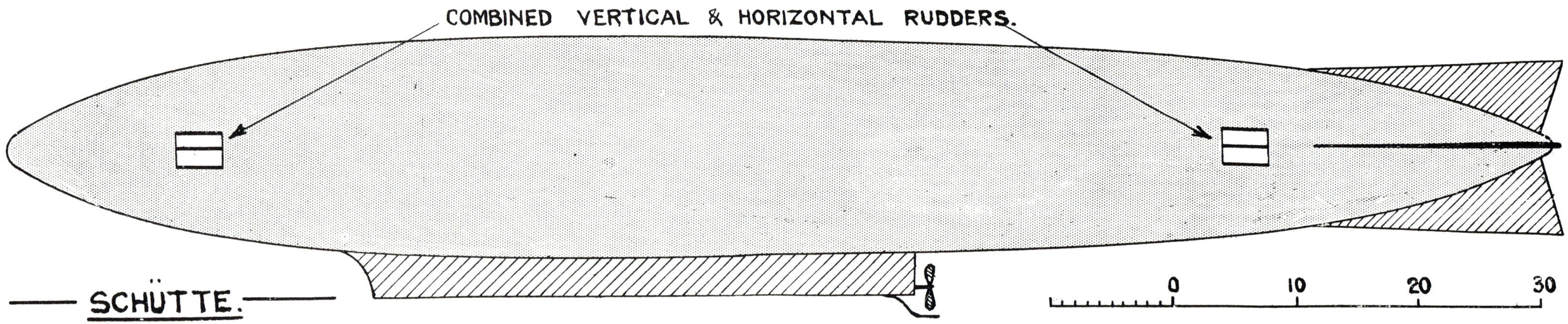

Maximum length, 420 feet / 128 m. **maximum diameter,** 59 feet / 18 m. **volume,** 671,020 c. feet / 19,000 m³.

Total lift.—*About* 20 tons (20,000 kgs.) **Useful lift.**—*About* 5 tons (5,000 kgs.)

Gas bags.—These are of great strength and of unusual shapes, made to fit the interior, which is encumbered with cross stays. All but two of the bags are always full, and when the gas expands it flows into the remaining two, which are nearly empty at sea level, and full at 6500 feet (2,000 m.) A centrifugal pump is used for distributing the gas.

Motors.—2, of 250-300 h.p. each. The propellers are at the ends of the car, driven through 1 set of reduction gear. Daimler motor.

Speed.—38-43 m.p.h.

Propellers.—2, aft. Also 1 with its axis vertical.

Steering.—Vertical and horizontal rudders at both ends of the ship.

Remarks.—Two of these ships are under construction, and probably one will be presented and one sold to the German government. The hull is built of special 3-ply wood made of Russian white fir; this wood is pressed into channel bars, angle bars, and all other requisite shapes. The strength of the hull is such that it can be supported at the ends without damage; its lightness is such that although the ship is nearly half as large again as *Zeppelin II*, yet the hull weighs about 3 tons less. One car is fitted, 131 feet (40 m.) long, and this contains the cabin and all other accommodation.

GERMAN DIRIGIBLES.

GERMAN MILITARY CLASS—GROSS (Semi-Rigid).

Up to date, these vessels have been designed by Major Gross.

The utmost secrecy is observed as to their details.

The system of employing 2 ballonets has been borrowed from the Parseval type, and presumably the Parseval system of working the automatic valves has also been adopted.

In all other features, these ships appear to resemble the French Lebaudy type, the shape of the hulls being rather better.

GROSS I.

(Gross Type. Semi-rigid.)

Closely resembles *II.*

Purely experimental. Designed for the War Office by Major Gross. Now used as an instructional ship.

Length,	**maximum diameter,**	**capacity,**	**total lift,**
131 feet.	39$\frac{1}{3}$ feet.	63,576 c. feet.	4,350 lbs.
40 m,	12 m.	1,800 m^3.	1,980 kg.

Gas bag.—Rubber-proofed material. Parallel body, with spherical coned ends. The marked dip in the centre is due to the weight of the car being borne by a short length.

Ballonet.—One-fifth of the total capacity.

Motor.—4-cylinder 20-24 h.p. Daimler-Mercedes, driving 2 propellers at 800 revs. close up under the gas bag.

Propellers.—2-bladed, 8·2 feet. / 2·58 m. diam.

Speed.—15 m.p.h. = 24 km. per hour.

Planes and rudders.—A large vertical plane under the after end of the gas bag, terminating in the vertical rudder.

Car.—Boat shaped. 13 feet below the gas bag.

Remarks.—Flight of 8 hours on October 28th, 1907. This ship is too small to be of any real value except as an instructional vessel. Built at Siemen's works, Berlin.

GROSS II and III.

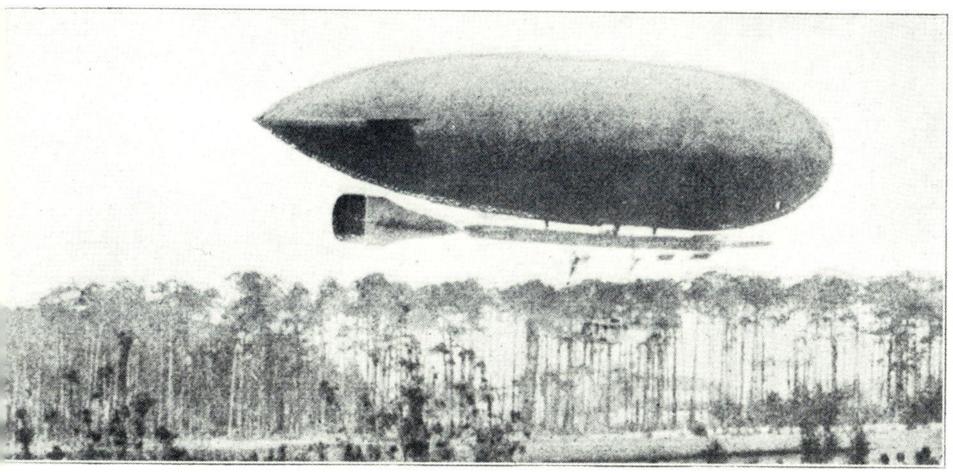

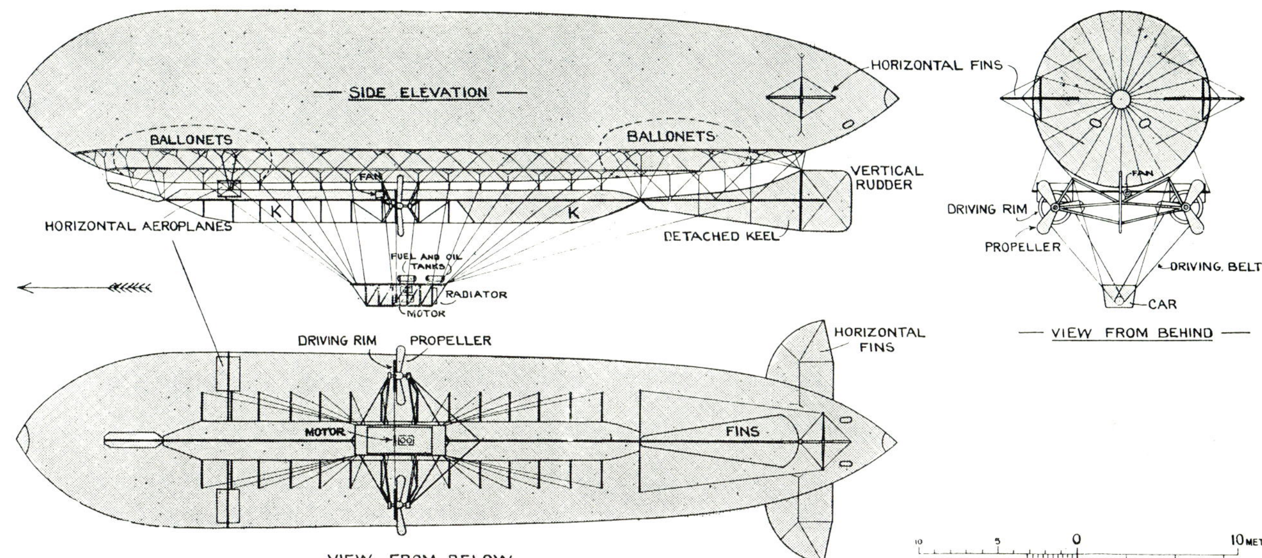

Length, feet m **maximum diameter,** feet m **capacity,** 176,000 cubic feet m³ **total lift,** lbs. kg.

Gas bag.—Continental rubber cloth, diagonal thread. Tapering shape.

Ballonet.—One-fifth of total volume.

Motor.—2—75 h.p. Daimler. 2 propellers, with 3 aluminium blades.

Remarks—Sept. 11th, 1908:—Covered 176 miles in 13 hours, returning to the starting point.

GROSS IV.

Building.

Maximum length, feet / m. **maximum diameter,** feet / m. **volume,** c. feet / m³.
Total lift.— lbs. (kgs.) **Useful lift,** lbs. (kgs.)
Gas bags.—
Motors.—
Speed.—
Propellers.—
Steering.—
Remarks.—Believed to be practically a sister to *III.*

GROSS V.

Building or preparing to build.

Maximum length, feet / m. **maximum diameter,** feet / m. **volume,** c. feet / m³.
Total lift.— lbs. (kgs.) **Useful lift,** lbs. (kgs.)
Gas bags.—
Motors.—
Speed.—
Propellers.—
Steering.—
Remarks.—

GERMAN DIRIGIBLES

PARSEVAL CLASS (Non-Rigid).

When the "Motorluftschiff Studiengesellschaft" was formed at the instigation of the German Emperor, a committee was formed to acquire an experimental airship of the most promising type. Major Von Parseval's first airship was selected, and since that time the above company has confined itself to improving this type, and to making exhaustive and costly researches, all of which have been embodied in successive ships.

The characteristic feature of every one of these craft is its unequalled portability. Almost all other so-called Non-Rigid vessels distribute the load by means of a long girder which also serves as a car. This girder is awkward to pack up and transport. Parseval uses a comparatively small car, and distributes the weight by hanging it further below the balloon than usual, and also by using 2 ballonets which are placed one near each end of the gas bag.

These two ballonets enable the ship to be trimmed by merely pumping air into either at the expense of the other.

Another essential feature of the type is the system by which the valves are worked automatically. At the present time there is no other system of valve working so reliable as this.

A third essential feature of the class is the use of a swinging car, in such a manner that pitching, due to alterations of propeller thrust, is automatically checked by an alteration of the position of the centre of gravity.

A fourth feature is the use of limp propeller blades. A propeller of this type is very easily packed up.

The shape of these vessels is in accordance with the experiments of Professor Prandtl.

PARSEVAL I was small, purely experimental, and bluff ended.
PARSEVAL II was better shaped, larger, and improved in details.
PARSEVAL III and IV are merely larger models of the same.

PARSEVAL I.

(Parseval Class.)

Bought by "Motor-Luftschiff-Studien-Gesellschaft" ("Society of the Study of Motor Ballooning,") and taken over by the German War Office.

Length, 170·6 feet (52 m.) **max. diam.** 29·2 feet (8·9 m.) **capacity,** 113,000 c. feet (3,200 m³.) **lifting power,** 7,800 lbs. (3,583 kgs.)

Gas bag.—Cylindrical, with semi-conical front. Of rubber-proofed material in longitudinal strips. Pressure in ballonets and gas bag, 1·18 inches (30 mm.) of water

Motor.—8-cylinder. 100 h.p. N.A.G.

Fuel.—700 lbs. (325 kg.) 88 gallons (400 litres)

Speed.—20 m.p.h.

This vessel was somewhat altered on being bought by the Society. Her essential principle is that she can be taken to pieces in a few minutes, and carried in a truck. Her main feature is that she has a ballonet at each end. This is described in the case of *Parseval II*. This class rise with the forward ballonet empty, and inclined up by the bow. The propellor is similar to that of *Parseval II*. The car also is mounted on wire runners, as in *Parseval II*. The first ship of a class that displays even more originality than the *Zeppelins*.

PARSEVAL II (or Type A).

Built by the "Society for the Study of Motor Air Ships," and taken over by the German War Office.

Length, 190 feet (58 m) **maximum diameter,** 31 feet ($9\frac{1}{2}$ m) **capacity,** 134,200 cubic feet (3,800 cubic m) **lifting power,** 9,200 lbs. (4,180 kgs.)

Gas bag.—Front end semi-ellipsoidal with semi-axes 15·4 feet (4·7 m.) and 11·8 feet, (3·6 m.) from which it increases to its maximum diameter.

This is maintained for about two-thirds of its length, when it begins to taper to a point at the stern.

Made of 2 layers of transverse strips of rubber proofed material, crossing each other diagonally. Fitted with a tearing strip.

Ballonets.—One at each end, together amounting to about one-quarter of the total capacity.

Owing to this disposition, the trim can be altered, and steering effected in the vertical plane by filling either more than the other.

Pressure in the ballonets and gas bag, ·79 inches (20 mm.) of water pressure.

Motor.—4-cylinder 100 h.p. Daimler placed at one side of the car to give more room. 1,000 to 1,200 r.p.m.

Propellors.—$12\frac{1}{3}$ feet (3·75 m.) diameter. 250 to 300 r.p.m. 4-bladed, the blades being of peculiar construction. When stopped, the fabric of which they are made hangs down limply; when running, these flaps fly out under centrifugal force.

Speed.—$33\frac{1}{2}$ miles per hour. kilometers per hour.

Car.—Length, $22\frac{1}{3}$ feet (6·8 m.) Width 4·1 feet. (1·22 m.)

Made of nickel steel, U bars, screwed together so as to take the pieces rapidly.

The sides are lattice girders. The whole is boat shaped and covered with canvas.

Contains motor, chart table; trail rope 480 feet (146 m.) long, weighing 220 lbs. (100 kg.)

Wheel for horizontal steering at the bow.

110 gal. (500 litre) cask of benzine on the girders at the after point.

41 feet ($12\frac{1}{2}$ m.) below the gas bag. It is capable of swinging horizontally on wires running over rollers.

Whereas without this device a forward swing of the car would lift the nose to a possibly dangerous extent, the free motion of the car shifts the centre of the gravity forward and so preserves stability.

Steering.—In vertical plane, by altering the trim.

In horizontal plane, by a rudder of 80·7 sq. feet ($7\frac{1}{2}$ m^2) immediately behind the vertical plane.

Two fixed horizontal planes are placed at the rear end of the gas bag above the central line.

Table of weights:—

Gas bag	1,653 lbs.
Cordage	220·5 ,,
Trail rope	220 ,,
Car and motor	529 ,,
Fuel	770 ,,
Oil	160 ,,
Oil and fuel tanks, instruments, miscellaneous...	1,637 ,,
Crew, passengers, ballast...	1,654 ,,
TOTAL ...	6,834 lbs.

Note.—This remarkably successful ship has performed a continuous flight of $11\frac{1}{2}$ hours. She also remained at a height of 4,800 feet (1,500 m) for 1 hour. She can be transported in 1 railway truck or 2 pair horse wagons, and be assembled and filled ready for ascent within 3 hours of arrival by train.

PARSEVAL III and IV.

(Type B.)

Appearance exactly as *II*.

Length 226 feet. / 69 m. **maximum diameter,** 36 feet. / 11 m. **capacity,** 197,700 c. feet. / 5,600 m³.

Gas bag.—Continental material. Total weight, (1,100 kg.). Shape similar to *Parseval II*.

Ballonets.—2, arranged as in *Parseval II*.

Motors.—Two 100 h.p. "N.A.G," started by compressed air. 6-cylinder motors. 150 m/m. bore × 130 m/m. stroke. Placed one each side of the car. Weight, 882 lbs. (400 kg.) each.

Propellers,—Two 4-bladed, "Parseval" pattern. Diam. feet (4 m.) when revolving.

Car.—Length, feet (9 m.) breadth, feet (1·8 m.) Similar in most respects to the car of *Parseval II*, and with the same device allowing it to swing.

Petrol.—500 litres (gallons).

PARSEVAL V and VI.

Building.

Maximum length, feet / m. **maximum diameter,** feet / m. **capacity,** c. feet / m^3.

Gas bag.—

Ballonets.—

Motor.—

Propellers.—

Car.—

Petrol.—

CLOUTH.

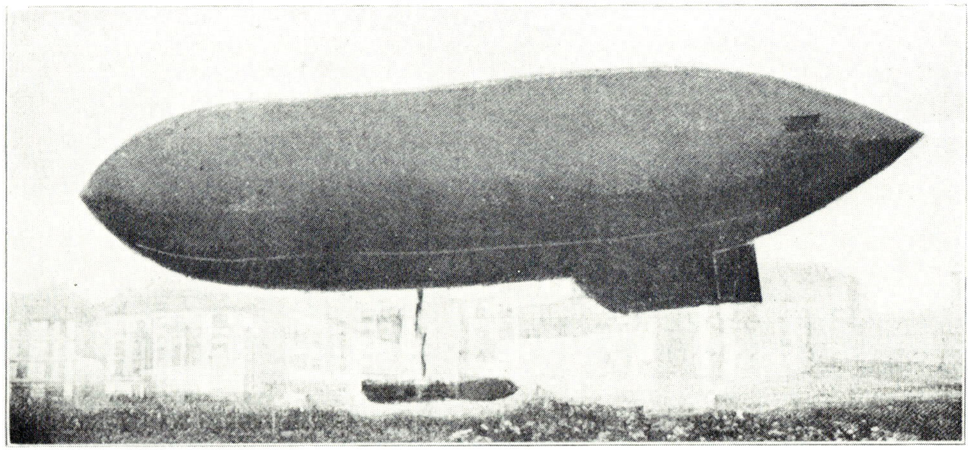

Maximum length, 138 feet / 42 m. **maximum diameter,** $26\frac{1}{4}$ feet / 8 m. **volume,** 60,040 c. feet / 1,700 m³.

Total lift.—862 lbs. (1,900 kgs.) **Useful lift,** lbs. (kgs.)

Gas bags.—Continental rubbered fabric.

Motor.—50 h.p. Clouth motor.

Speed.—22 m.p.h. (35 km.)

Propellers.—2, built of wood, arranged as in the Parseval type, one each side of and above the car.

Steering.—

Remarks.—

RUTHENBERG.

Maximum length, feet / m. **maximum diameter,** feet / m. **volume,** c. feet / m³.

Total lift.— lbs. (kgs.) **Useful lift.**— lbs. (kgs.)

Gas bags.—

Motor.—

Speed.—

Propellers.—

Steering.—

Remarks.—Built at Frankfurt. Like the *Clouth* apparently inspired by the *Parseval.*

?HEINISCH-WESTFÄLISCE MOTORLUFTSCHIFF GESELLSCHEFT.

Building.

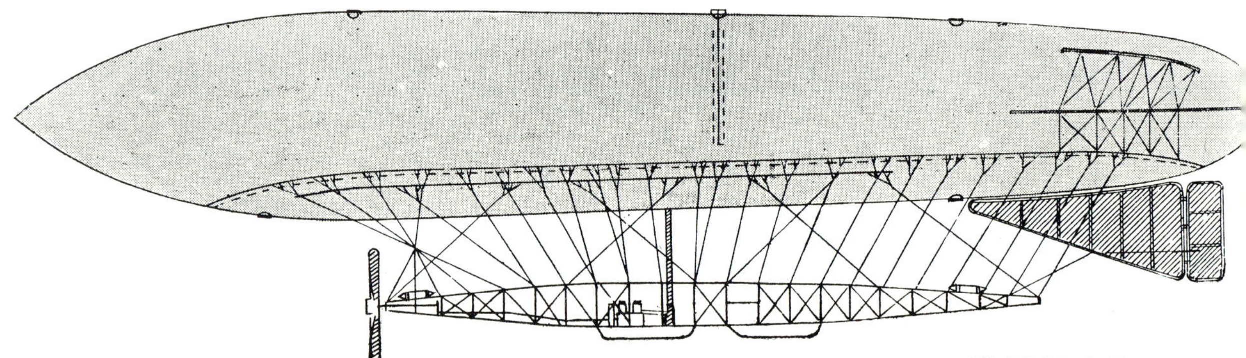

—R.W.M.G.—

ⁱpprox. maximum length, 255 feet / 77·70 m. **approx. maximum diameter,** 42 feet / 13 m. **volume,** 310,800 c. feet / 8,800 m³.

'otal lift.—21,164·lbs. (9,600 kgs.) **Useful lift,** lbs. (kgs.)

'as bags.—Continental rubber proofed material.

Iotors.—4-cylinder 110 h.p. Benz.

peed.—25 m.p.h. (40 km. p.h.)

ropellers.—1, in front of the girder. Connected by cardan shaft and reduction gearing to the motor.

teering.—Vertical rudder in rear of the vertical keel under the gas bag. No horizontal rudders, but the ship is steered upwards or downwards by being inclined up or down by the bow. This is effected by pumping water to and fro between 2 trimming tanks which are placed one at each end of the girder. The pump is connected through a friction clutch to the main motor.

Remarks. The girder which carries the car is built of ash, with junction pieces of aluminium alloy. ; is made to take into 7 separate pieces.

HILDEBRANDT.

(or BALDWIN II.)

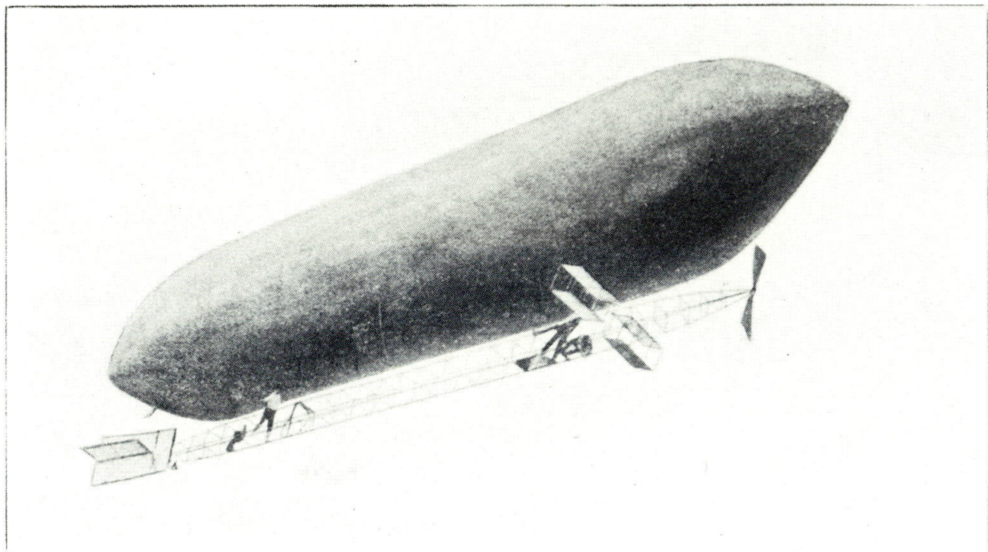

Length, 96 feet (29·26 m.) **diameter,** $19\frac{1}{2}$ feet (5·93 m.) **volume,** 20,000 c. feet (580 m³.)

Total lift.—1,370 lbs. (620 kgs.) **Useful lift** (crew, fuel, ballast, etc.) 500 lbs. (227 kgs.)

Ballonet.—2,800 c. feet = 79 m³.

Gas bag material.—2 layers of Japanese silk, with a layer of vulcanised rubber in between them.

Car.—A framework of square section, made of spruce wood, 66 feet long, $2\frac{1}{2}$ feet wide, $2\frac{1}{2}$ feet high. The car is slung only 5 feet (1·5 m.) below the gas bag by means of about 80 wires.

Motor.—20 h.p., 4-cylinder Curtis' water-cooled.

Speed.—Max. 19·61 m.p.h. (31 km.p.h.)

Propeller.—At the front end of the car, built up of spruce, connected to motor by a 20 feet (6 m.) steel shaft. Diameter, 10 feet, 8 inches (5·1 m.) Pitch, 11 feet (3·35 m.) Revs. 450 per minute.

Planes and rudders.—Fixed vertical plane at rear end. Vertical rudder at rear end with fixed horizontal plane attached to it. Wide moveable biplane, 13 feet (4 m.) from fore end of car, just before the motor. This is operated by a lever. Planes, 22 feet (6·4 m.) wide; 3 feet (·9 m.) deep; and 3 feet (·9 m.) apart. Carries two men only.

Remarks.—Built by Baldwin, and practically an exact copy of the *Baldwin* (U. S. Dirigible I). Bought by Captain Hildebrandt.

SIEMENS-SCHUCKERT.

(*Alias* SIEMENS).

Building.

Maximum length, 328 feet (100 m.) **maximum diameter,** $42\frac{1}{2}$ feet (13 m.) **volume,** 423,800 c. feet (12,000 m³.)

Total lift.— lbs. (kgs.) **Useful lift,** lbs. (kgs.)

Gas bag.—Of great strength. Consists of 2 layers of cotton, with a layer of rubber between them, and on both sides, divided into a number of separate compartments.

Motors.—500 h.p.

Speed.—

Propellers.—

Steering.—

Remarks.—

GREEK.

General Notes.—Nothing, so far, appears to have been done in Greece, but at least one Greek has been following recent trials in France, and some standard French machine may possibly be on order.

ITALIAN.

General Note.—Flying is now paid considerable attention to in Italy, and several Italian makes of machines and aerial motors are on the home market. *Miller* and *Moncher*, despite their non-Italian names, are both of them entirely Italian.

In dirigibles, Italy has evolved two distinct novelties.

Aerial Societies:—

Società Aeronautica Italiana.
L'Aeronauta.
Cadore Club.

Aerial Journals:—

Bolletino della Società Aeronautica Italiana.

Flying Grounds:—

Rome.
Milan (*projected*).
Brescia.

FACIELLI. Triplane.

(Signor Facielli of Turin.)
(Alias **S.P.A.)**

Maximum length, feet m. **maximum width of planes,** 32·80 feet 10 m. **supporting surface,** sq. feet m²

Total Weight.—1,102 lbs. (500 kgs.)—*not including* aeronaut.

Body.—

Planes.—The planes are 6·56 feet (2 m.) from front to rear. The middle plane is interrupted in the centre to give room for the aeronaut. Two stabilizing planes at outer ends of the middle plane.

Motor.—75 h.p.; 4 cylinder horizontal with 2 pistons to each cylinder, the pistons on each side driving a crank shaft which drives the propellers direct. Weight of motor 441 lbs. (200 kgs.) S.P.A. motor.

Speed.—

Propellers.—Two.

Steering.—Tail in rear with vertical rudder.

Remarks.—The special motor is the feature of this machine.

GEMMA. Aerocurvo.

Maximum length, feet 9 m. **maximum breadth,** feet 7·50 m. **supporting surface,** sq. feet m².

Total weight.—

Body.—

Wings.—

Motor.—Anzani. 1,500 r.p.m. Weight, lbs. (91 kgs.)

Speed.—

Propellers.—

Steering.—

Remarks.—Reported on trial so long ago as December, 1908, but not since been heard of.

MONCHER. Monoplane.

Maximum length, feet / m. **maximum breadth,** feet / m. **supporting surface,** sq. feet / m².

Total weight.—

Body.—

Wings.—

Motor.—Rebus. 120 × 130 m/m. Weight, 242½ lbs. (110 kgs.)

Speed.—

Propellers.—

Steering.—

Remarks.—

MILLER. Monoplanes.

Maximum length, 23 feet (7 m.) **maximum breadth,** 23 feet (7 m.) **supporting surface,** 236 sq. feet (22 m^2.)

Total weight.—551 lbs. (250 kgs.)

Body.—On 3 wheels; two forward, one aft; fitted with special Miller elastic suspension.

Wings.—Span of 23 feet (7 m.) Curved surfaces designed to give the maximum of longitudinal and latitudinal stability (see *Remarks*).

Motor.—55 h.p. Miller. Vertical 4-cylinder. 120×120 m/m. 1,400 r.p.m. Air-cooled by 2 fans. Overhead valves and cam shaft. 2 ignitions—H. T. magneto and accumulators independent of each other. Forced lubrication. Weight *complete*, $165\frac{1}{3}$ lbs. (75 kgs.) = 3 lbs. per h.p. (1·36 kgs. per h.p.) Motor is placed as low as possible.

Speed.—

Propellers.—1 tractor of steel. Diameter, $6\frac{1}{2}$ feet (2 m.) Chain driven.

Steering.—Single universal jointed rudder aft; controlled by a lever on the right side of driver.

Remarks.—Built by Franz Miller, of Turin. On the market for £500 (12,500 lire). The wings are very easily detached, and the fuselage can then be driven like an ordinary motor car.

MILLER. Biplane.
(Also known as COBIANCHI).

Maximum length, feet / m. **maximum breadth,** feet / m. **supporting surface,** 430 sq. feet / 40 m².

Total weight.—882 lbs. (400 kgs.) *including* driver.

Body.—Light frame mounted on 4 wheels.

Wings.—Upper surface curved and interrupted altogether in centre. Flexible wing tips. Lower deck has less span. Vertical fins rise from its extremities half way up to upper deck. Stabilising planes near ends between the decks.

Motor.—100 h.p. 9-cylinder Miller. Weight complete, 286½ lbs. (130 kgs.) = 2·8 lbs. per h.p. (1·3 kgs.)

Speed.—

Propellers.—1, in rear.

Steering.—Single plane elevator both forward and aft. 2 vertical rudders between the planes.

Remarks.—Built by Franz Miller, of Turin. Property of Mario Cobianchi. The type is on the market.

RADICE. Biplane.

Maximum length, $29\frac{1}{2}$ feet (9 m.) **maximum breadth,** 36 feet (11 m.) **supporting surface,** sq. feet (m².)

Total weight.—882 lbs. (400 kgs.)

Body.—

Wings.—Span, 36 feet (11 m.)

Motor.—35 h.p. 4-cylinder Rebus automobile engine.

Speed.—

Propellers.—2. Diameter, $8\frac{1}{2}$ feet (2·60 m.) Pitch, $8\frac{3}{4}$ feet (2·70 m.), geared to 450 r.p.m.

Steering.—Elevator of special design forward.

Remarks.—Somewhat resembles a Wright.

BOSSI. Biplane.
(Also known as F.I.A.M.)

Rebuilding

Maximum length, feet / m. **maximum breadth,** 46 feet / 14 m. **supporting surface,** sq. feet / m^2.

Total weight.—

Body.—On 2 wheels in centre line.

Wings.—Span, 46 feet (14 m.) 2 ailerons in rear amidships.

Motor.—35 h.p. Züst.

Speed.—

Propellers.—2. Diameter, 9·8 feet (3 m.)

Steering.—Biplane elevator forward. Vertical rudder aft. Special stability device.

Remarks.—This machine was built by the Faebrica Italiana Aeroplani in Milano (F.I.A.M.) to the designs of Signor Bossi. Wrecked September, 1909.

MONCHER. Helicoplane.

Maximum length, feet } m. } **maximum breadth,** feet } m. } **supporting surface,** sq. feet } m^2. }

Total weight.—

Body.—

Wings.—

Motor.—35 h.p. Rebus.

Speed.—

Propellers.—

Steering.—No vertical rudder.

Remarks.—

BERTELLI. Helicopter.

Building.

Note.—This may be one of the Howard Wright machines.

Maximum length, feet / m. **maximum breadth,** feet / m. **supporting surface,** sq. feet / m^2.

Total weight.—

Body.—

Wings.—

Motor.—

Speed.—

Propellers.—

Steering.—

Remarks.—No details.

FUSERI.

Building.

Maximum length, feet / m. **maximum breadth,** feet / m. **supporting surface,** sq. feet / m².

Total weight.—

Body.—

Wings.—

Motor.—

Speed.—

Propellers.—

Steering.—

Remarks.—No details procurable.

HOWARD WRIGHT No. I.

(Combined helicopter and glider.)

Maximum length, 28 feet (8·53 m.) **Maximum span,** (including fans) 50 feet (15·23 m.) **supporting surface,** sq. feet (m².)

Total weight.—1,250 lbs. (568 kgs.) *including* aeronaut.

Body—On 4 wheels. Frame of steel tubes welded without sockets.

Planes.—Main plane 30 feet (9·14 m.) span. Fans at each extremity 20 feet (6 m.) diameter. 2-bladed pitch variable. Action mostly vertical, slightly forward also. Total lift, 650 lbs. (295 kgs.)

Motors.—50 h.p. Antoinette (water cooled).

Speed.—Soars at 20 m.p.h. (32 km.)

Propellers.—One.

Steering.—Vertical rudder on same plane as main plane. 2 small horizontal rudders also on same plane.

Remarks—Built in London. Has made several successful flights, but the fans proved to be too small. As it was impracticable to increase their size, No. 2 was built.

HOWARD WRIGHT, No. II bis.

(Combined helicopter and glider.)

No photo available.

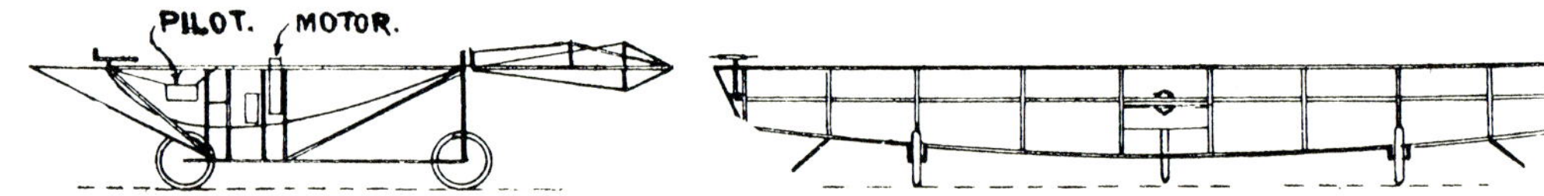

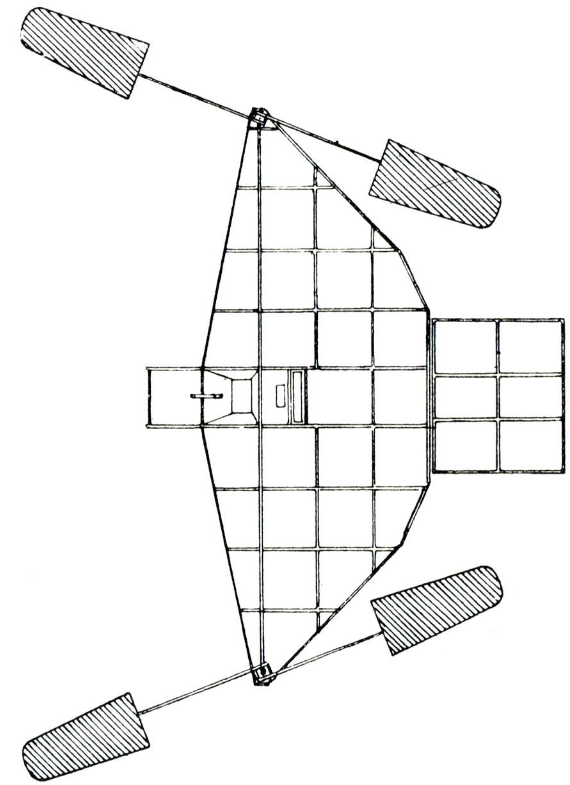

— HOWARD WRIGHT II bis. —

Maximum length, 28 ft. / 8·53 m. **maximum span** (including fans tip to tip), 62 ft. / 18·89 m. **depth of main plane,** 22 ft. / 6·70 m.

Total weight.—600 lbs. (272 kgs.), *not including* aeronaut.

Body.—Mounted on 4 wheels with spring forks. Track 20 feet (6 m.) Frame of steel tubes welded without sockets. Frame weighs 120 lbs. (54 43 kgs.

Planes.—Main plane is 30 feet (9·14 m.) span. At either end are fans, with vertical spindles 26 feet (7·92 m.) in diameter, running at 90-100 r.p.m. These fans are two-bladed. Pitch is varied during each revolution, so that they have an angle of incidence only when reversed. The fans are connected and produce a vertical action, also a slight forward action. The object is to fall by reducing speed of fans. Lift of fans, 33 lbs. per h.p. These fans weigh only 40 lbs. (18·14 kgs.) each. The blades are 6½ feet (1·97 m.) long by 3 feet (·91 m.) wide.

Motors.—30 h.p. 7 cylinder R.E.P., air cooled by propeller. Weight *complete*, 130 lbs. (58·96 kgs.)

Speed.—Soars at 15 m.p.h. (24 km.)

Propellers.—

Steering.—Vertical rudder on the same plane as main planes. 2 small horizontal rudders also on same plane.

Remarks.—Built in London. Embodies marked improvements on *No. 1* (page 213). The motion of this machine is in a series of curves, rising and falling, thus requiring comparatively little power to keep aloft.

VOISIN Biplanes. (*See France*).

The Avis Co. has the licence for Voisins in Italy. Normally an Anzani motor is fitted.

Machines of this type, built, building, or on order in Italy:—

Model 1908-9.

1 Florio*
2 Cagno
3 Anzani
4 *Order*
5 ,,
6 ,,
7 ,,

Notes.—*1. Has been alluded to as Signor Gloria. His is a standard older type Voisin.

2. Itala motor.

WRIGHT BROS. Biplanes. (*see U.S.A.*)

Italian Concessionaire:—

Of this make there are in Italy, built, building, or to order:—

Model 1908-9.

1 Lieut. Calderara
*(for Italian Government)
2 ———
3 Italian Government
4 Cadore Club

* Badly damaged in a fall at Rome: since repaired. Now fitted with an 80 h.p. engine.

ITALIA.

(Alias DA SCHIO).

Maximum length,	**maximum diameter,**	**volume,**
128 feet / 39 m.	$19\frac{3}{4}$ feet / 6 m.	42,663 c. feet / 1,208 m³.

Total lift.—2,976 lbs. (1,350 kgs.) **Useful lift,** lbs. (kgs.)

Gas bags.—Of rubbered cloth, except along the bottom, where a length of pure rubber is let in.

Motor.—40-50 h.p. 8-cylinder Antoinette.

Speed.—25 m.p.h. (40 km.)

Propellers.—1, above the motor, between the car and the gas bag, 15 feet (4·5 m.) diameter. The blades can be turned on their axes for going astern.

Steering.—A horizontal rudder each side of the car, which is of steel; large vertical and horizontal rudders at the rear end of the car, and a very large vertical rudder.

Remarks.—Designed and built in 1905 by Count Almorico da Schio. Since repeatedly reconstructed and altered. Finally, in 1909, the gas bag was destroyed, consequent on the air-ship becoming out of control when the motor had broke down. A new gas bag is being constructed. The unique feature of this ship is her "belly" of Para rubber. The use of this obviates the use of a ballonet. As a ballonet, on leaving the ground, usually occupies about a sixth of the total volume, this modification must be considered as increasing the lifting power by a sixth. If the system should prove successful when tried on a larger scale it should be very valuable to all semi and non-rigid air-ships.

RICALDONI I & II.

*(Alias **ITALIAN MILITARY AIRSHIP No. 1 & 2.**)*

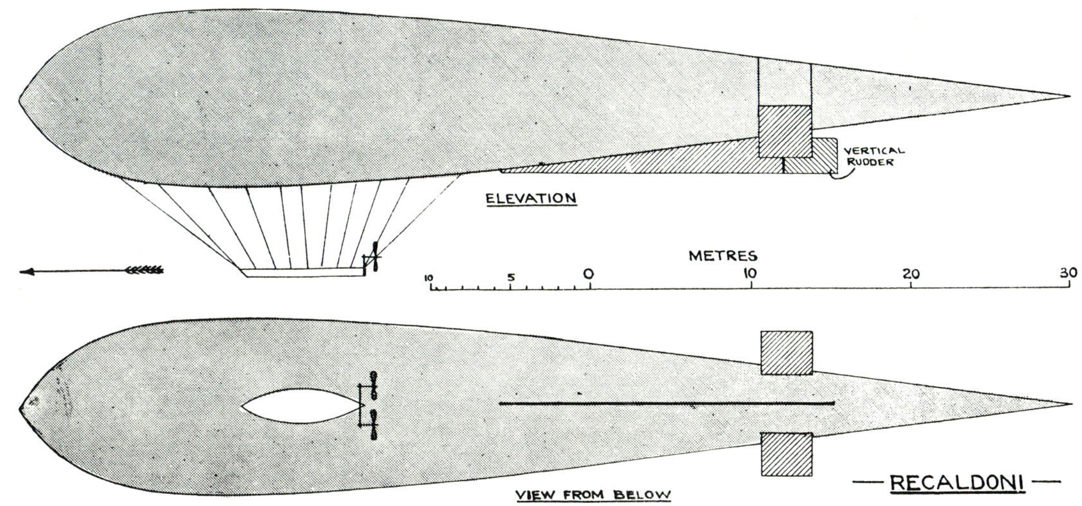

(Dimensions approximately only).

Maximum length, 216 feet / 66 m. **maximum diameter,** 36 feet / 11 m. **volume,** c. feet / m^3.

Total lift.— lbs. (kgs.) **Useful lift,** lbs. (kgs.)

Gas bags.—Of rubbered fabric, with ballonet. The shape is unique, the maximum diameter being very far towards the front, and the after part tailing off into a fine point.

Motor.—70 h.p. Bayard-Clement.

Speed.—*About* 27 m.p.h. (45 km.p.h.)

Propellers.—Two, 2-bladed, mounted on lattice girders, one on either side of the nacelle aft and a few feet above it.

Steering.—A box kite arrangement, containing moving planes, is fitted over the tail and acts as a stabiliser. The vertical rudder is at the rear of a long vertical fin under the gas bag.

Remarks.– Designed by Captains Crocco and Ricaldoni. This ship appears to have been very successful. Her lines are finer than those of any other type. No details have at any time been published. A second vessel of the same type has recently commenced her trials, and it is reported that a very large number of sister ships are under manufacture. The car is a beautifully shaped wooden boat, hung under the greatest section. In no other existing air-ship have such pains been taken to reduce air resistance to a minimum. It is, however, doubtful whether the shape of the gas bag is not unnecessarily fine.

FORLANINI (now FORLANINI bis).

Maximum length, 131¼ feet / 40 m. **maximum diameter,** 46 feet / 14 m. **volume,** 121,842 c. feet / 3,450 m³.

Total lift.—1,315 lbs. (2,900 kgs.) **Useful lift.**— lbs. (kgs.)

Gas bags.—Rubbered fabric. Fitted with 1 ballonet. Divided into 7 compartments.

Motors.—A 40 h.p. motor for propulsion, and also a 4 h.p. for auxiliary purposes.

Speed.—

Propellers.—

Steering.—Horizontal steering by means of a vertical rudder at the rear of the vertical fin. Vertical steering by altering the trim by means of moving the various weights.

Remarks.—The lower portion of the apparent hull is separated from the upper by a diaphragm, and serves as a ballonet and living space, and contains the motor. This lower portion contains a girder, making the ship Semi-Rigid. Under the rear two-thirds of the hull is a vertical keel connected to the internal girder. The keel terminates at the front end in a look-out place and captain's conning tower. The internal girder is 49¼ feet (15 m.) long by 5 feet (1·50 m.) wide by 5 feet (1·50 m,) high, and the motor is placed in this girder. The shaft runs through the middle of this girder and emerges from the ballonet near the tail, driving by means of gearing the propeller which is mounted at the tail point. First ascent, July 22nd, 1909. Recently reconstructed, as here given.

JAPANESE.

General Note.—Next to nothing is known as to aerial progress in Japan. One or two Wright machines are reported to have been ordered, but there is no confirmation obtainable. Two other machines are recorded. The "*Yamada* dirigible," despite very detailed accounts, is shrouded with a great deal of mystery. It appears to be nothing but a standard Louis Godard "kite balloon" of considerable dimensions. It was used at Port Arthur in the war for observation, but not for any other purpose.

Probably very little has yet been done, as there is very small public interest in the question as yet.

Aerial Societies:—

One reported forming.

Aerial Journals:—

Flying Grounds:—

Near Yokohama.

HARRISON. Biplane.

Maximum length, feet / m. **maximum breadth,** feet / m. **supporting surface,** sq. feet / m^2.

Total weight.—

Body.—Peculiar. Described as "like a street car with the sides knocked out and replaced by slender rods."

Wings.—Apparently reverse all accepted practice and placed fore and aft.

Motor.—Steam.

Speed.—

Propellers.—

Steering.—

Remarks.—Designed by the American steel capitalist, John W. Harrison. All rights purchased by Japan. The machine has flown, but without anyone on board. This machine has immense stability and obeys its rudder without regard to winds.

YAMADA. *Biplane.*

Maximum length, feet / m. **maximum breadth,** feet / m. **supporting surface,** sq. feet / m^2.

Total weight.—
Body.—
Wings.—
Motor.—
Speed.—
Propellers.—
Steering.—

Remarks.—Understood to resemble a *Wright* in its main features. Reported to be completely successful.

NORWEGIAN.

General Note.—So far as can be ascertained there are no aeroplanes in Norway. The German Wright Flying Machine Co. controls Norway for that machine, so a demonstration machine may shortly be sent.

Aerial Societies:—

Aerial Journals:—

None.

Flying Grounds:—

PERUVIAN.

Maximum length, feet } m. **maximum breadth,** feet } m. **supporting surface,** sq. feet } m^2.

Total weight.—

Body.—

Wings.—

Motor.—

Speed.—

Propellers.—

Steering.—

Remarks.—Nothing known.

PORTUGUESE.

General Note.—An aeroplane, of which no details are known, reported building at Lisbon.

There are also rumours of negotiations for the purchase by a Portuguese subject of a Pischoff-Koechlin aeroplane.

Aerial Societies:—

One reported *projected.*

Aerial Journals:—

Flying Grounds:—

RUSSIAN.

General Note.—Aviation has been studied in Russia for a considerable while and much quiet progress made. So long ago as 1903, the late Lieut. Pavloff, of the Navy, flew a small machine at Sevastopol, which is believed to be embodied in the present Baronovski machines. Man-carrying kites were in use in the Navy in 1900.

Russian investigation is almost entirely in the direction of helicopters.

Aerial Societies:—

Russian Aeronautical Society.
Odessa Aero Club.

Aerial Journals:—

Aeronautical Journal of St. Petersburg.
Vozdookhoplaratel.

Flying Grounds:—

Gatchina.
Odessa.

BARANOVSKI. Monoplane-helicopter (Nos. 1-5).
(Russian Government).

Maximum length, feet / m. **maximum breadth,** feet / m. **supporting surface,** sq. feet / m^2.

Total weight.—

Body.—Bird shaped.

Wings.—1 on either side of the engine.

Motor.—

Speed.—

Propellers.—1, mounted behind the planes.

Steering.—

Remarks.—On trial one of these left the ground immediately without any run. The type was officially selected for that reason.

ZATOFF. Biplane.

Maximum length, feet / m. **maximum breadth,** feet / m. **supporting surface,** sq. feet / m^2.

Total weight.—

Body.—

Wings.—

Motor.—Engine uses a special fuel—said to be a form of glycerine.

Speed.—

Propellers.—

Steering.—

Remarks.—

LOUTZKY (Boris). *Biplane-helicopter.*

Maximum length, feet / m. **maximum breadth,** 46 feet / 14 m. **supporting surface,** sq. feet / m^2.

Total weight.—*About* 2,200 lbs. (998 kgs.)

Body.—Very strongly built.

Wings.—Span of 46 feet (14 m.)

Motors.—2 of 50 h.p. each.

Speed.—

Propellers.—

Steering.—

Remarks.—Built in Berlin. Designed to rise vertically in any wind, and to carry 50% extra load.

LEBEDEFF. Biplane.

Maximum length, feet / m. **maximum breadth,** feet / m. **supporting surface,** sq. feet / m^2.

Total weight.—

Body.—

Wings.—

Motor.—

Speed.—

Propellers.—

Steering.—

Remarks.—Said to have flown. No data. He also either owns a Voisin or flies it for the Odessa Aero Club, and this may be the same machine, and not one of his own invention, as generally reported. No reply from him to date of going to press.

H. FARMAN Biplanes.

Of this type there are in Russia, built, building, or to order:—

Model 1908-9.

1 ——— (*Built*)

PISCHOFF-KOECHLIN Biplanes. (*See France*).

Concessionaire:—

Of this make there are in Russia, built, building, or to order:—

Model 1908-9.

1 *Building* in France.

VOISIN Biplanes. *(See France).*

Of this make there are in Russia, built, building, or on order:—

Model 1908-9.

1 Graf de Malynski (*ex* Delagrange)
2 ——— Odessa Aero Club
3 ———

WRIGHT Biplanes. *(See U.S.A.)*

Of this make there are in Russia, built, building, or on order:—

Model 1908-9.

1, 2, 3, 4, 5, 6 } Russian Government
7 ———

TATOUROFF

Building.

Maximum length, feet / m. **maximum breadth,** feet / m. **supporting surface,** sq. feet / m^2.

Total weight.—

Body.—

Wings.—

Motor.—

Speed.—

Propellers.—

Steering.—

Remarks.—Nothing known ; except that he is subventioned by the Government.

BOLATOFF (Prince). Triplane.

Maximum length, feet / m. **maximum breadth,** feet / m. **supporting surface,** sq. feet / m^2

Total weight.—
Body.—
Wings.—
Motor.—100 h.p.
Speed.—
Propellers.—
Steering.—

Remarks.—Built by Voisin (France). This machine is being tried at Reigate, England.

Sister Ship to LA FLANDRE.

(Astra class.)

Maximum length, 246 feet (75 m.) **maximum diameter,** $48\frac{1}{4}$ feet (14·7 m.) **volume,** 69,914 c. feet (6,500 m^3.)

Total lift.—15,763 lbs. (7,150 kgs.) **Useful lift,** lbs. (kgs.)

Gas bags.—Continental rubbered fabric, yellow. Ballonet, 16,146 c. feet (1,500 m^3.)

Motors.—2 Pipe motors of 110 h.p. each, placed in line with each other in the fore and aft line, and with clutches and the necessary gearing in between them.

Speed.—35 m.p.h.

Propellers.—3, namely, one at the fore end, driven by the two motors when coupled together, and two placed above and on either side of the centre of the car, for use when only one motor is running. Made of wood, and of the "Integrale" type.

Steering.—Vertical steering by means of a large double aeroplane fixed above the car, about a third from the front. Horizontal steering by means of a double vertical rudder above the rear end of the car. Stability is secured by the usual Astra pear shaped stabilising gas bags, with fins of rubbered cloth spread between the inner edges of these shapes.

Remarks.—The distinctive feature of this ship is the arrangement of the propellers. Both motors can be coupled either on to the front propeller or on to the two rear propellers, or on to all three together, but they are actually intended only to drive the front one. On stopping either motor the other is connected to the two rear propellers, which are designed for a slower speed of translation than the front one, with the result that the running motor does not find itself overloaded as it would if the same propeller had to serve both for one and for two motors.

BAYARD-CLÉMENT. Astra Type.

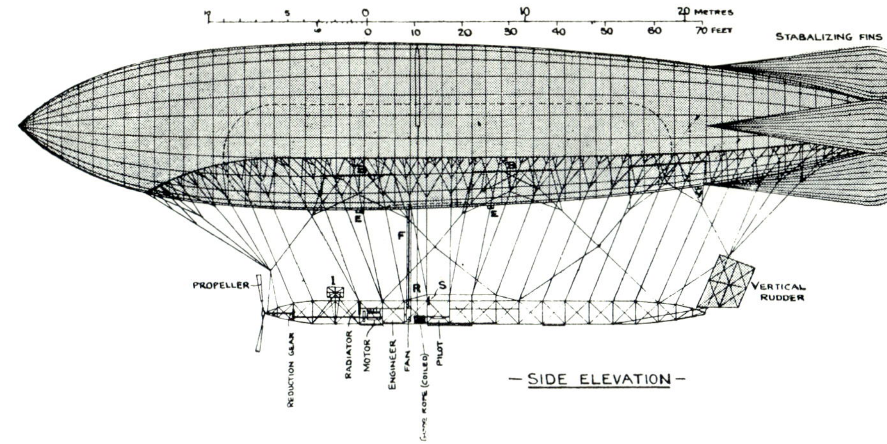

Maximum length, 184 feet. / 56 m., 25. **maximum diameter,** $34\frac{3}{4}$ feet. / 10 m., 58. **volume,** 123,608 c. feet. / 3,500 m³

Total lift.—8,488 lbs. (3,850 kgs.) **Useful lift,** lbs. (kgs.)

Gas bags.—Of Continental balloon material, surface 23,214 sq. feet (2,250 m².) Weight, lbs. (805 kg.) The stabilising fins at the rear end connect with the interior of the gas bag through small holes. Ballonet, 38,844 c. feet (1,100 m³), divided in the middle into 2 airtight compartments.

Motor.—1 Bayard-Clément motor, 105 h.p. Bedded on springs.

Speed.—$33\frac{1}{2}$ m.p.h. (54 km. p.h.)

Propellers.—1 in front of the car, $16\frac{1}{2}$ feet (5 m.) diameter. Driven by reduction gearing at 380 r.p.m.

Steering.—Large double vertical rudder of $193\frac{3}{4}$ sq. feet (18 m².) in rear of and just above the car. Vertical steering by means of a treble horizontal aeroplane just above the car and before the motor; area, 172 sq. feet (16 m²).

Remarks.—Built for the private use of M. Clément. Bought by the Russian Government. The car is a girder built of steel tubes, $93\frac{3}{4}$ feet (28 m , 50) long, and of square section, with feet (1 m., 50) sides. It is raised in the centre to form a better shelter, and covered with aluminium plate and waterproof material. The unusually large size of the ballonet admits of unusual heights being attained, up to 9,000 feet. The gas bag has 2 automatic valves under the stern, opening under 40 mm. of internal pressure. The ballonet also has 2 automatic valves, opening under 1·18 inches (30 mm.) of internal pressure; these latter can also be worked by hand.

KEY TO PLAN.

B = Ballonets.
V = Gas valve.
E = Air valves.
F = Pipe from fan to ballonet.
I = Horizontal rudder.
R = Water tank.
S = Steering wheel.

LEBED.

(*ex* La Russie.)

[To all intents and purposes a sister of the ill-fated *Republique* (see page 161).]

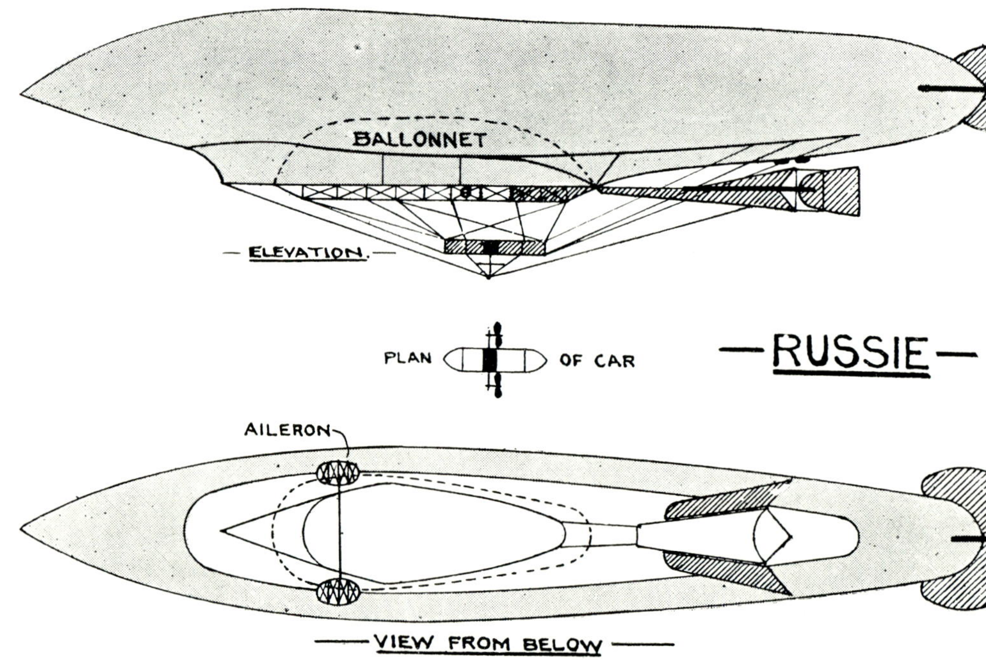

OUTCHEBNY.

(Russian Government.)

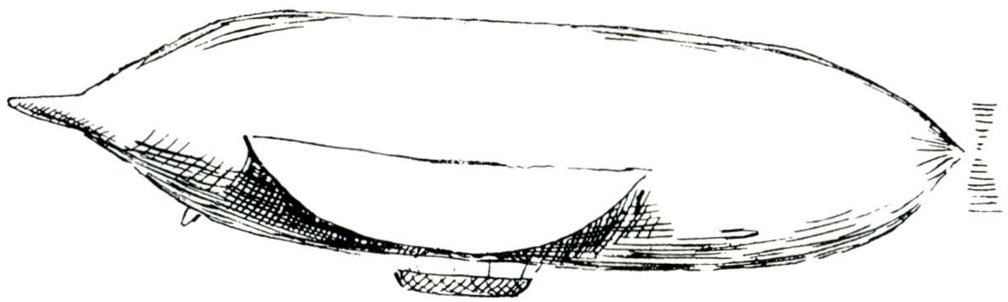

Maximum length, feet } m. } **maximum diameter,** feet } m. } **volume,** c. feet } m³. }

Total lift.— lbs. (kgs.) **Useful lift,** lbs. (kgs.)

Gas bags.—Composed of 2 kite balloons, cut up and joined together.

Motor.—

Speed.—

Propellers.—

Steering.—

Remarks.—This is a very primitive "home-made" article.
Reported to be a failure.
Details unknown.

SOUTH AMERICAN REPUBLICS.

General Note.—There are rumours of aeroplanes building in South America—especially in Brazil, but nothing can be authenticated.

It is probable, however, that one or two standard machines are on order for South America.

Santos Dumont, one of the earliest pioneers, is a Brazilian; but his machines, being always in France, will be found recorded under the head of "French."

SPANISH.

General Note.—Spain has produced a few aeroplanes, also a dirigible possessing novel features.

Aerial Societies:—

El Real Aëro Club de España.
La Asociacion de Locomocion Aérea.

Aerial Journals:—

Barcelona Aero Club Journal.
Boletin Oficial de la Asociacion de Locomocion Aérea.

Flying Grounds:—

AMA. Monoplane.

(Ametzoy, Magica and Azeona.)

Building.

Maximum length, feet / m. **maximum breadth,** feet / m. **supporting surface,** sq. feet / m^3.

Total weight.—

Body.—

Wings.—

Motor.—

Speed.—

Propellers.—

Steering.—

Remarks.—

CANELLA. Biplane.

Building.

Maximum length, feet / m. **maximum breadth,** feet / m. **supporting surface,** sq. feet / m^2.

Total weight.—

Body.—

Wings.—

Motor.—

Speed.—

Propellers.—

Steering.—

Remarks.—Building near Valladolid. Reported to be a sort of Wright Bros.

VOISINS. *(see France.)*

Concessionaire :

Of this type there are in Spain :—

Model 1908-9.

1 Barker

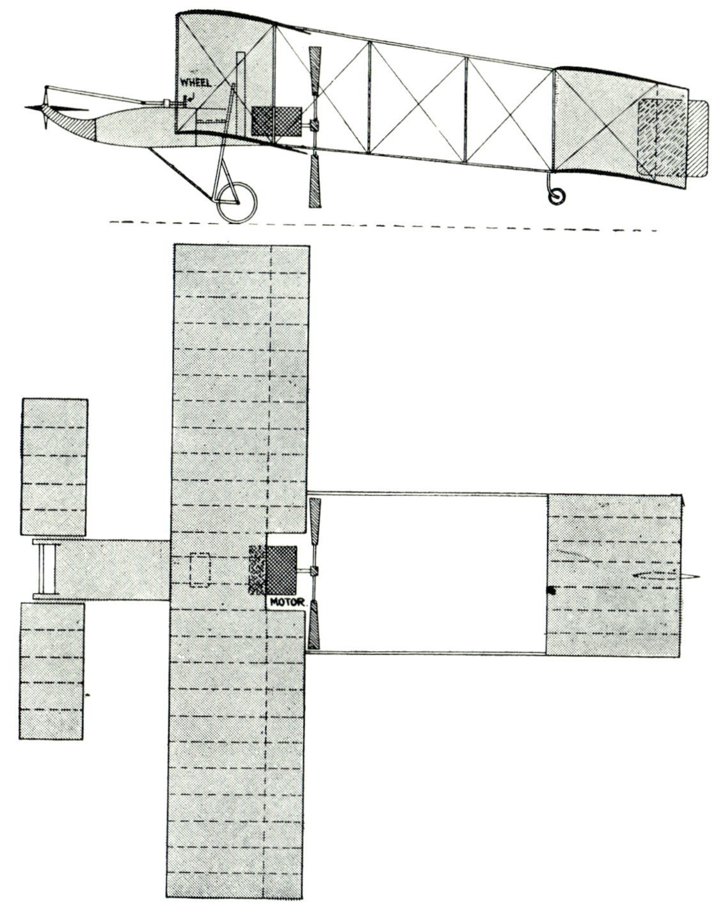

VOISIN type.

SANCHEZ. Biplane.
(Spanish Government.)

Building.

Maximum length, feet m. **maximum breadth,** 39½ feet 12 m. **supporting surface,** 430 sq. feet 40 m^2.

Total weight.—

Body.—On 4 wheels.

Wings.—Span, 39½ feet (12 m.) Depth, 6 feet (1·85 m.)

Motor.—

Speed.—

Propellers.—2, chain driven; abaft the main planes.

Steering.—

Remarks.—Closely resembles a Wright in its main features. Said to be the invention of Lieut. Sanchez, of the Spanish Army.

SALAMANCA.

Building.

Maximum length, feet / m. **maximum breadth,** feet / m. **supporting surface,** sq. feet / m².

Total weight.—
Body.—
Wings.—
Motor.—
Speed.—
Propellers.—
Steering.—
Remarks.—

WRIGHT BROS. Biplanes. (*See U.S.A.*)

Concessionaire :—

Of this type there are in Spain, built, building, or to order :—

Model 1908-9.
1 ——
2 ——

WRIGHT type.

TORRÈS-QUEVEDO.

Maximum length, 118 feet } 36 m. } **maximum diameter,** feet } m. } **volume,** 33,550 c. feet } 960 m³. }

Total lift.— lbs. (kgs.) **Useful lift,** lbs. (kgs.)

Gas bags.—The arrangement of the gas bag is unique and is best seen from the accompanying sketch The object of this arrangement has not been revealed.

Motors.—Two 24 h.p., each Antoinette, placed side by side.

Speed.—

Propellers.—2 propellers, one for each motor, 5 feet (1·5 m.) diameter.

Steering.—

Remarks.—This vessel is only experimental, preparatory to building one of 141,270 c. feet (4,000 m³.) volume. It was designed by Captain Kindelan and Engineer Torres Quevedo.

ASTRA class.

Maximum length, 197 feet / 60 m. **maximum diameter,** 35$\frac{1}{3}$ feet / 10·75 m. **volume,** 43,057 c. feet / 4,000 m³.

Total lift.—9,700 lbs. (4,400 kgs.) **Useful lift.** lbs. (kgs.)

Gas bags.—Yellow coloured rubber proofed Continental fabric.

Motor.—One 100 h.p. 4-cylinder Panhard.

Speed.—29 m.p.h.

Propellers.—1, at the front end of the car, of wood, "Integrale" type.

Steering.—As in *Clement Bayard I* and *Ville de Nancy*.

Remarks.—The two side stabilising shapes are duplicated, as they are in the *Ville de Paris*. A webbing stretched on steel tubes is introduced between the inner edges of the 4 main stabilising shapes to provide extra stabilising surface.

SWEDISH.

General Note.—Two aeroplanes of Swedish design are possibly in an early stage of construction.

A *Wright* machine is building for Sweden by the German Wright Flying Machine Co. (which controls this biplane for Sweden), and the Swedish Government is reported to have ordered two *Zeppelins* and several *Wrights*.

Aerial Societies:—

Svenska Aëronautiska Sallskapet.

Aerial Journals:—

None.

Flying Grounds:—

VOISIN type. *(see France.)*

Of this make there are in Sweden:—

1 Dr. Folmer Hansen.
2 *On order.*

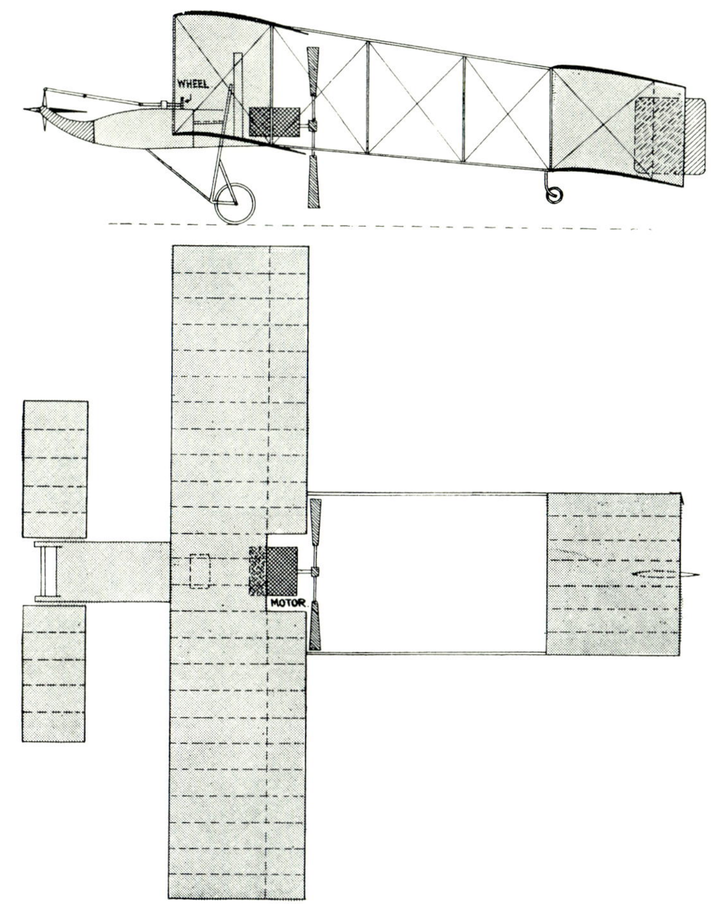

VOISIN type.

SWISS.

General Note.—Switzerland is keenly interested in the German dirigibles, and already making preparations for an aerial service.

Aerial Societies:—

Aëro Club Suisse, Berne.

Aerial Journals:—

Flying Grounds (with hangars):—

Lucerne (with 60 acre park), *building*.
Zurich, *building*.

COMISETTI AND GUIGNET.

Building.

Maximum length, feet / m. **maximum breadth,** feet / m. **supporting surface,** sq. feet / m^2.

Total weight.—
Body.—
Wings.—
Motor.—130 h.p.
Speed.—
Propellers.—
Steering.—

Remarks.—Building at Geneva.

PLAZZERIAUD. Triplane.

Building.

Maximum length, feet / m. **maximum breadth,** feet / m. **supporting surface,** sq. feet / m^2.

Total weight.—
Body.—
Wings.—
Motor.—
Speed.—
Propellers.—
Steering.—Elevator placed forward. Vertical rudder aft.

Remarks.—Constructed at Zurich. *Special features*: Extremely low centre of gravity.

GENÈVE.

Building.

Maximum length, feet / m. **maximum diameter,** feet / m. **volume,** 123,600 c. feet / 3,500 m^3.

Total lift, 8,380 lbs. (3,800 kgs.) **Useful lift.**— lbs. (kgs.)

Gas bags.—
Motor.—125 h.p.
Speed.—
Propellers.—
Steering.—

Remarks.—Built at Vernier, to the designs of Engineer Liwenthal. The construction of this vessel has been much criticised and much delayed. At one time her construction was reported to have been abandoned; later it was reported to have been resumed. No details are available.

TURKISH.

General Note.—The German Wright Flying Machine Co. having secured all rights for their machines in Turkey, there is presumably an expected market; but so far as can be ascertained, there are at present no aviators in Turkey.

The late Sultan ordered a dirigible in France, and negotiations for a couple of *Zeppelins* have more recently been in progress.

U. S. A.

General Note.—In the early nineties, Professor Langley and the Bros. Wright were experimenting with heavier-than-air machines, but general interest in the subject is quite recent. Though some dirigibles exist, American attention is mainly devoted to aeroplanes of various types, chiefly (except the Wright Bros. and Curtis) triplanes and helicopters. A number of experimental helicopters, of which there is no record, are believed to be building or ordered.

It is stated that there are no less than *eight thousand* people in the U. S. A. who have flying machine designs in some stage or other. For "8,000," 800 is the more probable figure. Of these many are either schemes for fanciful machines never likely to be built, or if built, extremely unlikely to rise from the ground; or else home-made copies of *Wright* or *Curtis* machines.

In the following pages an attempt has been made to include every original American machine actually in hand.

Aerial Societies:—

Aerial Experiment Association (*now disbanded*).
Aeronautic Society.
International School of Aeronautics.
Junior Aero Club.
Woman's Aero Club.
Aero Club of America, New York.
Aero Club of New England, Boston.
Aero Club of Philadelphia.
Philadelphia Aeronautical Recreation Society.
Aero Club of Ohio.
Aero Club of St. Louis.
Milwaukee Aero Club.
Ben Franklin Aeronautical Society.
North Adams Aero Club.
Pittsfield Aero Club.
Aero Club of Chicago.
Aeronautique Club of Chicago.
Aero Club of San Antonio.
Pacific Aero Club.
International Aeroplane Club, Dayton.

Aerial Journals:—

American Aeronaut.
Aeronautics.
Fly.

Flying Grounds:—

Fort Myer.
Washington.
Morris Park (Aeronautic Society).
Fort Sill Reservation, Oklahoma (Government Station, *building*).
Dayton.
Near San Francisco.
Hemsted Plain, Long Island.
Westburg, Long Island (600 acres).

BERLINER & WILLIAMS I. Helicopter.

Completing.

Maximum length, feet / m. **maximum breadth,** feet / m. **supporting surface,** sq. feet / m^2.

Total weight.—

Total lift.—610 lbs. (277 kgs.)

Body.—

Wings.—

Motor.—2 Adams-Farwell.

Speed.—

Propellers.—2 superposed 2-bladed helices on concentric shafts revolving in opposite directions.

Steering.—

Remarks.—Parachute carried for emergencies.

BERLINER & WILLIAMS II. Helicopter.

Maximum length, feet / m. **maximum breadth,** feet / m. **supporting surface,** sq. feet / m^2.

Total weight.—

Total lift.—780 lbs. (354 kgs.)

Body.—Tubular frame.

Wings.—*See* Propellers.

Motor.—2 Adams Farwell.

Speed.—

Propellers.—2 large 2-bladed superposed on concentric shaft. Revolve in opposite directions.

Steering.—

Remarks.—To all intents and purposes merely a larger edition of *Berliner & Williams I.*

BOLAND. Monoplane.

Completing. Being altered.

Maximum length, feet / m. **maximum breadth,** feet / m. **supporting surface,** sq. feet / m^2.

Total weight.—

Body.—

Wings.—

Motor.—

Speed.—

Propellers.—

Steering.—

Remarks.—Built at Rahway, N.J. All details kept most strictly confidential. The machine has been seen making short flights.

BOURDIN. Monoplane.

Building.

Maximum length, feet / m. **maximum breadth,** feet / m. **supporting surface,** sq. feet / m^2.

Total weight.—

Body.—

Wings.—Fixed planes.

Motor.—Mounted to shift.

Speed.—

Propeller.—1, direct driven.

Steering.—Shifting motor is pivoted in front of the machine, and the idea is to use this for steering.

Remarks.—Built at S. Francisco.

BEACH-WILLARD. Monoplane.

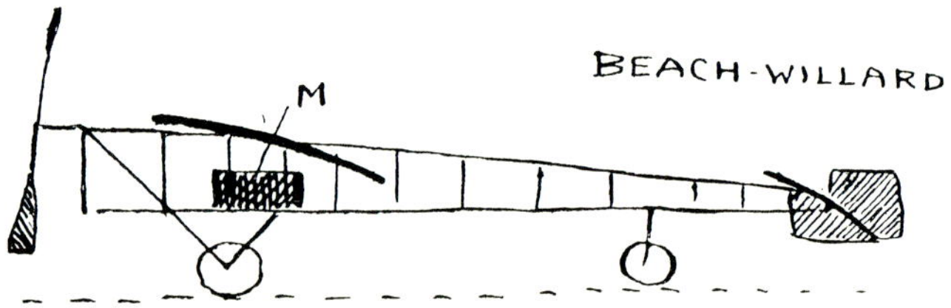

Maximum length, 36 feet / 11 m. **maximum breadth,** $39\frac{1}{4}$ feet / 12 m. **supporting surface,** sq. feet / m².

Total weight.—*About* 750 lbs. (340 kgs.) *including* aviator.

Body.—Square body frame in front, triangular in rear. It is *about* $3\frac{1}{4}$ feet (1·00 m.) high in front, tapering down to *about* $1\frac{1}{2}$ feet (0·50 m.) in rear. Length of frame, 33 feet (10 m.)

Wings.—38 × 8 feet (11·60 × 2·50 m.) maximum dimensions. At the tips the wings are 6 feet (1·83 m.) wide. Triangular wing tips fitted. These are controlled by the aviator swaying his body.

Motor.—50 h.p. 4-cylinder, water-cooled, high compression engine of special design (Willard) 1,000 r.p.m. Concentric valves. Dimensions, 100 × 105 m/m. Weight complete, 350 lbs. (159 kgs.) = 7 lbs. per h.p. (*about* 3 kgs. per h.p.) Automobile type. The motor is placed in the bottom of the body.

Speed.—

Propellers.—1, chain and sprocket drive. Ratio of revolutions engine to propeller, $2\frac{1}{2}$-1. Diameter of propeller, 9 feet (2·74 m.) Pitch, 9 feet (2·74 m.)

Steering.—There is a horizontal surface at the end of the body frame 5 × 6 feet (1·50 × 1·80 m.) This has warping wing tips 2 feet square. These are operated by a single lever on the aviator's left. Vertical rudder combined with the after supporting wheel. Both operated by a single lever on aviator's right.

Remarks.—Completed 31st May, 1909. It can be transported in a motor car (this has been actually done several times over rough roads). Designed by Stanley Y. Beach, Aeronautical Editor, *Scientific American*, and Charles F. Willard. At first glance it suggests a cross between a *Bleriot* and an *Antoinette*; but on inspection is found to contain many original features, carefully designed and thought out. It is still somewhat in the trial and experiment stage; but promises to become an extremely successful design.

HENDRICK. Monoplane.

Completing.

Maximum length, 20 feet / 6 m. **maximum breadth,** feet / m. **supporting surface,** sq. feet / m².

Total weight.—400 lbs. (181·50 kgs.)

Body.—

Wings.—

Motor.—2-cylinder Curtiss.

Speed.—

Propellers.—

Steering.—

Remarks.—Built by C. J. Hendrick, Middletown, N.J. The machine closely resembles Santos Dumont's *Demoiselle*.

V. L. OCHOA. Flapper.

(JERSEY MOSQUITO.)

Maximum length, feet / m. **maximum breadth,** feet / m. **supporting surface,** sq. feet / m².

Total weight.—

Body.—

Wings.—2 very large steel flapping wings.

Motor.—

Speed.—

Propellers.—

Steering.—

Remarks.—Was on show at Arlington, May, 1909, but did not fly.

LATE AERIAL EXPERIMENT ASSOCIATION.

Biplanes: (1) JUNE BUG, (2) RED WING, (3) WHITE WING.

Photo, C. Malcuit.

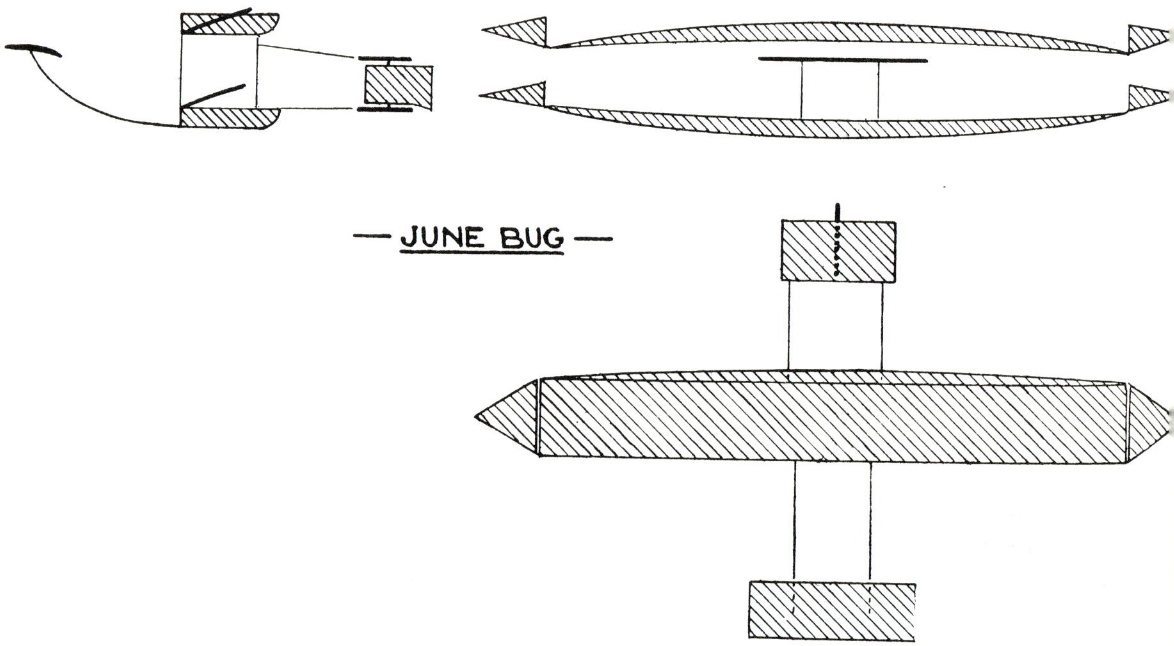

Maximum length, feet. / m. **maximum breadth,** 36¼ feet. / 11 m. **supporting surface,** sq. feet. / m².

Total Weight.—600 lbs. (272 kgs.)

Body.—Built of wood, covered with rubbered fabric. Mounted on 3 wheels.

Wings.—Planes are 30 feet wide (9·1 m.) by 4 feet (1·2 m.) from front to rear. Stabilising triangular planes at the ends of the planes. An upward in rear of planes.

Motor.—

Speed.—

Propellers.—One of 5½ feet (1·67 m.)

Steering.—Single horizontal rudder in front. Box tail in rear; Vertical rudder inside it.

Remarks.—There is an original method of keeping automatic stability. *Special feature*: Planes are curved towards each other.

RICHARDSON. Biplane.

Completing.

Maximum length, feet, m. **maximum breadth,** 38 feet (11·58 m.) **supporting surface,** sq. feet, m².

Total weight.—

Body.—

Wings.—Upper and lower planes form dihedral angles and touch at the extremities. Span of planes, 34 feet (10·36 m.) At extremities are ailerons, 2 × 8 feet (0·60 × 2·43 m.) Rear edges flexible. In rear of planes, two surfaces at a dihedral angle, 26 feet long × 8 feet wide (7·92 × 2·43 m.)

Motor.—

Speed.—

Propellers.—

Steering.—Biplane tail, universally jointed, 3 × 8 feet (0·91 × 2·43 m.) Distance between surfaces, 2 feet (0·60 m.)

Remarks.—Control by 2 levers.

Built by H. C. Richardson, Othello, Washington.

CUTHBERTSON II. Monoplane.

Completing.

Maximum length, feet / m. **maximum breadth,** 30 feet / 9·14 m. **supporting surface,** sq. feet / m².

Total weight.—

Body.—

Wings.—Fixed planes with a considerable upward inclination.

Motor.—16 h.p.

Speed.—

Propellers.—1. Diameter 5½ feet (1·67 m.) Chain driven at 600 r.p.m.

Steering.—Single plane elevator, forward (lever controlled). Single vertical rudder aft (wheel controlled).

Remarks.—Built to the inventor, designs by the Michigan Steel Boat Co., Detroit. *C.I.* was smashed on its first trials. This one is an exact duplicate.

DE HAVEN & WATKINS. Monoplane.

Building.

Maximum length, feet / m. **maximum breadth,** feet / m. **supporting surface,** sq. feet / m².

Total weight.—

Body.—

Wings.—

Motor.—

Speed.—

Propellers.—

Steering.—

Remarks.—Building by Claude de Haven at San Francisco. To have A. C. Watkins patent equilibrium device.

BABCOCK, ROBINSON & GLEASON. Biplane.

Building.

Maximum length, feet / m. **maximum breadth,** feet / m. **supporting surface,** sq. feet / m^2.

Total weight.—

Body.—

Wings.—About a quarter of the surface of each wing is set at a dihedral angle.

Motor.—

Speed.—

Propellers.—

Steering.—Horizontal rudder in front. Combined horizontal and vertical rudder in tail.

Remarks.—

LOOSE. Monoplane.

On trials.

Maximum length, feet / m. **maximum breadth,** 26 feet / 8 m. **supporting surface,** 268 sq. feet / 25 m^2.

Total weight.—750 lbs. (340 kgs.)

Body.—

Wings.—Span 26 feet (8 m.)

Motor.—20 h.p.

Speed.—

Propeller.—1. Diameter, 8 feet (2·44 m.) Pitch, $6\frac{1}{2}$ feet (2 m.)

Steering.—

Remarks.—Designed by G. H. Loose, of Redwood City, California. Somewhat on Pischoff-Koechlin lines. Engine tried August, 1909.

BEACH. Biplane.
(*Alias* **WHITEHEAD.**)

Maximum length, feet / m. **maximum breadth,** 40 feet / 12 m. **supporting surface,** sq. feet. / m².

Total weight.—1,200 lbs. (544 kgs.), *including* pilot.

Body.—Boat shaped. Designed to give longitudinal stability and assist lateral stability. Mounted on 4 wheels, the rear pair connected with the steering wheel.

Wings.—Main planes are 40 × 8 feet (12 × 2·43 m.)

Motor.—50 h.p. water cooled, 5 cylinder Whitehead, with automatic inlet valves. Bore, 127 m/m. Stroke, 140 m/m. Fitted with supplementary exhausts. The front of the body acts as a combined tank and radiator, the water being sprayed against it internally. Engine carried underneath the body.

Propellers.—2. Diameter 8 feet (2·43 m), belt driven, mounted in front of the planes, one on either side of the body.

Steering.—By ordinary motor car wheel for the vertical rudder and rear landing wheels. The horizontal rudder is worked by a special lever.

Remarks.—In general features this machine somewhat suggests a Bleriot monoplane turned into a biplane. Its own special feature is the very low centre of gravity and considerable stability. Built to the designs of Stanley Y. Beach by G. Whitehead, Bridgeport, Conn.

CURTIS. Biplane.

Photo, Benoit.

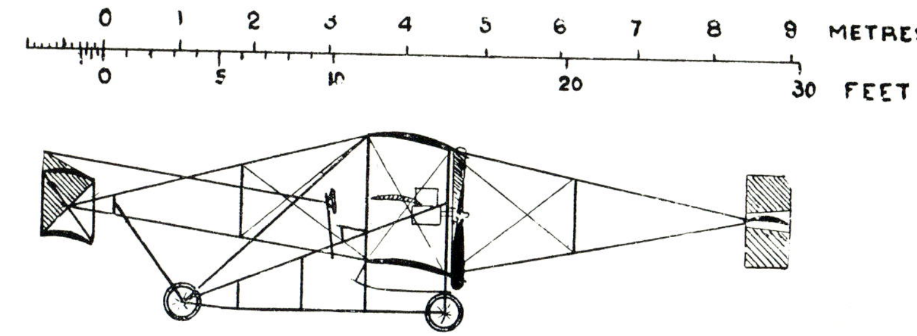

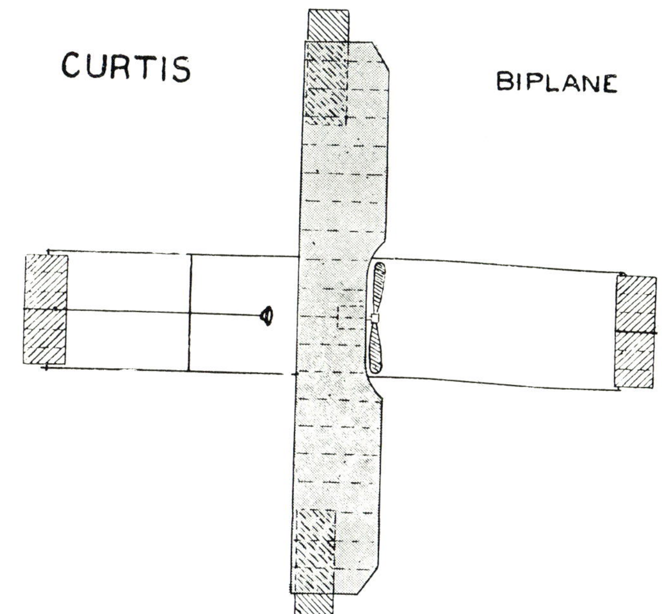

Maximum length, 33½ feet (10·20 m.) **maximum breadth,** 32 feet (9·75 m.) **supporting surface,** 272 sq. feet (25 m².)

Total weight.—550 lbs. (249·5 kgs.) *not* including aeronaut.

Body.—Plane frames, oregon spruce, bamboo for the general framework.

Wings.—Maximum span, 29¾ feet (9 m.) Depth, 4½ feet (1·40 m.) Planes spaced apart, 4½ feet (1·40 m.) Surfaces have a curvature of about 1 in 9. Angle of inclination, 6°. Material: Baldwin rubber coated silk. Ailerons, 6×2 feet (1·82×0·60 m.) at the ends, between the planes, operated by the driver swaying his body.

Motor.—30 h.p. 4-cylinder Curtis, 90×100 m/m, 1,200 r.p.m. Water cooled, copper water jackets. Bosch H.T. magneto. Weight of engine, 97 lbs. (44 kgs.)=about 3 lbs. per h.p. (1·36 kgs.) Total power plant, 192 lbs. (87 kgs.) Engine carried amidships.

Speed.—45 m.p.h. (72 km.) Rises at 25 m.p.h. (40 km.)

Propellers.—One direct drive, placed amidships just behind engine. Diameter, 6½ feet (2 m.) Pitch 6 feet (1·8 m.) Blades, 5 inches (126 m/m) in width. Thrust 225 lbs. (102 kgs.)

Steering.—Horizontal rudder, double surface, 6×2 feet (1·82×0·60 m.) carried 10 feet (3 m.) in front of the main planes, with large triangular steadier in centre. 10 feet (3 m.) behind the main planes is a single adjustable horizontal plane, 6×2 feet (1·82×0·60 m.) in the centre of which is the vertical rudder, 2½ sq. feet (0·19 m².) Control is by a steering wheel which is pushed or pulled for rising and falling. For horizontal steering this wheel is employed in conjunction with movements of the pilot's body. Rising and falling is also controlable by varying the engine power.

Remarks.—Built for the Aeronautic Society of America by Messrs. Curtis & Herring. It is an improved *June Bug (q.v.)*

HERRING CURTISS Biplanes. *(see Curtis)* of Hammondsport, N.Y.

Messrs. Herring and Curtiss were formerly associated with Dr. Graham Bell, Messrs. F. S. Baldwin and J. A. D. McCurdy in producing the *June Bug*, *Silver Dart*, etc. The Herring-Curtiss wing is now a separate Company.

Herring-Curtiss machines, built. building, or to order:—

Model 1908-9.

1 G. H. Curtiss (*q.v.*) *bought.*

2 Cortland Field Bishop.

3 Signal Corps, U. S. Army, *building*

4 *Private order. Building.*

5 — Willard

6 G. H. Curtis

7

8

9

10

11

GREENE (Dr.) Biplane.

Completing.

Maximum length, feet / m. **maximum breadth,** 40 feet / 12·19 m. **supporting surface,** sq. feet / m².

Total weight.—

Body.—

Wings.—The planes measure 40×6½ feet (12·19×2 m.) Distance between planes 6 feet (1·82 m.) Automatic stabilising planes above and below the main planes at either extremity. Size of these 10×5 feet (3×1·50 m.)

Motor.—

Speed.—

Propellers.—3. Two in front and one astern.

Steering.—Entirely by the tail, which can be operated for both vertical and horizontal steering.

Remarks.—Built and owned by Dr. William Greene, treasurer of the U.S.A Aeronautic Society.

HENDRICKSON. Biplane.

On trials.

Maximum length, feet / m. **maximum breadth,** 40 feet / 12·12 m. **supporting surface,** sq. feet / m².

Total weight.—

Body.—

Wings.—40 × 7 feet (12·12 × 2·13 m.)

Motor.—60 h.p.

Speed.—

Propellers.—

Steering.—

Remarks.—

GOULD, CONTENT AND LOENING. Biplane.

Building.

Maximum length, 40 feet (12·20 m.) **maximum breadth,** 26 feet (7·90 m.) **supporting surface,** sq. feet (m².)

Total weight.—

Body.—Boat like; intended to rise from water.

Wings.—Planes have 26 feet (7·90 m.) span, Special rubber treated glazed cloth. Bamboo frames. Pointed ends to planes.

Motor.—

Speed.—

Propellers.—

Steering.—

Remarks.—Building on the Hudson by aeronautical students of Columbia University.

HEZOG (R.D.) Biplane.

Maximum length, feet m. **maximum breadth,** 48 feet 14·63 m. **supporting surface,** 600 sq. feet 55 m².

Total weight.—773 lbs. (350·62 kgs.)

Body.—On 3 wheels.

Wings.—Bi-curved. Planes 48 × 6½ feet (14·63 × 2 m.) Ailerons on the lower plane. Skids are fitted under the lower plane to assist steady landing.

Motor.—21 h.p. air-cooled.

Speed.—

Propellers.—

Steering.—Horizontal rudder in front, consisting of a single plane 16 × 2¼ feet (4·87 × 0·68 m.)

Remarks.—Built at Harvard, Neb.

KERRISON. Monoplane.

Building.

Maximum length, 57 feet 17·37 m. **maximum breadth,** feet m. **supporting surface,** sq. feet m².

Total weight.—

Body.—

Wings.—

Motor.—

Speed.—

Propellers.—

Steering.—Elevator is placed 4 feet (1·21 m.) ahead of main planes and moved forward or backward to rise and fall. Vertical rudders aft, 16 feet (4·87 m.) abaft main planes. Single lever control to both.

Remarks.—Building by Dr. Davenport Kerrison, of Jacksonville, Fla.

JOHNSON. Biplane.

Completing.

Maximum length, feet / m. **maximum breadth,** feet / m. **supporting surface,** 500 sq. feet / 46·50 m².

Total weight.—

Body.—Weighs 120 lbs. Both decks above it. Suspended from two points on the front edge of the lower plane and to one point of the rear edge.

Wings.—Planes are 3 feet (0·90 m.) apart, crescent shaped with convex edge to the front. Upper plane has only about one-third the area of the lower plane.

Motor.—

Speed.—

Propellers.—

Steering.—Large tail hinged at rear suspension. A single lever depresses tail and elevates rear edge of lower plane, rotating it round front edge, and vice vêrsa.

Remarks.—There is an arrangement for lateral stability, whereby, a side of the machine being tilted up, it automatically receives the whole weight of driver and engines.

EICHENFELDT. Biplanes.

Model.

Maximum length, feet / m. **maximum breadth,** feet / m. **supporting surface,** sq. feet / m².

Total weight.—510 lbs. (231 kgs.) *not including* aviator.

Body.—Steel tubing.

Wings.—Sheet aluminium. The upper plane is semi-circular, bowed to meet the main wings. Both planes are rigid.

Motor.—

Speed.—

Propeller.—1, forward.

Steering.—Single elevator forward. Vertical rudder aft.

Remarks.—Invented by E. I. Eichenfeldt, of Minneapolis. Put on the market by H. J. Nice & Co. Price, about £210 or so ($1,000).

Several are building, but none yet complete. Models have been very successful, and remarkably stable.

KIMBALL. Biplane.
(*alias* **New York I.**)

Maximum length, feet m. **maximum breadth,** feet m. **supporting surface,** sq. feet m^2.

Total weight.—

Body.—On 3 wheels. One (steering) in front, two behind.

Wings.—Distance between planes about 4 feet (1·20 m.)

Motor.—50 h.p., 2-stroke, 4-cylinder.

Speed.—

Propellers.—8 between the planes. Diameter of each, 4 feet (1·20 m.)

Steering.—Biplane elevator forward.

Remarks.—First test 18th May, 1909. Christened by Miss Vera Held, the famous American actress. Damaged slightly on trials as the propellers would not all work. Wrecked July, 1909.

LAWRENCE III. Biplane.

Completing.

Maximum length, feet / m.	**maximum breadth,** 44 feet / 13·41 m.	**supporting surface,** *about* 600 sq. feet / 55 m².

Total weight.—*About* 1000 lbs. (453 kgs.), *including* pilot.

Body.—Chassis is 28 feet (8·53 m.) long. Frame, steel tube, keyed with cast bronze bolts. Mounted on runners, filled in with canvas, to act as a keel.

Wings.—Planes are 44 × 6½ feet (13·41 × 2 m.). 4 rigid keels under each plane. Ailerons in rear between the planes. These are worked electrically or by pedal.

Motor.—Motor car engine.

Propellers.—2 steel, shaft and bevel geared. Diameter, 7 feet (2·13 m.) Pitch, 6 feet (1·82 m.)

Steering.—2 horizontal in front, 2 others behind. 1 vertical rudder in front, 2 others behind.

Remarks.—This machine is generally on *Silver Dart* lines. It is chiefly remarkable for extreme lateral stability.

LINDSAY. Biplane.

Maximum length, feet / m. **maximum breadth,** feet / m. **supporting surface,** sq. feet / m².

Total weight.—

Body.—On 4 wheels.

Wings.—In three sections, outer section adjustable.

Motor.—

Speed.—

Propellers.—

Steering.—Elevator forward, single plane.

Remarks.—

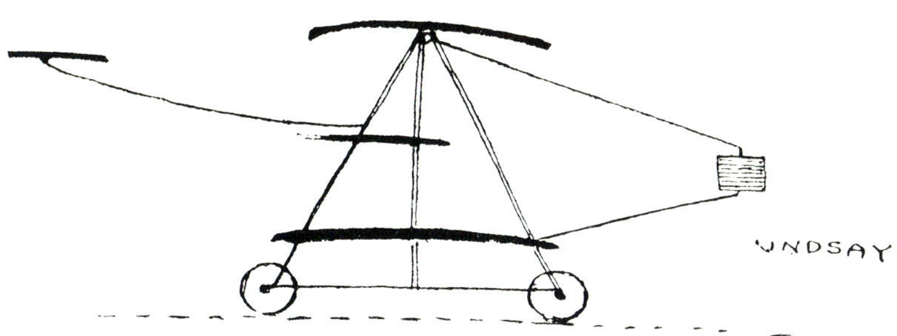

JEAN. Helicopter.

Building.

Maximum length, feet / m. **maximum breadth,** feet / m. **supporting surface,** sq. feet / m².

Total weight.—

Body.—

Wings.—Rectangular frames with helices inside.

Motor.—2-cylinder.

Speed.—

Propellers.—

Steering.—

Remarks.—Company formed to exploit this machine, June, 1909, but little known about it.

PRENTICE NEWMAN. Biplane.

(Wright Model).

See *Wright.*

Maximum length, feet / m. **maximum breadth,** feet / m. **supporting surface,** 533 sq. feet / 50 m^2.

Total weight.—

Body.—

Wings.—

Motor.—

Speed.—

Propellers.—

Steering.—

(All details closely resemble a *Wright, q.v.*)

Remarks.—Has flown 12 miles (19 km.) at a height of 150 feet (46 m.) and been up for 30 minutes.

BRADLEY. Biplane.

Building.

Maximum length, feet / m. **maximum breadth,** feet / m. **supporting surface,** sq. feet / m^2.

Total weight.—

Body.—

Wings.—

Motor.—

Speed.—

Propellers.—

Steering.—

Remarks.—Building by Chas. C. Bradley, Vice-President of the Pacific Aero Club. Commenced to build August, 1909. No details.

ORME. Biplane.

Completing.

Maximum length, feet m. **maximum breadth,** feet m. **supporting surface,** 174 sq. feet 16 m².

Total weight.—Only *about* 100 lbs. (45 kgs.) *not including* driver.

Body.—Frame of spruce, covered with canvas.

Wings.—The planes are 18 feet (5·50 m.) span. They are rigidly trussed with steel wire.

Motor.—8 h.p. air-cooled Belgian make. Weight, 45 lbs. (20·41 kgs.) = 5·62 lbs. per h.p. (2·55 kgs. per h.p.) H. T. magneto. 1,800 r.p.m.

Speed.—

Propellers.—2 chain driven, geared down 2 to 1. Variable pitch.

Steering.—Horizontal rudder in front which bends in the middle to act as a brake. Vertical rudder aft. This rudder can be converted into a flat plane at will. On *top* of the planes is a mushroom shaped plane, on springs, capable of any distortion, intended to act as a bird's tail does.

Remarks.—Glider models have been very satisfactory indeed. Built by Harry A. Orme, Washington, D.C.

RICKMAN. ***Miscellaneous.***

Maximum length, feet } m. } **maximum breadth,** feet } m. } **supporting surface,** sq. feet } m². }

Total weight.—
Body.—
Wings.—
Motor.—
Speed.—
Propellers.—
Steering.—
Remarks.—

SHNEIDER II. Biplane.

Building.

Maximum length, feet / m. **maximum breadth,** feet / m. **supporting surface,** 537 sq. feet / 49 m².

Total weight.—

Body.—

Wings.—A keel fixed longitudinally below the main upper plane and above the engine. Flexible extensions in rear at extremities of planes.

Motor.—30 h.p.

Speed.—

Propellers.—2, chain driven, as Wright.

Steering.—As Wright.

Remarks.—Differs little from a Wright, except for the details mentioned.
A *Shneider I*, similar, smashed completely, 12th July, 1909.

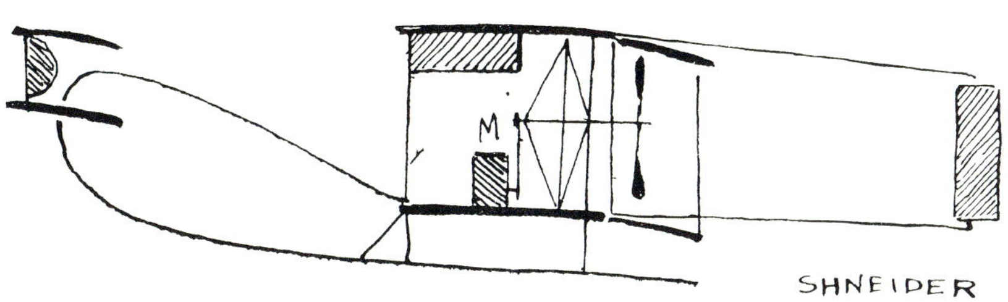

RAICHE I & II. Biplane.

Building.

Maximum length, 28½ feet / 8·70 m. **maximum breadth,** 33 feet / 10 m. **supporting surface,** sq. feet / m².

Total weight.—497 lbs. (225 kgs.), *including* aviator.

Body.—On 3 wheels. Keel to fore wheel.

Wings.—Planes in 3 sections. Escalloped in rear. Stability planes between main planes at ends.

Motor.—Special Crout 28-32 h.p. 4-cylinder. 1,100 r.p.m. Weight, 130 lbs.

Speed.—

Propeller.—1. Chauviere model.

Steering.—Special biplane elevator, 6 × 2 feet (1·80 × 0·61 m.) and 1½ feet (0·46 m.) apart. Single bamboo support. Diamond shaped vertical fin between. Rear rudder a single horizontal plane (adjustable), vertical plane intercepting it.

Remarks.—

M

TELFER. Multiplane.

Building.

Maximum length, feet / m. **maximum breadth,** feet / m. **supporting surface,** sq. feet / m^2

Total weight.—

Body.—

Wings.—

Motor.—

Speed.—

Propellers.—

Steering.—The top plane is used as a horizontal rudder.

Remarks.—Building at Richmond, Va. All details confidential. It is claimed for this machine that it cannot capsize.

HUDSON AND O'BRIEN. Biplane.

Building. Ready about December, 1909.

Maximum length, 36 feet / 10·97 m. **maximum breadth,** 40 feet / 12 m. **supporting surface,** sq. feet / m^2.

Total weight.—

Body.—On 4 wheels.

Wings.—Planes, 40×6 feet (12×1·80 m.) Distance between planes, 6 feet (1·80 m.)

Motor.—Special by designers. 35-37 h.p. Details secret.

Speed.—

Propellers.—2. Diameter, 6½ feet (2 m.) Chain driven.

Steering.—As *Wright*.

Remarks.—Designed by John W. Hudson and Clifton O'Brien, of the Pacific Aero Club. Building at San Francisco. Special automatic balance.

THOMAS.

Building. Not far advanced.

Maximum length, feet / m. **maximum breadth,** feet / m. **supporting surface,** sq. feet / m².

Total weight.—

Body.—

Wings.—

Motor.—

Speed.—

Propellers.—

Steering.—

Remarks.—Designed for E. L. Thomas, by Myers, of Hammondsport. Said to incorporate many very novel features.

A. L. SMITH. Biplane.

Building.

Maximum length, feet / m. **maximum breadth,** 40 feet / 12·12 m. **supporting surface,** sq. feet / m².

Total weight.—

Body.—On 2 wheels forward, and runners aft.

Wings.—2 fixed planes. Vertical rudders between them act as stabilisers. Width of planes, 7 feet (2·13 m.) Space between planes, 6 feet (1·80 m.)

Motor.—60 h.p.

Speed.—

Propellers.—2, abaft planes, chain driven.

Steering.—Single plane elevator forward, another in rear. Vertical rudders at end between main planes.

Remarks.—This is the same A. L. Smith as the "gasless air-ship" Smith. Building in California.

WRIGHT BROS. Biplane.

Photo, C. Malcuit.

Maximum span, 41 feet / 12·5 m. **maximum depth,** $6\frac{1}{2}$ feet / 2 m. **supporting surface,** 538 sq. feet / 50 m². slightly reduced in later models.

Total weight.—$826\frac{3}{4}$ lbs. (375 kgs.) or with two passengers about $1,157\frac{1}{2}$ lbs. (525 kgs.) Weight with Wilbur Wright is about 992 lbs. (450 kgs.)

Body.—Body mounted on runners. Flight is attained by a special apparatus, but Mr. Wilbur Wright has risen on wet grass without this.

Planes.—Continental aeroplane sheeting in the more recent machines. Distance between planes 5·8 feet (1·80 m.) Hickory wood frame, curved (see sketch plan).

Motor.—22-27 h.p. Barringuard (Wright design), 4 cylinder, water cooled, 1,650 revs. Automatic valves. Magneto ignition. Bore, 105 mm.; stroke, 100 mm. (*about*). Engine weighs about 7·9 lbs (3·6 kgs.) per h.p.

Speed.—32 m.p.h. (52 km.)

Propellers.—Two 2-bladed, each 9 feet 4 inches (2·8 m.) in diameter; driving in opposite directions at 450 r.p.m.

Steering.—Vertical rudder (double), aft for horizontal steering; horizontal rudder forward for rising and falling. Tips of planes controlled to rectify tilting during turning.

Remarks.—This is the pattern now in existence. The latest patent shows a single vertical rudder in rear, a fixed vertical plane in front of main planes, with a vertical rudder before it; and ahead again a single plane elevator in line with lower deck.

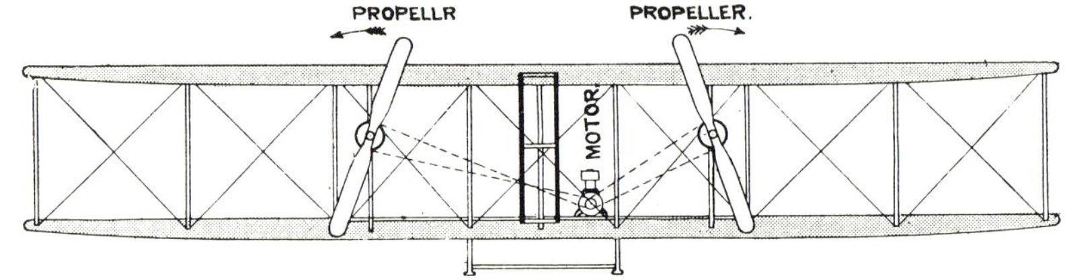

VIEW FROM BEHIND.

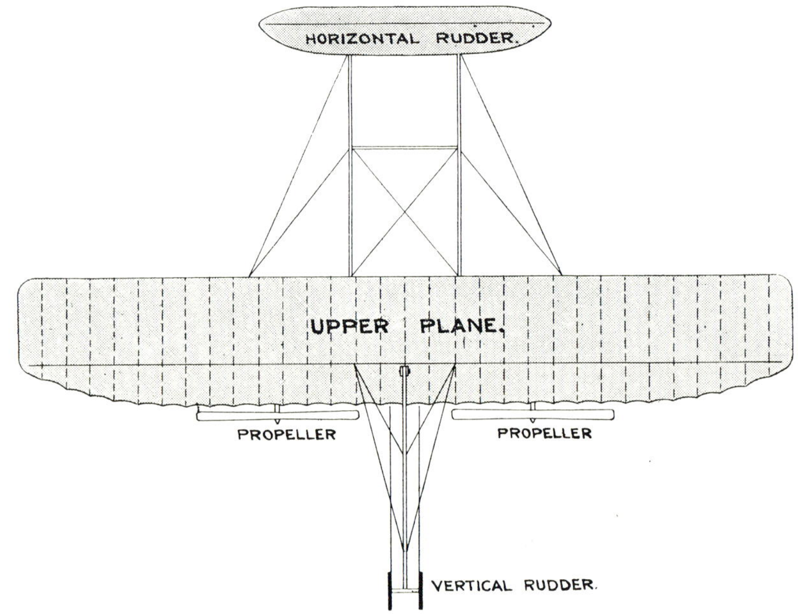

VIEW FROM ABOVE.

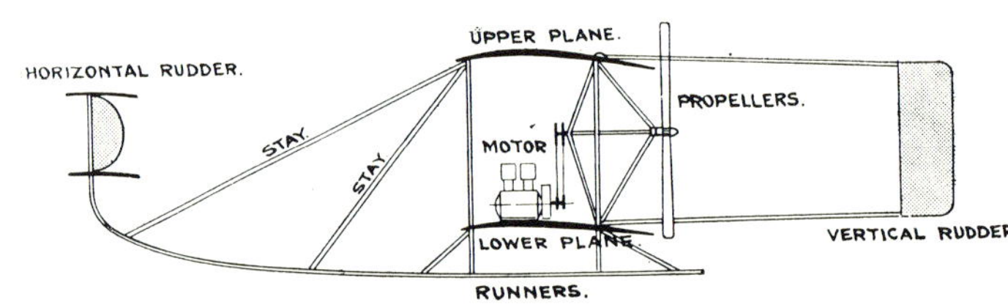

SIDE VIEW.

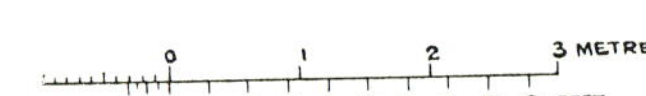

WRIGHT BROS. Biplanes.

Private machines built, building or on order:—

Model 1908-9.	Model 1908-9.	Model 1908-9.	Model 1908-9.
1 Orville Wright	22 E. R. Thomas	43	64
2 U. S. Government	23	44	65
3 ,,	24	45	66
4 Brothers Alger	25	46	67
5 F. M. Barnes (Chicago)	26	47	68
6 E. Green (Texas)	27	48	69
7 — Pitman	28	49	70
8 Howard Colby	29	50	71
9 E. S. F. Randolph	30	51	72
10 Dayton Aeroplane Club	31	52	73
11	32	53	74
12	33	54	75
13	34	55	76
14	35	56	77
15	36	57	78
16 } *Building.* (11–21)	37	58	79
17	38	59	80
18	39	60	81
19	40	61	82
20	41	62	
21	42	63	

(Most of these cannot be supplied till new works have been erected).

WALDEN (Dr. H. W.) Biplane.

Building.

Maximum length, feet / m. **maximum breadth,** feet / m. **supporting surface,** 533 sq. feet / 50 m^2.

Total weight.—

Body.—

Wings.—

Motor.—

Speed.—

Propellers.—

Steering.—

Remarks.—Building. The machine resembles a Wright in general principles, but embodies a variety of innovations in detail. No particulars available.

ROGERS. Monoplane.

Building.

Maximum length, feet / m. **maximum breadth,** feet / m. **supporting surface,** sq. feet / m^2.

Total weight.—

Body.—Circular tube.

Wings.—One plane on either side of the tube.

Motor.—

Speed.—

Propellers.—

Steering.—

Remarks.—Building at Hyatsville, Md.

ZORNES. Biplane.

Not quite complete yet.

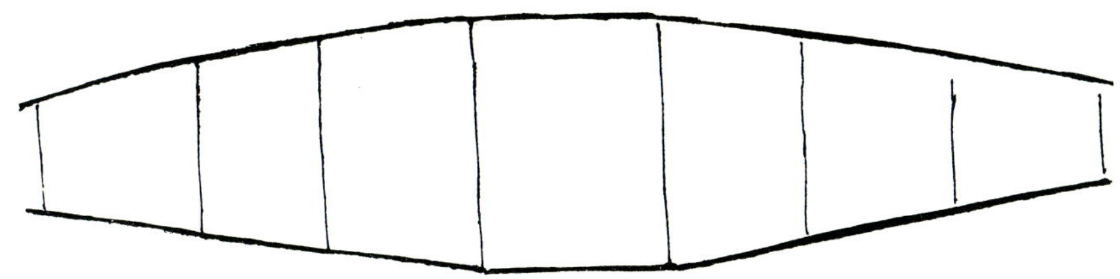

ZORNES - FRONT VIEW

Maximum length, 41 feet (12·50 m.) **maximum breadth,** 37 feet (11·27 m.) **supporting surface,** 533 sq. feet (50 m^2.)

Total weight.—660 lbs. (299 kgs.), *not including* aviator.

Body.—Hickory body. Bamboo frame, much trussed with piano wire. *Voisin* type bow. Fuselage is *about* 10 × 3 feet (3 × 0·90 m.)

Wings.—Novel. The lower planes have a slight upward tilt towards the edges. The upper planes curve downward towards the outer edge considerably. Ailerons. Span, 37 feet (11·27 m.) Depth, 6 feet (1·80 m.)

Motor.—*About* 25-35 h.p.

Speed.—

Propellers.—2, chain driven.

Steering.—Horizontal rudder in front, vertical behind.

Remarks.—Designed by C. A. Zornes. Built at Lind, Wash. Property of the Zornes Co., which is putting these machines on the market. First trials, July, 1909.

BOKOR (M). Triplane.

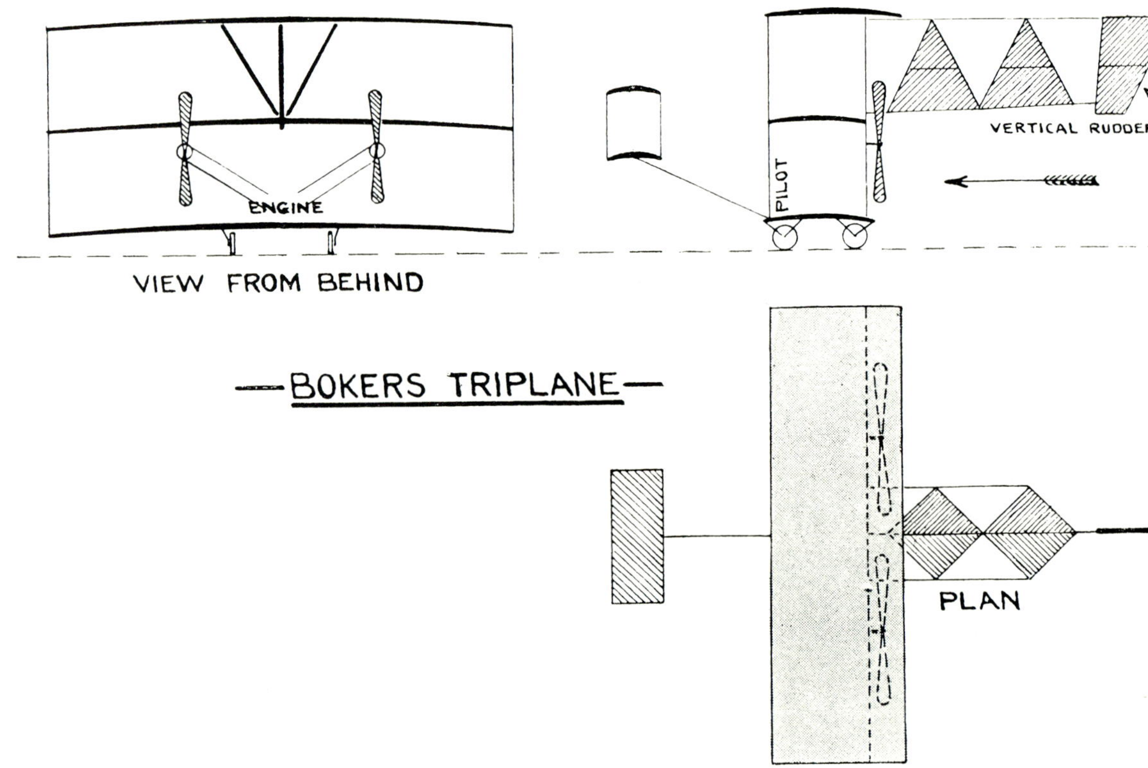

Maximum length, feet / m. **maximum breadth,** 26 feet / 8 m. **supporting surface,** 580 sq. feet. / 54 m².

Total weight.—1,181 lbs. (536 kgs.) *including* aviator.

Body.—On 4 wheels, but these are occasionally replaced by runners.

Wings.—3 surfaces of rather small span, the two lower ones curved, the upper one flat. Small flat planes in rear of bottom planes. There is a considerable tail, consisting of two pairs of triangular planes set at a dihedral angle, on a frame extending from the middle plane and from about 2 feet below the top plane. Top to middle plane, 5 feet (1·5 m.) Middle to lower, 6 feet (1·80 m.) Flexible rear edges to main planes. Width of planes, 6½ feet (2 m.)

Motor.—34 h.p. Thomas, with a number of special fittings. 4-cylinder. Bore, 100 m/m. Livingstone radiator.

Speed.—

Propellers.—2, chain driven. Placed behind central plane. Diameter, 8 feet (2·43 m.)

Steering.—Biplane elevator in front, 14×2½ feet (4·26×0·75 m.) Vertical rudder (shape see photograph) behind the tail.

Remarks.—This was the third American type to leave the ground. It has made several short flights but none have been attended with conspicuous success. The troubles are, however, being gradually overcome.

STADTLER. Triplane.

Building.

Maximum length, 40 feet / 12 m. **maximum breadth,** 25 feet / 7·61 m. **supporting surface,** sq. feet / m²

Total weight.—810 lbs. (367 kgs.)

Body.—

Wings.—3 in front, 3 behind. Span in each case, 25 feet (7·61 m.)

Motor.—

Speed.—

Propellers.—2, chain driven, placed between the forward and after plane groups. These propellers have a variable pitch.

Steering.—

Remarks.—Not yet completed.

SCOTT. Triplane.

Building.

Maximum length, feet / m. **maximum breadth,** 26¼ feet / 8 m. **supporting surface,** 560 sq. feet / 52 m².

Total weight.—650 lbs. (295 kgs.)

Body.—

Wings.—The 2 front planes are 144 sq. feet (13·37 m²); there are 3 rear ones of 414 sq. feet (38·46 m²) Front planes are divided in the centre to allow of independent warping. Rear planes adjustable to any angle to suit speed.

Motor.—

Speed.—

Propellers.—2, chain driven, as Wright.

Steering.—

Remarks.—Built at Laurenceburg, Ind.

STEFFAN. Re-inforced Biplane.

Building.

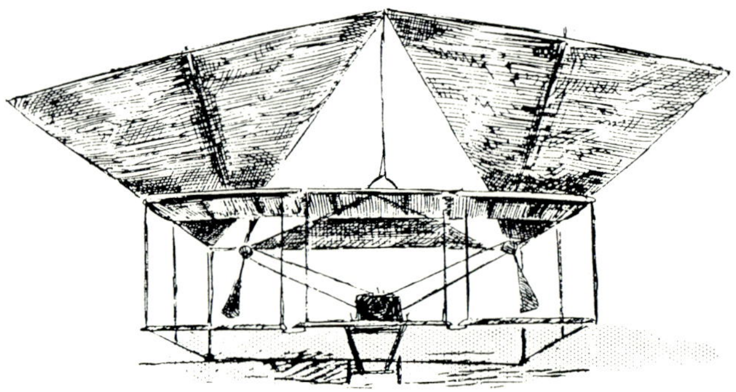

Maximum length, feet / m. **maximum breadth,** feet / m. **supporting surface,** sq. feet / m².

Total weight.—

Body.—Frame in form of a square.

Wings.—Main planes divided into 3 equal sections, central one fixed, side ones for use as elevators. Above the upper plane 2 large triangular planes side by side, each with its apex to the rear. A third triangular plane, with apex forward in rear of upper main plane.

Motor.—Curtis.

Speed.—

Propellers.—Two 2-bladed in rear of planes, chain driven.

Steering.—

Remarks.—Building at Los Angeles, California, by Frank Steffan.

TRIACA (A.C.) Biplane.

Building.

Maximum length, feet / m. **maximum breadth,** feet / m. **supporting surface,** sq. feet / m^2

Total weight.—

Body.—

Wings.—

Motor.—40-50 h.p., water-cooled.

Speed.—

Propeller.—

Steering.—

Remarks.—Building by the International School of Aeronautics. Messrs. Luttgen & Beckert are connected with the design of this machine.

MYERS.

Maximum length, feet / m. **maximum breadth,** feet / m. **supporting surface,** sq. feet / m^2.

Total weight.—

Body.—

Wings.—

Motor.—

Speed.—

Propellers.—

Steering.—

Remarks.—No details.

DOWNER. Quintuplane.

Building.

Maximum length, feet / m. **maximum breadth,** feet / m. **supporting surface,** sq. feet / m².

Total weight.—

Body.—

Wings.—There are 5 planes (see Steering).

Motor.—

Speed.—

Propellers.—

Steering.—Horizontal rudder in front. No vertical rudder, moveable wing tips being relied on instead.

Remarks.—It is claimed for this machine that it cannot capsize. Built by the owner, C. L. Downer, of Salt Lake City, Utah. Details kept strictly confidential at present.

CLARKE. Biplane.

Building.

Maximum length, feet / m. **maximum breadth,** feet / m. **supporting surface,** sq. feet / m².

Total weight.—

Body.—

Wings.—

Motor.—

Speed.—

Propellers.—

Steering.—

Remarks.—Understood to be something like a *Wright* in its main feature. Built by J. Clarke, of Chicago.

ADAMS.

Maximum length, feet / m. **maximum breadth,** feet / m. **supporting surface,** sq. feet / m^2.

Total weight.—

Body.—

Wings.—

Motor.—

Speed.—

Propellers.—

Steering.—

Remarks.—No details.

CALIFORNIA AERIAL MFG. CO. Biplane.

Completing

Maximum length, feet / m. **maximum breadth,** feet / m. **supporting surface,** sq. feet / m^2.

Total weight.—

Body.—

Wings.—

Motor.—

Speed.—

Propellers.—

Steering.—

Remarks.—A species of Wright Bros. type; built at Los Angeles.

BATES I (of Chicago).

Maximum length, feet / m. **maximum breadth,** feet / m. **supporting surface,** sq. feet / m².

Total weight.—

Body.—

Wings.—

Motor.—

Speed.—

Propellers.—

Steering.—

Remarks.—Has flown on Daytona Beach, Fla. Experimental only. See *Bates II.*

BATES II. Biplane.

Building.

Maximum length, feet / m. **maximum breadth,** feet / m. **supporting surface,** 463 sq. feet / 43 m².

Total weight.—1400 lbs. (635 kgs.) *including* aeronaut.

Body.—McAdamite patent castings for quick stowage.

Wings.—Planes 42 × 6 feet (12·80 × 1·80 m.) Covered with Baldwin's vulcanised rubber silk.

Motor.—Of special Bates design. 40 h.p. water-cooled. Livingston radiator.

Speed.—

Propellers.—

Steering.—1 elevator in front; 1 vertical rudder aft. Special steering features (*secret*).

Remarks.—Designed by Carl Bates, of Chicago. An improved Bates I. special lateral balance device. Special braking device. To carry 2.

DIXON.

Building.

Maximum length, feet (m.) **maximum breadth,** feet (m.) **supporting surface,** sq. feet (m^2.)

Total weight.—

Body.—

Wings.—

Motor.—

Speed.—

Propellers.—

Steering.—

Remarks.—Cromwell Dixon, of Columbus, is a boy inventor who exhibited a machine to large crowds at Indianola Park, May, 1909. The machine now being built is entered for the New York-Albany $10,000 prize.

POSADAS. Biplane.

Completing.

Maximum length, feet (m.) **maximum breadth,** 35 feet (10·66 m.) **supporting surface,** 430 sq. feet (40 m^2.)

Total weight.—

Body.—

Wings.—Planes are 35 × 4 feet (10·66 × 1·22 m.)

Motor.—7 h.p. only (see propeller).

Speed.—

Propellers.—One of special design. It is claimed that this has such a high efficiency that more h.p. is unnecessary.

Steering.—As Wright.

Remarks.—Designed by J. Zenon Posadas, Jr., of San Francisco.

CALL (H.L.) Biplane.

Reconstructing.

Maximum length, feet m. **maximum breadth,** feet m. **supporting surface,** 533 sq. feet 50 m^2.

Total weight.—830 lbs. (376 kgs.)

Body.—

Wings.—Planes 6 feet (1·82 m.) apart. Span, 41 feet (12·50 m.)

Motor.—Curtis.

Speed.—

Propellers.—Two 2-bladed.

Steering.—Horizontal and vertical rudder, as *Wright* (*q.v.*)

Remarks.—Closely resembles a *Wright*. Recently smashed. Re-building.

WILSON. Biplane.

Building.

Maximum length, feet m. **maximum breadth,** feet m. **supporting surface,** sq. feet m^2.

Total weight.—

Body.—

Wings.—The planes incline bodily round a central transverse rib.

Motor.—

Speed.—

Propellers.—2, in rear of planes, chain driven and superposed. Vertical rudder in rear of planes between the propellers. Main planes as elevator.

Steering.—

Remarks.—Built by John H. Wilson, Middlesex, Pa.

WILLOUGHBY (H.L.) Biplane.

Building.

Maximum length, feet / m. **maximum breadth,** feet / m. **supporting surface,** sq. feet / m^2.

Total weight.—

Body.—

Wings.—

Motor.—

Speed.—

Propellers.—

Steering.—

Remarks.—Building at Newport. Delayed for engine. No details known; but believed to be a biplane.

BULASK-HIDALGO.

Building.

Maximum length, feet / m. **maximum breadth,** feet / m. **supporting surface,** 355 sq. feet / 32 m^2.

Total weight.—

Body.—

Wings.—

Motor.—

Speed.—

Propellers.—

Steering.—

Remarks.—Designed by H. C. Bulask and Professor Hidalgo. Building at San Francisco. No details procurable.

BAPTISTE. Helicopter.

Completing.

Maximum length, feet / m. **maximum breadth,** feet / m. **supporting surface,** sq. feet / m².

Total weight.—

Body.—

Wings.—

Motor.—

Speed.—

Propellers.—

Steering.—

Remarks.—Completing at St. Louis. Nothing known about it.

POTTS. Biplane Helicopter.

Building.

Maximum length, feet / m. **maximum breadth,** feet / m. **supporting surface,** sq. feet / m².

Total weight.—

Body.—

Wings.—

Motor.—30 h.p. Placed well below lower plane.

Speed.—

Propellers.—4. Two on either side of motor. Variable.

Steering.—All by helices.

Remarks.—Designed by John Potts, Winchester, O.

BULTZING-SLÖWEN. Biplane-Helicopter.

Building.

Maximum length, $12\frac{2}{3}$ feet (3·80 m.) **maximum breadth,** 15 feet (4·57 m.) **supporting surface,** 520 sq. feet (48 m^2.)

Total weight.—750 lbs. (340 kgs.)

Body.—Tubular frame mounted on four 18 inch bicycle wheels (0·45 m.), which can be geared to the engine if required.

Wings.—2 sets of adjustable planes. Maximum span, 15 feet (4·57 m.) Depth, 6 feet (1·82 m.) Planes, $3\frac{1}{2}$ feet (1·06 m.) apart. Under the planes are 2 adjustable helicopters, 6 feet (1·82 m.) in diameter, running at 120 r.p.m. Calculated lift, 500 lbs. (228 kgs.)

Motor.—36 h.p. placed immediately over the frame and underneath the aviator.

Speed.—

Propellers.—Two 3-bladed, placed aft. Diameter, 6 feet (1·82 m.)

Steering.—2 vertical rudders in front. Planes used as elevators.

Remarks.—Building in New York.

DORLAND. Helicopter.

Building.

Maximum length, feet (m.) **maximum breadth,** feet (m.) **supporting surface,** sq. feet (m^2.)

Total weight.—500 lbs. (227 kgs.) *not including* aviator.

Body.—

Wings.—Eight 12-bladed helices, each 2 feet (60 c/m.) in diameter. Each is enclosed in a separate well.

Motor.—40 h.p., 8-cylinder, Curtis.

Speed.—

Propellers.—(See *Wings*).

Steering.—

Remarks.—Building by L. S. Dorland, San Francisco.

SHAFFER.

Maximum length, feet / m. **maximum breadth,** 27 feet / 8·23 m. **supporting surface,** 323 sq. feet / 30 m².

Total weight.—

Body.—On runners, Wright Bros. style. Hickory and tubular steel.

Wings.—Fixed planes. Span, 27 feet (8·23 m.) Distance between planes, 5¾ feet (1·60 m.) Stabilising planes between main planes near the extremities.

Motor.—*Not yet fitted.*

Speed.—

Propellers.—

Steering.—Wright type elevator forward.

Remarks.—Under test June, 1909, towed flights.

GILLESPIE.

Completing.

Maximum length, feet / m. **maximum breadth,** feet / m. **supporting surface,** sq. feet / m².

Total weight.—

Body.—

Wings.—

Motor.—

Speed.—

Propellers.—

Steering.—

Remarks.—Built by G. Curtis Gillespie, Brooklyn, N.Y.

KLASSEN. Gyroplane.

Building.

Maximum length, feet / m. **maximum breadth,** feet / m. **supporting surface,** sq. feet / m^2.

Total weight.—

Body.—Girder framework on a runner forward and 2 wheels aft.

Wings.—Four 12-sided slightly mushroom planes, carried one on either side right forward and right aft.

Motor.—

Speed.—

Propellers.—Four 3-bladed, one under each plane, chain driven.

Steering.—By altering angle of planes.

Remarks.—Building at S. Francisco.

CULVER. Miscellaneous.

Building.

Maximum length, feet / m. **maximum breadth,** feet / m. **supporting surface,** sq. feet / m^2.

Total weight.—

Body.—On 4 wheels. Fusilage well above it.

Wings.—Several, in 2 sets; all connected and all variable.

Motor.—

Speed.—

Propeller.—1, in rear of planes.

Steering.—Elevator forward, vertical rudder aft. Variable planes.

Remarks.—Laid down May, 1909. Building. Designed by Chas. R. Culver, of Springfield, Mass.

LANE. Helicopter.

Maximum length, feet / m. **maximum breadth,** feet / m. **supporting surface,** sq. feet / m^2.

Total weight.—400 lbs. (181 kgs.)

Body.—

Wings.—There is a single plane, 40×20 feet (12×6 m.), in centre of which is a single horizontal screw, 16 feet in diameter (4·87 m.). The idea is that the screw sucks air from above, compresses it, and discharges it underneath from special cups at 250 times per minute.

Motor.—36 h.p.

Speed.—

Propellers.—

Steering.—

Remarks.—It is stated that Mr. Lane, using hand and foot power only, flew $1\frac{1}{2}$ miles on this machine on September 8th, 1908. The machine is the property of the Lane Automatous Air-ship Co., San Francisco.

TWINING. Helicopter.

Building.

Maximum length, feet / m. **maximum breadth,** feet / m. **supporting surface,** sq. feet / m^2.

Total weight.—

Body.—

Wings.—2 lifting screws.

Motor.—

Speed.—

Propellers.—

Steering.—

Remarks.—Designed by H. la V. Twining, President of the Aero Club of California.

LUYTIES OTTO. Helicopter.

Maximum length, feet / m. **maximum breadth,** feet / m. **supporting surface,** sq. feet / m^2.

Total weight.—

Body.—

Wings.—

Motor.—

Speed.—

Propellers.—2 large helices.

Steering.—

Remarks.—So far as can be gathered this machine is rather like the French *Bertin* in general appearance. It is stated to have risen on several occasions.

SNELL. Miscellaneous.

Building.

Maximum length, feet / m. **maximum breadth,** feet / m. **supporting surface,** sq. feet / m^2.

Total weight.—

Body.—

Wings.—Supporting surfaces, which rotate in a direction opposite to the direction of flight.

Motor.—

Speed.—

Propellers.—

Steering.—

Remarks.—Building by Harry B. Snell, of Toledo, O.

KUNOW. Flapper.

Building.

Maximum length, feet / m. **maximum breadth,** feet / m. **supporting surface,** sq. feet / m^2.

Total weight.—

Body.—

Wings.—

Motor.—

Speed.—

Propellers.—

Steering.—

Remarks.—Building. Models have flown successfully.

THOMPSON. Air Sucker.

Building.

Maximum length, feet / m. **maximum breadth,** feet / m. **supporting surface,** sq. feet / m^2.

Total weight.—

Body.—Ovoid.

Wings.—Monoplane wings. Vertical plane on top of body.

Motor.—

Speed.—

Propellers.—Placed inside the body. Suck air in and expel it to work the machine.

Steering.—Rudders aft.

Remarks.—Designed by George Thompson, of Kingston, Okla.

ROBINSON (A. J.) Monoplane.

Building.

Maximum length, $39\frac{1}{4}$ feet / 12 m. **maximum breadth,** $32\frac{3}{4}$ feet / 9 m. **supporting surface,** 269 sq. feet / 25 m^2.

Total weight.—1,234 lbs. (559 kgs.)

Body.—

Wings.—Warping wing tips.

Motor.—20 h.p. Curtis.

Speed.—

Propellers.—1 in front.

Steering.—Elevators each side of tail. Vertical rudder tail.

Remarks.—Designed by Arthur J. Robinson, Sheridan, Wyo.

SMIDLEY. Monoplane.

Maximum length, feet (m.) **maximum breadth,** 32 feet (9·75 m.) **supporting surface,** sq. feet (m^2.)

Total weight.—225 lbs. (102 kgs.) *not including* aviator.

Body.—Bamboo frame on 4 small wheels.

Wings.—Canvas. Main plane, 32 × 5 feet (9·75 × 1·52 m.) Rear plane, 17 feet (5·13 m.) span. The planes have their maximum curve in rear, with a knife edge entering.

Motor.—Duryea.

Speed.—

Propeller.—1 wooden 2-bladed, 8½ feet (2·60 m.) in diameter.

Steering. 2 single planes in front placed side by side. Used together these act as elevator; by the use of one only, steering in the horizontal plane is effected.

Remarks.—

ANDREAE. Multiplane.

Building.

ANDREAE MODEL

Maximum length, 30 feet (9·14 m.) **maximum breadth,** 30 feet (9·14 m.) **supporting surface,** 559 sq. feet (52 m².)

Total weight.—1,000 lbs. (453½ kgs.) *including* aviator and 14½ gallons of petrol.

Body.— Tubular steel construction, 1,118 feet (341 m.) being used altogether. Surfaces, vulcanised silk. Body on wheels.

Wings.—There are altogether no less than 19 planes, set at various curves and angles. In addition, about 66 sq. feet (6 m².) of vertical steadying fins.

Motors.—

Speed.—

Propellers.—1. Diameter, 6½ feet (2 m.) Chain driven at 1,000 r.p.m.

Steering.—70 sq. feet (6½ m².) of horizontal rudders, vertical, 35 sq. feet (3 m².)

Remarks.—Building by F. O. Andraea. Central Valley, N.Y., patent law expert of the journal *Aeronautics*. Based on a large number of models.

LAKE. Air Sucker.

Building.

Maximum length, feet / m. **maximum breadth,** feet / m. **supporting surface,** sq. feet / m^2.

Total weight.—

Body.—

Wings.—Hollow curved surfaces, on which "heated elastic fluid" is discharged.

Motors.—

Speed.—

Propellers.—

Steering.—

Remarks.—Building for C. J. Lake at Bridgeport, Con., by the Lake Submarine Co. Work was commenced May, 1909, and considerable progress has now been made.

WADE. Air Sucker.

Building.

Maximum length, feet / m. **maximum breadth,** feet / m. **supporting surface,** sq. feet / m².

Total weight.—

Body.—

Wings.—

Motor.—

Speed.—

Propellers.—

Steering.—

Remarks.—The invention of Mr. Wade, of Chicago. Extraordinary claims have been made for this machine and its applicability to an ordinary motor car, producing the opinion that it was a mere myth. It appears, however, to be in the ordinary category of "air suckers," collecting air in front and above, creating a partial vacuum, and expelling the air behind and below.

ENGLISH. Monoplane-Helicopter.

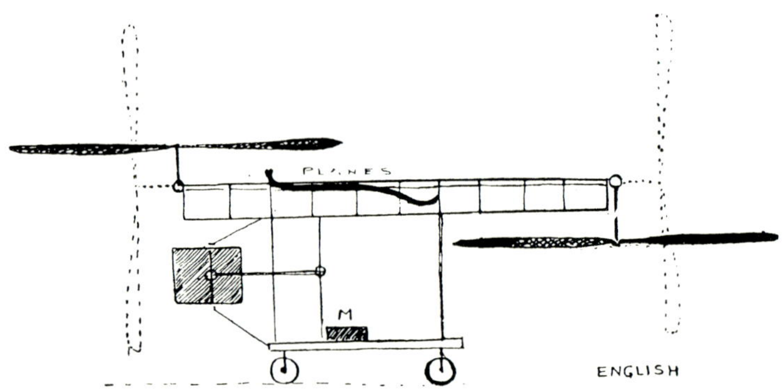

Maximum length, feet / m. **maximum breadth,** feet / m. **supporting surface,** 400 sq. feet / 37 m².

Total weight.—About 600 lbs.

Body.—Triangular steel tubular frame (like a U. S. dirigible). Square platform below mounted on 4 wheels.

Wings.—Planes curve up both in front and rear.

Motor.—60 h.p. 8-cylinder V type. Weight, 150 lbs. (kgs.)

Speed.—

Propellers.—Two 2-disced, one at either end. Diameter, 16 feet (4·87 m.) Bevel gear drive. Revolve in opposite directions. Act as helices. Lift claimed to be 1,660 lbs. (kgs.)

Steering.—Universally jointed rudder behind forward propeller.

Remarks.—Built at San Francisco. Considerable claims are made as to the efficiency of this machine. Tried in its shed July, 1909, it broke loose and damaged itself badly.

METCALF. Helicopter.

Building at Malden, Mass.

Maximum length, 22 feet / 6·70 m. **maximum breadth,** 48 feet / 14·60 m **supporting surface,** sq. feet / m².

Total weight.—

Body.—Directly under the helices.

Wings.—*See* propellers.

Motor.—

Speed.—

Propellers.—2, diameter, 20 feet (6 m.) placed 8 feet (2·43 m.) apart on the same shaft. Propellers rotate in opposite directions and tilt to all angles.

Steering.—By tilting the machine.

Remarks.—Claimed to fly in any direction.

SMITH. Gasless Airship.

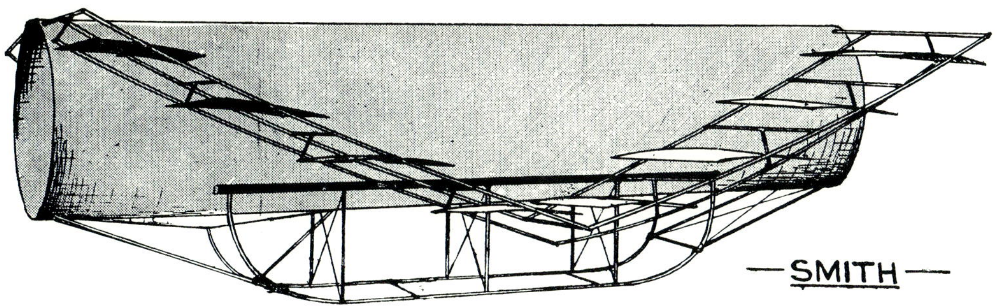

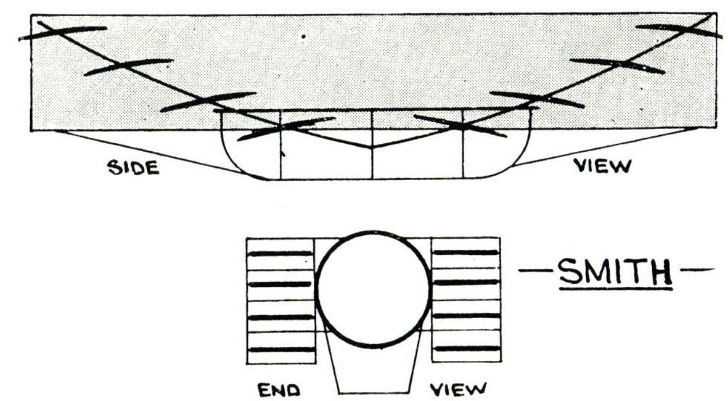

Maximum length, feet } m. } **maximum breadth,** feet } m. } **supporting surface,** sq. feet } m^2. }

Total weight.—

Body.—Steel tubular frame on runners.

Wings.—A large silk covered cylinder with 16 small elevating planes—eight on either side in two groups.

Motor.—

Speed.—

Propellers.—

Steering.—

Remarks.—Building.

PURCELL.

Building.

Maximum length, feet / m. **maximum breadth,** feet / m. **supporting surface,** sq. feet / m^2.

Total weight.—

Body.—

Wings.—

Motor.—40 h.p., 4-cylinder, 2-cycle Purcell. Weight, 80 lbs. (kgs.)

Speed.—

Propellers.—

Steering.—

Remarks.—Built by John D. Purcell, Chattanooga, Tenn.

Roshon. Multiplane.

Irvine.

C. W. Williams. Monoplane.

Kinek.

Williams. Heliocopter.

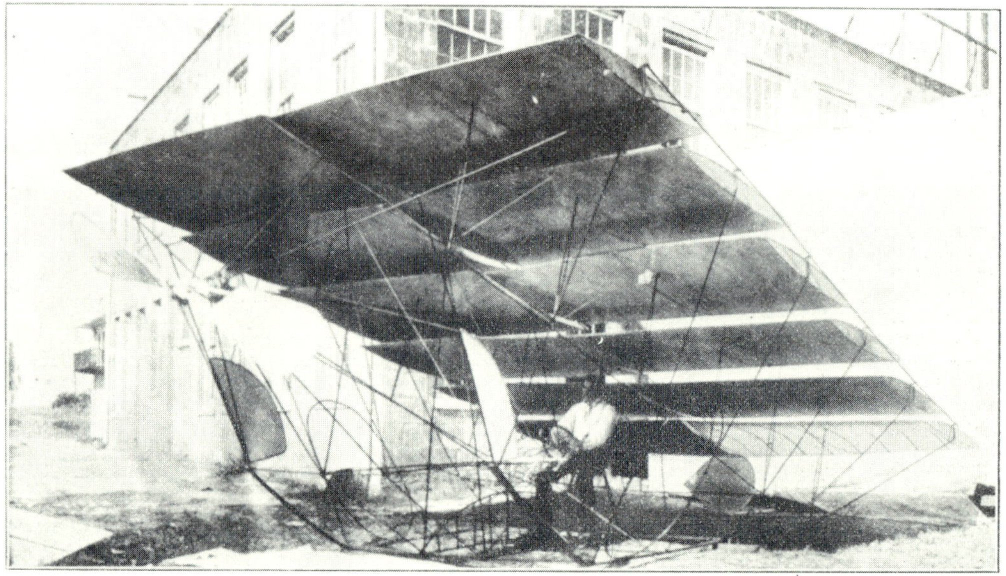

Zerbe.

DIRIGIBLE No. 1.

(Alias **BALDWIN.)**

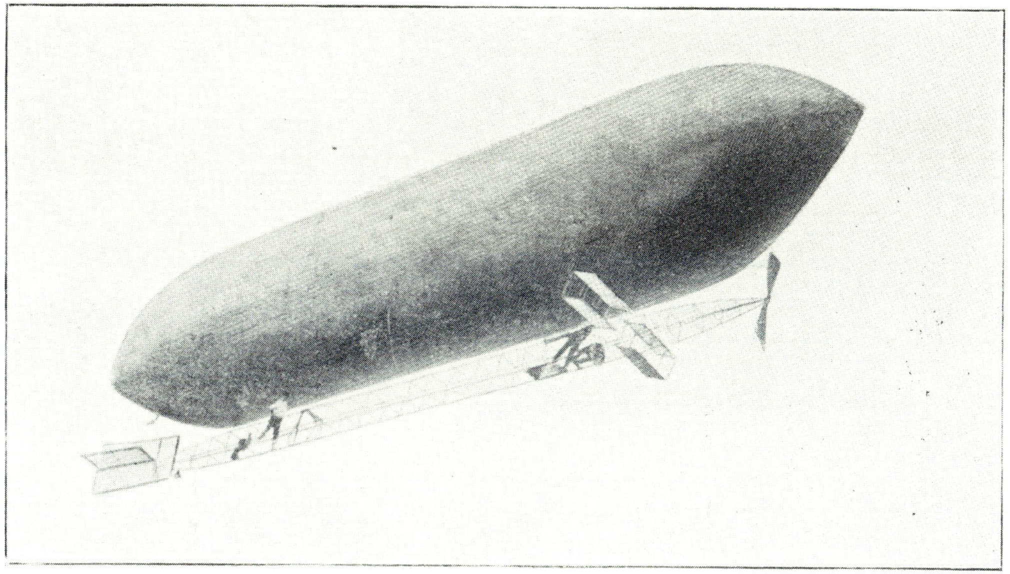

Belongs to American Government.
Built to specifications by Captain Baldwin.

Length, 96 feet. (29·26 m.) **diameter,** 19½ feet. (5·94 m.) **volume,** 20,000 cubic feet. (580· m³.) **total lift,** 1,370 lbs. (620 kg.) **useful lift** (crew, fuel, ballast, etc.) 500 lbs. (227 kg.)

Ballonet.—2,800 cubic feet=79 m³

Gas bag material.—2 layers of Japanese silk, with a layer of vulcanised rubber in between them.

Car.—A framework of square section, made of spruce wood, 66 feet long, 2½ feet wide, 2½ feet high. The car is slung only 5 feet (1·5 m.) below the gas bag by means of about 80 wires.

Motor.—20 h.p., 4 cylinder Curtis' watercooled.

Speed.—Max. 19·61 m.p.h. (31 km.p.h.) **during 2 hours run,** 14 m.p.h. (22 km.p.h.)

Propeller.—At the front end of the car, built up of spruce, connected to motor by a 20 feet. (6 m.) steel shaft. Diameter 10 feet 8 inches (5·1 m.) Pitch 11 feet (3·35 m.) Revs. 450 per minute.

Planes and rudders.—Fixed vertical plane at rear end. Vertical rudder at rear end with fixed horizontal plane attached to it. Wide moveable biplane 13 feet (4 m.) from fore end of car, just before the motor. This is operated by a lever. Planes 22 feet (6·4 m.) wide; 3 feet (·9 m.) deep, and 3 feet (·9 m.) apart. Carries 2 men only. Taken over by U.S. Signal Corps, in August, 1908.

KNABENSHUE.

Maximum length, 112 feet (34 m.) **maximum diameter,** 17½ feet (5·33 m.) **volume,** 22,000 c. feet (m³.)

Total lift.—1,540 lbs. (682 kgs.) **Useful lift,** lbs. (kgs.)

Gas bag.—A complete bag of varnished Japanese silk; over the upper half is a second layer of thinner varnished silk. Ribbons of Ponghee silk, 1 inch wide and spaced 4 inches apart, are passed over the gas bag; the ends of these ribbons each side are sewn to a strip of silk 90 feet long and 14 inches wide. To this is sewn a band of four thicknesses of Ponghee silk, from which the car is suspended. The whole is prevented from sliding over by 2 patches of silk sewn under the two ends of the gas bag, from which stays are led to the car. Ballonet, 3,500 c. feet.

Motor.—4 cylinder, 3½ inch bore by 4 inch stroke, 2 cycle, water cooled, 25 h.p., 1,000 revs. p.m. Weight, with water and cooling system, 145 lbs.

Speed.—25½ m.p.h. (40 km.)

Propellers.—1, at the front of the car, 9 feet diameter. Direct coupled.

Steering.—Vertical rudder, 50 sq. feet (4·6 m²), 10 feet by 5 ($3 \times 1·52$ m.) at rear end of car. This vertical rudder also carries a horizontal fin each side, 15 feet by 3½ feet ($4·57 \times 1·06$ m.), total 104 sq. feet (9·66 m²) Horizontal rudders, one each side of the front of the car, just abaft the propeller.

Car.—62 feet (18·89 m.) long, of triangular section, each side being 4 feet 4 inches (1·31 m.) Constructed of 1¼ inch (3 c/m.) spruce, strung with piano wire.

Remarks.—Has made several successful trips. The chief trouble seems to be lack of gas tightness.

AMERICA. Wellman Polar Expedition, U.S.A.

Maximum length, 184 feet. (56 m.) **maximum diameter,** 52 feet. (15.84 m.) **volume,** 300,000 c. feet. (m³.)

Total lift.—21,000 lbs. (kgs.) **Useful lift,** (kgs.)

Gas bags.—Formed of a triple layer of vulcanized rubber-proofed cotton and silk, cemented together by varnish. Strength, lbs. per foot run (3,000 kgs. p.m.)

Motors.—2 motors of 60 and 25 h.p. have been replaced by a single motor of 80 h.p., 4 cylinder. Water cooled. Hydrogen from the gas bag can be fed into the motor as fuel, thereby obviating the usual waste of hydrogen which takes place to compensate for the weight of petrol used.

Speed.—18 m.p.h. Petrol, 2 tons.

Propellers.—Formerly one at each end of the car. Recently replaced by one at each side of the centre of the car, gear driven, and supported on strong outrigger brackets.

Steering.—Large vertical rudder in rear of the keel.

Remarks.—Designed solely for Arctic exploration, this ship has undergone and is undergoing continuous improvement. The keel now consists of one long and very strong tubular petrol tank, extending two-thirds the length of the ship. From this upwards the space is covered in by thick waterproof canvas. Bunks are arranged near the motor, branches from the exhaust pipe being led close under the bunks. The exhaust heat is also used for cooking. The car is fitted to contain three explorers, a pack of dogs, two sleighs, and a very large store of food. The feature of using hydrogen as a fuel is unique; it seems likely to save one-sixth of the fuel consumption. The ballonet is small, only one-eighth of the total volume as it is intended to keep the ship close to the ice. The ship complete, with everything except ballast, only weighs about five tons, leaving 4½ tons available for emergencies, such as the weight of snow on the upper surface.

ANDERSON.

Maximum length, 300 feet / 92 m. **maximum diameter,** feet / m. **volume,** c. feet / m^3.

Total lift, lbs. (kgs.) **Useful lift,** lbs. (kgs.)

Gas bags.—In two halves, the whole resembling an egg cut in two halves lengthwise, air space between the spheroids. Each half is 40 feet high. The nacelle is hung under lower half.

Motor.—

Speed.—

Propellers.—

Steering.—

Remarks.—Building by the Anderson Air-ship Co., New York. Capital, $25,000.

CHARLEMAGNE SIRCH.

Building.

Maximum length, feet / m. **maximum diameter,** feet / m. **volume,** c. feet / m^3.

Total lift.— lbs. (kgs.) **Useful lift,** lbs. (kgs.)

Gas bags.—Depends on hot air for lift.

Motor.—

Speed.—

Propellers.—

Steering.—

Remarks.—Designed by Charlemagne Sirch, of Los Angeles.

EAST ST. LOUIS.

(*Alias* **ZELLER.**)
(*Alias* **ZELLER-SMITH.**)
(*Alias* **SMITH.**)
(*Alias* **AMERICAN EAGLE.**)

Completing.

Maximum length, feet / m. **maximum diameter,** feet / m. **volume,** c. feet / m³.
Total lift.— lbs. (kgs.) **Useful lift,** lbs. (kgs.)
Gas bags.—
Motor.—
Speed.—
Propellers.—
Steering.—
Remarks.—Built by W. J. Smith, of St. Louis.

GOODALE.

The "Start."

Maximum length, 58 feet / 17·67 m. **maximum diameter,** feet / m. **volume,** 8,000 c. feet / 226 m³.
Total lift.— lbs. (kgs.) **Useful lift,** very slight indeed.
Gas bags.—
Motor.—7 h.p.
Speed.—
Propellers.—One 3-bladed.
Steering.—
Remarks.—Frank B. Goodale, a boy, of Toledo, has made several spectacular flights in this air-ship. On June 11th, 1909, he circled over the U. S. cruiser, "New York." On July 12th, he went over New York and followed the track of Broadway from 130 Street to 42 Street, where he circled and returned on his track.

LINCOLN BEACHEY.

Maximum length, feet / m. **maximum diameter,** feet / m. **volume,** c. feet / m³.

Total lift.— lbs. (kgs.) **Useful lift,** lbs. (kgs.)

Gas bags.—

Motor.—

Speed.—

Propellers.—

Steering.—

Remarks.—Nothing known, beyond that he is entered for a race with *Baldwin*.

MYERS.

Maximum length, 55 feet / 16·76 m. **maximum diameter,** 18 feet / 5·50 m. **volume,** 7,500 c. feet / 212 m³.

Total lift.— lbs. (kgs.) **Useful lift,** lbs. (kgs.)

Gas bags.—Cotton machine fabric, varnished seven times and a final weather coat in addition. The whole bag, except the extreme ends, covered with close fitting net.

Frame consists of two gunwhale and a keel, forming a triangle in cross section. Length, 39 feet (11·88m.) Weight, 36 lbs. (16·33 kgs.) Rectangular car, placed amidships on the frame.

Motors.—7 h.p. 2-cylinder Curtis, air-cooled.

Speed.—

Propellers.—One 2-bladed, direct driven. Diameter, 4 feet (1·22 m.) Placed right forward.

Steering.—Small vertical emergency rudder just behind propeller. Ordinary vertical rudder aft. A horizontal plane on either side of the aeronaut's seat. Elevating also done by shifting ballast or the operator shifting his position.

Remarks.—Only 200 lbs. of lift available for aeronaut and ballast.

The balloon is constructed with an eye to very rapid dismantling and packing.

Has been bought by Aeronautic Society, but only the lighter members can do much in the way of flying her.

REKAR.

Maximum length, feet (m.) **maximum diameter,** feet (m.) **volume,** c. feet (m³.)

Total lift, lbs. (kgs.) **Useful lift,** lbs. (kgs.)

Gas bag.—Cylindrical trussed gas bag with ends converging to vertical edges.

Motor.—

Speed.—

Propellers.—In horizontal and vertical planes, front and rear, at corners of the gas bag.

Steering.—

Remarks.—Invented by John J. Rekar, of San Francisco. Building by the Rekar Airship Construction Co., Portland, Ore. Capital, $150,000. Does not sound particularly convincing, but no further details are procurable.

RIGG & RICE.

(*Alias* **AMERICAN EAGLE.**)

Maximum length, 100 feet (30·47 m.) **maximum diameter,** feet (m.) **volume,** c. feet (m³.)

Total lift.— lbs. (kgs.) **Useful lift,** lbs. (kgs.)

Gas bags.—

Motor.—

Speed.—

Propellers.—

Steering.—

Remarks.—Building for J. A. Riggs and Joel A. Rice, of Hot Springs, Ark., by A. L. Stevens, at Morris Park.

SHUTT.

Completing.

Maximum length, feet / m. **maximum diameter,** feet / m. **volume,** c. feet / m^3.

Total lift.— lbs. (kgs.) **Useful lift,** lbs. (kgs.)

Gas bags.—

Motors.—

Speed.—

Propellers.—

Steering.—

Remarks.—Built by Daniel C. Shutt, of Chattanooga, Tenn. Very like *Knabenshue*. Was to have been completed in July, 1909.

SILHOUETTES OF THE WORLD'S DIRIGIBLES.

AUSTRIAN AIR-SHIPS.

PARSEVAL

BELGIAN AIR-SHIPS.

LAMBERT

LA BELGIQUE

LA FLANDRE

BABY

CLEMENT BAYARD II.

LEBAUDY (MORNING POST)

FRENCH AIR-SHIPS.

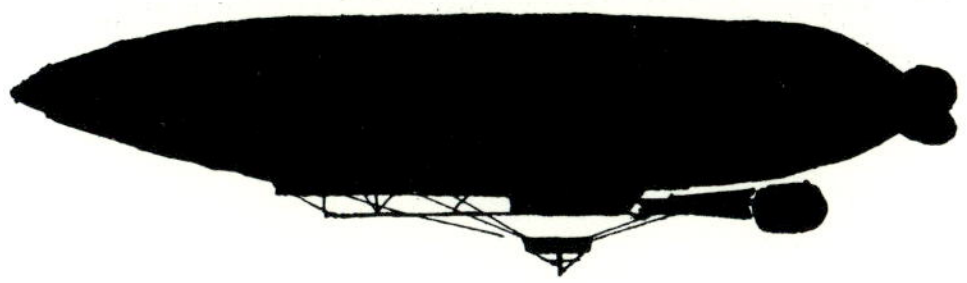

Lebaudy

Republique

Liberte

Ville de Paris

Ville de Nancy

Ville de Bordeaux

Identical

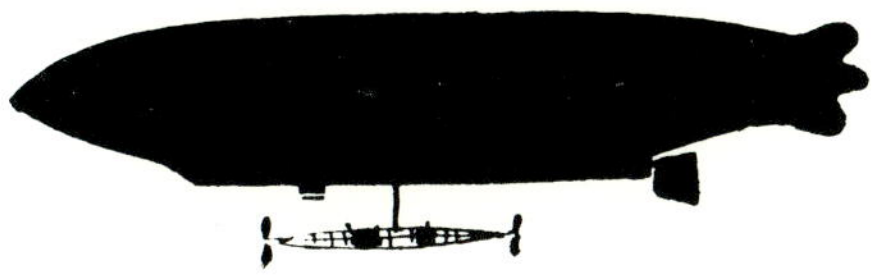

Colonel Renard

Malecot

Marcay

Faure

Zodiac I.

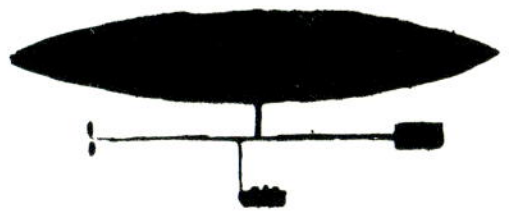

Zodiac II.

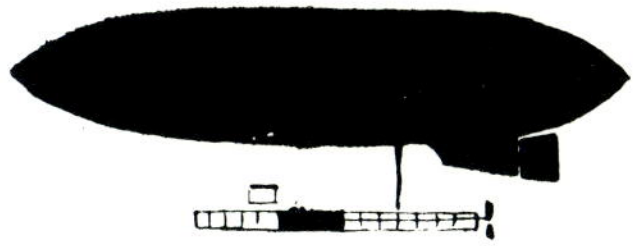

Zodiac III.

SWISS AIRSHIPS.

La Geneve

ITALIAN AIR-SHIPS.

Military I.

Military II.

Identical

Forlanini

Italia

Zeppelin I.

Zeppelin II.

Zeppelin III.

Parseval I.

Parseval II.

Parseval III.

Parseval IV.

Gross I.

Gross II.

Gross III.

Gross IV.

Clouth

Ruthenberg

R. G. W. G.

Hildebrandt

RUSSIAN AIR-SHIPS.

Lebed

Outchebny

Clement-Bayard

SPANISH AIR-SHIPS.

Torres Quevedo

Bayard-Clement
(Identical with Russian *Clement-Bayard.*)

Myers

Knabenshue

Baldwin Military

Wellmann

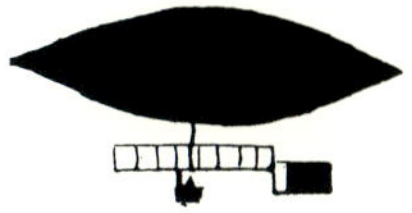

Early Baldwin

PART II.

AERIAL WARFARE,

By Vice-Admiral SIR PERCY SCOTT, K.C.V.O., C.B., LL.D.

THE progress recently made in aviation and the existence of so many comparatively practical machines compels attention from every thinking man. The performances of the *Zeppelins* are sufficiently satisfactory to indicate that the time has arrived when the flying warship is a factor to be seriously reckoned with, but when I am asked to forecast the aerial warfare of the future, I am confronted by a double difficulty. In the first place, I cannot claim sufficient technical acquaintance with the subject of flying to warrant discussing the matter closely. And in the second place, the details of any ideas that we may have on the subject of destroying airships are naturally confidential.

As an adjunct to H.M. Navy, the useful function of an airship or aeroplane would appear to be in gaining information of the locality, strength, and disposition of the enemy's fleet, and so possibly unmask his strategy. In this direction an airship's services would be invaluable, for it might not be possible to obtain the information in any other way.

If it be allowed that an airship is of value as a Scout to acquire information, then airships or aeroplanes we must have, but as the enemy will use similar appliances to watch our strategical operations, secrecy can only be arrived at by the destruction of his observers, and the method of aerial warfare becomes a subject for serious consideration.

The heretofore only traversers of the air use beak and talon to destroy one another, the human aviator having neither beak nor talon, must be provided with some means of offence, it may be a gun, if it is, then the aviator will realise that his safety depends upon whether the projectile out of his gun hits the mark aimed at or not, and accurate gunnery, that is *quick-hitting*, will in the air be as important as it is on land or on the sea in deciding a final issue. Whatever the weapons used are, practice with them will be necessary, and we may live to see two airships each towing a suitable target carrying out a test of their efficiency in *quick-hitting*.

PERCY SCOTT,

VICE-ADMIRAL.

THE FUTURE OF DIRIGIBLES,

By "RIPPING PANEL."

ONE of the greatest of the pioneers of Mechanical Aviation once asked the question: "Would you willingly go to sea, for an extended trip along a rocky and dangerous lee coast, in a ship which you knew to be full of leaks, yet unprovided with pumps, to be so frail that a touch spelt disaster; to be propelled by engines so unreliable that you could almost count on their breaking down at a speed well below that of the currents you might have to encounter; and to be unprovided with a ship's great resource: an efficient anchor?"

The argument against dirigibles could hardly be better put; and it would be difficult to find a better starting point for a reasoned consideration of the probable development and future of this type of vessel.

Dirigibles are usually divided into Rigids, Semi-Rigids, Non-Rigids; but this division is very arbitrary; one type is, in fact, as rigid as another. It would be more accurate to class them as non-collapsible, collapsible, and quickly collapsible; but as the former designation are generally accepted, they will be adhered to.

Every dirigible consists of two main parts, an elongated gas bag, and a car or cars containing motors and men, with all their respective appurtenances; both these main divisions differ in an endless number of details in the various ships of to-day; but only in details; the great distinction between the different types lies in the method of attaching these two main parts to one another. This has been and still is the principal problem confronting all designers, all other matters are details. The designer uses the best engine he can find, and the best material available at the time; but since dirigibles first began, he has had a free hand to choose any method of attachment he preferred, and as certain methods became general, so the ships in which they were practised came to be divided into classes.

The problem of attachment divides itself naturally into two parts: how to distribute the load along the necessary length of gas bag, and how to make connection to any given section.

A very simple calculation shows how much the load must be distributed. There will always be a tendency for the gas bag to turn up its nose and collapse. If this is not to take place, then the internal pressure "P" multiplied by the area of the cross section "A," multiplied by the radius "R," must exceed the lift "L," of the portion of gas bag under consideration, multiplied by the distance "D" between the centre of lift of this portion and the particular section, or:

$$P \times A \times R = F \times L \times D,$$

where F is a factor of safety.

Now if we know our general dimensions, we know the pressure we can allow ourselves to maintain in the gas bag by means of the balloonet; so that a simple calculation shows us how far we must extend our support.

Having decided the length of gas bag from which support is to be derived, the next matter is, how to obtain it.

The old and simple method of putting a net over the gas bag was generally abandoned on account of the great resistance offered to passage through the air by the network; gradually more direct methods were adopted of sewing strips along the belly of the gas bag, and securing cordage to these.

Still the difficulty remained that if the car were small and short the lines going towards the two ends required to have great tension in them in order to provide the vertical loads required by our first calculation. This meant bringing unfair stresses at the points where these lines are secured to the gas bags, and also producing a compression along the underneath side of the gas bag, which would have the effect of slackening off all the cordage secured to these parts, and so still further increasing the load on the end lines which, owing to their inclinations, are least fitted to stand them.

So Lebaudy introduced an oval steel frame close up to the belly of the gas bag; this frame he hung from the gas bag by very numerous short vertical strings, and from this stiff frame he led all his cordage to the car. Thus the frame took all the compression; the stresses in the end lines now exerted their efforts on connections to the steel, instead of tending to tear the threads out of the sensitive gas bag fabric; and the so-called "semi-rigid" type developed.

But the obvious question arises: why have this great frame close up to the gas bag, when it would serve its purpose equally well much lower down, with the additional advantage that it could then form part and parcel of a more extended car? This was done in the *Ville de Paris* type: the Lebaudy girder, much altered, but still the same in purpose, is combined with the car into one long frame hung well below the gas bag, dividing the load along the required length, avoiding undue stresses in any part, and at the same time providing a long and comfortable place for the crew to move in. Merely because the girder is a little lower down, this type has somehow come to be known as "Non Rigid."

To account for the growth of the "Rigid" type is more difficult; it is often thought that a vessel with a "gas bag" nose is precluded from attaining great speeds by the liability of the nose to collapse or dent in; but experiment and theory have both proved conclusively that this danger is unreal, and that certainly up to 50 miles per hour no caving-in takes place; so this argument, often used by the non-expert in support of the rigid type falls to the ground. Another view is that the resistance of the suspension in the case of the first two types is so great that it becomes worth while to adopt a rigid framework in order to enable the cars to be fixed close up to the gas bag. But this could be effected in either of the other types, and probably would be, were it not for the danger of fire originating in the car and spreading to the gas bag.

Lastly, there are a number of small advantages accrueing to the "rigid" type, such as the possibility of putting the propellers in line with the centre of resistance; the possibility of doing away with the ballonet and its pump; the possibility of having a large number of separate compartments in the gas holder; the possibility of using lighter material for the gas bags.

Probably it was all these together which led to the adoption of the rigid type. Looking at existing specimens, and forgetting for the moment the liability of the rigid ship to damage from occurrences which would probably not injure the other types, it seems as if:—

(1) The successes of the *Zeppelin* type are due exclusively to the large size of these vessels, enabling them to carry motors that are powerful and reliable, and enough petrol for extended voyages.

(2) Justification for the existence of this type will only be found when the dimensions are such that the hull weight of a rigid ship is less than the hull weight of any other type.

It is easy to see that this latter condition may be attained; the strength, and hence the weight per square foot, of the material used for making gas bags, must vary directly as the diameter, so long as the same internal pressure is used; the area of material used varies as the second power of the linear dimensions; so that the total weight of gas bag, for a Semi or Non-rigid ship, increases as the third power of the linear dimensions. But in the Rigid type, where there need be no internal pressure in the gas bags, the weight of gas bag and the weight of hull structure need not increase faster than the second power of the dimensions; indeed the weight of hull structure need not increase so fast, seeing that, regarding the hull as a long tubular girder, the depth of the girder is increasing with the dimensions.

Consequently there is a certain size beyond which it becomes advantageous from the point of view of weight alone to use the rigid type, and incidentally to gain its several advantages. It is difficult, impossible even, to say at what point this size is reached. This point could be fixed approximately if only the researches on the best shape of hull were sufficiently conclusive; but they are not. We know that Professor Prandtl has carried out researches, and that their results are seen in the latest *Gross* and *Parseval*; so also has Count Zeppelin, yet his shape is totally different. We can only say that if Professor Prandtl is right and Count Zeppelin wrong, then the diameter of the *Zeppelin* is smaller than it should be, when he would be on the borderland of those dimensions above which the rigid hull becomes desirable. If, however, the shape of the *Zeppelin* is the best, then it is doubtful whether a rigid ship of such moderate dimensions is justifiable.

Evidence is against the *Zeppelin* shape; the *Schütte* form seems to be more in accordance with the results of researches; but this we can say with absolute certainty, that whoever sets himself the task of ultimately producing a type of vessel that can keep the air in all weathers, reach its destination in time and with certainty, and undertake the work of the modern liner, must start with the rigid type, or he will have to change in due course, and his previous experiments and experience will be wasted. So we can foresee the rigid type growing and growing till it becomes supreme in a certain field; and we see it eliminated from among dirigibles of moderate cost and dimensions. But all alike are still in the state described by the question alluded to at the beginning of this article.

How are these drawbacks to be overcome, and with what result? "The ship is full of leaks, and there are no pumps!" Rubber is the leak stopper of to-day, but it has reached its limits. There are only two courses open, namely, to stop the leaks, or to provide the equivalent of "pumps."

The former can only be done by adopting a substitute for rubbered fabric. In the case of large and rigid dirigibles, this is not far to seek; Schwartz pointed the way in his ill-fated attempt, and only the lack of a good method of joining thin aluminium plates stands between us and realization; a rigid airship with a complete skin of aluminium plate is a possibility of to-day; the weight compares favourably with that of the present systems; so does the cost; and once adopted, "leakage" would be abolished. The last problem left to face would then be that of replacing the weight of the petrol used; this is not insuperable, and even if it should defy solution for many years, this will be but a slight impediment to the general use of dirigibles: 1% off the dividends perhaps.

Aluminium is not the only material for the structure or for the skin. A wooden structure is already in use in the *Schütte* ship; a wooden skin is already contemplated in Germany; German experts have even produced new materials, better than either wood or metal, which can doubtless be turned to account. But the principle adopted by Schwartz would seem to be the one to which all rigid airships must ultimately come.

In the case of other types, a substitute for rubber must be found, and there is every reason to believe that it very shortly will be; the number of varnishes, oils, and miscellaneous processes now being produced is great, and it seems probable, almost to certainty, that another year will see the required process commercially obtainable.

Then as to "pumps," *i.e.*, a method of compensating for losses of buoyancy, whether due to voluntary or involuntary losses of gas. In Germany experiments have been made, though on a very small scale, with liquid hyrogen; and it has even been taken up in a balloon and used to prolong the voyage. At the same time it has been suggested that liquefying plant should be installed in connection with the various alkali works which produce pure hydrogen as a by-product. The movement is progressing, and there can be little doubt that if there is a demand, *i.e.*, if the above two problems are not speedily solved, airships will take up in a liquid state enough gas to make good all losses for a week or a month, according to particular requirements.

The charge of frailty is hardly worth refuting; ships of the rigid type will be able to rely on reaching their destinations like ocean liners; if they join issue with the "rocks" half-way, they will suffer. The other types are wonderfully safe against such accidents. A ship of the *Lebaudy* type can hardly come to any harm, unless by some extraordinary bad fortune she should tear her gas bag; even then, the repair is not an expensive one; there is no other craft in existence which enjoys an equal impunity when it is driven on to a lee shore.

The engines are unreliable it is true; but we have seen almost the last of "one motor" airships; and this reduces the risk very greatly; add to this the fact that reliability is increasing hour by hour. For the small ships, whose two motors will both necessarily be light, and which will not attempt ocean voyages, there are no more dangers or inconveniences in store than there are for the motor car with its liability to breakdown: less in fact, for no motor cars have two separate engines. For the large and powerful aerial liners the risk attendant on disablement is greater, yet no greater than for a waterborne ship, to the reliability of which they must necessarily attain.

But lack of speed, this is the fatal drawback to these craft. Consider the latest *Clement Bayard*, 440 horsepower, and yet with a speed of only 35 miles an hour. Notoriously, the speed of the wind frequently exceeds this figure, and obviously the speed will only increase as the cube root of the horse power. There is little probability of the weight per horsepower being decreased for many years to come; the demand for increased reliability will be so strong, and even if it were decreased, the weight of the petrol would not thereby be affected. So that there are only two alternatives, namely to increase the size of the ship, when we can afford to add horse-power faster than the resistance increases, since lift varies as the cube of the dimensions and resistance as the square, or to find ways of decreasing the resistance or applying the power more efficiently. While undoubtedly the former is the easier way of achieving results, the latter is the better. Doubtless dimensions will grow by leaps and bounds, till the present *Zeppelin* appears a mere canoe by comparison with the craft of two or three years hence. For it must be borne in mind that size is only relative; in the case of ships we have two limits, and we stop in deference to whichever we encounter first. One is the quantity of cargo or passengers that will be available for our ship when built, the other is the size of the harbours into which she will have to fit. Applying these considerations to airships, and assuming the craft to be adapted in other respects for passenger or cargo traffic, it is clear that we should have to build vessels of dimensions that seem to our present ideas absurd before we reached the first of these two limits. As to the second limitation, it applies, but in a very modified way: depth of water is the decisive factor with regard to a harbour; an airship has no such trouble; it merely becomes a question of how great an area of flat land a nation or a company can afford to possess in a suitable place. Judged by these standards, there seems no reason why the proposals of Albert Wetzel should not be carried out to-day, and airships built, as he suggests, a thousand feet long, or even double this length, carrying 300 tons of passengers if required, at any speed which may be thought most lucrative from the commercial point of view.

But while this is the probable line of development in the immediate future for the large craft, it is out of the question for vessels of the *Zodiac* type, for medium sized military scouts, for "club ships" belonging to various societies, and for family yachts which will undoubtedly largely take the place of motor cars. For all these, mere hulking size and brute force must give way to refinements in the shape and in the application of power.

There is probably still room for a little improvement in the shape of hull; undoubtedly too, the car will come closer and closer up, thus reducing resistance by about a third.

Still considering the *Clement Bayard*, we might say that an improved shape would perhaps reduce the power required for her present speed from 440 to 350 horse-power; bringing the car close up, which can be done little by little as all danger of fire and of pitching is overcome, should reduce this figure to 250.

What next? The only direction in which at this present moment there seems to be a glimmering of hope is that indicated by Monsieur Marcay, (vide page 165). It is quite obvious that his propeller is too far forward, but forgetting this for a moment, let us consider the theory of the method.

A modern airship with one propeller leaves behind her two separate air currents, one in rear of the propeller going in a direction opposite to that of the ship, and one directly in rear of the gas bag, caused by the skin friction and partly by the hull form going in the same direction as the ship. Now it is clear that with the ideal shape of hull the skin friction will be equal, or nearly so, to the "form loss" or resistance which would still exist if there were no skin friction.

Since the energy imparted to the propeller current is equal to the sum of these two, it follows that by placing our propeller suitably we can regain some of the energy at present wasted, and that if we can only leave the whole of the air in the track of the ship at rest, we shall require very little power for propulsion at high speeds. This fact is clearly brought out in all standard works (Lanchester and Maxim for example), and needs no further explanation here; but it means that by adopting the Marcay system, and placing the propeller near the rear end of the ship, we could recover something more than a half of the energy at present wasted, namely, all that is not expended in creating heat or setting up eddies, and all that is used instead in creating a moving stream.

As the nature of a wake has not been thoroughly investigated, it is impossible to say how great an economy could hereby be effected. But, to take the lowest estimate, we can safely say that when this system has been brought into practical operation the power of our *Clement Bayard* which we are considering could be further reduced from 250 to 150 horse-power.

In justice to Monsieur Marcay, it must be added that his vessel appears to be merely an experiment on a very large scale, and that the propeller was probably placed in its present position simply because in such a small vessel the trim would have been upset by placing it at the end.

While, during the next few years, power is being economised along these lines, doubtless refinements of gas bag manufacture will be introduced, allowing of a further saving of weight; refinements in the manufacture of cars, etc., etc.; so that the near future should see any given craft of to-day replaced by one of equal accommodation and speed, but with a smaller gas bag, and with motors of between a quarter and a fifth of the present power; or, putting it another way, there is every reason to anticipate that in a short time vessels of the present sizes will be able to proceed at nearly double their present speeds on the same horse-powers as at present, and, seeing that the present speeds are amply sufficient over the half of any closed circuit, they will be able to cover distances many times greater than at present, and remain in the air for periods of as many days as they at present do hours.

It would appear from this that there are likely to be two general types, namely the small type, of which the *Zodiac* is the forerunner, and the large type, of which the *Zeppelin* is a very diminutive specimen; between these two will be numbers of types for special purposes. Doubtless, aeroplanes will develop rapidly, and most special purposes will be met by them: postal services, express passenger services, military services, etc., etc. But there seems no prospect in the immediate future of the aeroplane being able to replace the Atlantic Liner or the Naval Aerial Scout, or for it to take the place of the motor car for the man of moderate means with a family. Can any better sport or amusement be imagined that could be obtained with an airship of the *Zodiac* type, endowed with a speed of 40 miles an hour for four hours, or 20 miles an hour for eight times this period, and so on in cubic proportion?

Always able to reach a desired goal, but with the ever changing wind to add an element of interest to the journey; free from dust and the dangers of the road; always able to stop and enjoy still air. An airship of this type would combine the delights of a motor car, a balloon, a sailing yacht, an aeroplane, with the dangers of none. A study of all the facts can leave little doubt that such a vessel is a possibility, and that it could be produced to carry four or five, and at a cost of about £500. But not till, one by one, the many problems have been solved which have here been alluded to.

It is perhaps worth while contrasting such a vessel with an aeroplane designed for the same purpose: condemned to be rushing through the air every moment of its time; never slowing, never pausing while its occupants look down on the mountain tops, or eat a quiet meal; unable to come down except where the ground has been specially prepared.*

It would seem that it is to the dirigible that the ordinary family man must look for his aerial source of health and daily pleasure, and he will not be disappointed.

But last in the indictment comes the charge of being unprovided with an efficient anchor.

This is one of those things which the airship has not yet grown to require, but without which it can never develop to the full. Wanted! an anchor which will hold the ship by sea or by land, and which can be instantly and easily weighed; wanted also a method of attachment such that the airship can at all times ride like a balloon kite; then when a tempest overtakes her, she has only got to drop her anchor and wait till it is past; on every service, on every pleasure trip, an anchor will be wanted, will be in fact essential if the most is to be made of the ship, but as yet there are no indications as to what it will be like. It is just one of those things which will develop quickly when it is required, and not before, but it will be the final touch which will render airships a success.

Then in a few short years we shall be able to alter the question and ask "Won't you come for a cruise over beautiful country in a staunch ship, with powerful pumps, with two reliable engines, at a speed as great as you could wish, and provided with a faithful anchor by means of which you can at all times pass a night in peace, or ride the fiercest storms?" And the whole world will answer "Yes."

*Some of these difficulties would perhaps be solved by practical helicopters.

THE POLITICAL ASPECT OF AVIATION,

By L. CECIL JANE, M.A.

(Editor of the "Chronicle of Jocelin of Brakelond," Asser's "Life of Alfred," etc.)

"HOW much the greatest event of the century, and how much the best!" The exclamation, with which Charles James Fox greeted the fall of the Bastille, was in effect repeated by a section of the daily press, when M. Blériot crossed the Channel from Sandgate to Dover. His achievement was hailed as the dawn of a new era, as the prelude to the realisation of the Christmas message of peace and good will. It was urged that, as the new means of communication is common to all peoples, the boundaries between States would disappear, and the essential identity of the races of mankind be admitted. It was to be the end of war. Universal free trade must follow, and the consequent cessation of all commercial rivalry remove the ultimate purpose for which fleets and armies have been created, the securing or holding of a market. All peoples, united in a new fellowship, would labour together for the good of the human race, for the advancement of civilisation. And many of those, who were too pessimistic to believe in the likelihood of such a change in the nature of mankind, were yet disposed to agree that the armaments of the present day would become valueless before the prospect of bombs of precision falling on helpless ships and regiments from invulnerable and almost invisible airships, poised in middle air.

But viewed in a more sober spirit, and not under the influence of the excitement naturally aroused by the first definite advances towards the conquest of the air, the chances of so complete a revolution seem to be somewhat remote. The great airship fleets of the sensational novelists, such as George Griffith, or the "Clipper of the Clouds," of Jules Verne, are at least likely to remain dreams for the present, and it is improbable that either inclination or necessity will lead to the immediate disbanding of armies and to the hasty "scrapping" of fleets. None the less, it is on all sides admitted that the advent of the airship, whether dirigible or aeroplane, must lead to some change in the art of war, must be a factor in the determination of the course of political evolution. The question is not whether there will be any change, but as to the nature and magnitude of the change. The present article aspires to be a contribution towards the answering of this question. More it could not be, for it is obvious that, in the present state of aviation, suggestion as to its possibilities is alone admissible, dogmatism an absurdity.

Yet even so, it may safely be asserted that the more sensational conjectures as to the extent of the change are unsound. Quite apart from the fact that an airship, capable of carrying either guns or a large quantity of bombs or military stores in bulk, is as yet a thing of the future, the influence of the new factor may much more easily be exaggerated than minimised. There would appear to be three ways in which airships might revolutionise future wars by acting as a directly offensive weapon. They might cruise above an army and drop bombs upon it; they might do the same to a fleet; they might appear above a city and either destroy it or hold it to ransom. (i) Such have been the suggested rôles for the new factor; whether it would play any of them with even passable success is questionable.

The lessons of recent wars have tended rather to falsify the expectation, once entertained, of the possibility of the infliction of serious damage in this particular way. An army, operating in open order, becomes, when viewed from above, little more than a loosely united aggregation of individuals, and would present so unsatisfactory a target, that a few expert riflemen might do more damage to the airship than the army would sustain from it. Even less danger would be run by a fleet. An armoured upper deck and the possession of quick-firing guns capable of the necessary elevation would probably give the ship a

(i.) Attemps to destroy dockyards, and bases in general, might be mentioned. But our lessons which may be drawn from the bombardment of towns apply to this form of attack also, while there would be even more opportunity for defensive precautions, when once the necessity for them had been realised.

considerable advantage. And the experience of the ineffectiveness of prolonged bombardments of towns in recent wars seems to be enough to dispel the idea that a city could be seriously injured by bombs dropped from an airship. There would be nothing like the same opportunity for taking careful aim as was enjoyed, for example, by the Boer artillerymen before Ladysmith. (ii)

No doubt, for a while, there would be the possibility of profound moral effect. It may well be that an army would be panic-stricken at the appearance of an airship above it, that a fleet would be disorganised, that a town would offer all its wealth to escape destruction. But the first successful resistance to the new method of attack, the first demonstration of any vulnerability and relative powerlessness in the new enemy, would be enough to restore discipline and to destroy the moral superiority of the aviator. And it is worth remembering that the aviators themselves would be risking certain death, since one lucky shot would infallibly destroy an aeroplane and might easily cripple even a dirigible. It is, perhaps, unlikely that moral effect would make the airship master of fleets and armies for more than the briefest period, and it may be suggested that with every increase of knowledge of the limitations on the powers of the new vessel, the probability of such moral supremacy being established at all is greatly reduced.

The function of the airship, then in ordinary war, as a weapon of offence, as a direct agent of attack, may be regarded as distinctly limited. Its province as an indirect agent, a negative factor, may be admitted, but it is not the purpose of this article to consider the extent of its influence as an adjunct of Intelligence Departments. However great its value in this connection may be, it will not effect any great change in the art of war. Hostilities will not the less be waged by fleets and armies, because the strategical operations prior to an actual engagement were assisted or impeded by the scouting exploits of airships. Those exploits might well have a profound effect on the preliminaries of a battle, but not on the battle itself, not on the art of war or on the political development of mankind. That the country, which possessed an overwhelming airship fleet, would necessarily be victorious would seem to be no more an axiom than that victory always awaits the State, whose Intelligence Department is the most efficient. "Air-power" can hardly be more than one of many factors in deciding the issue of future wars. (iii)

This limitation of the sphere of the aviator is in direct accord with what history might have led us to anticipate. It has been the general fate of new discoveries in military or naval science. The difficulty of a frontal attack and the value of turning movements had been abundantly illustrated at Thermopylae, before the operations on the Modder River. The idea of breaking the line had occurred to Phormio, before it was adopted by Rodney. Destruction of the enemy in detail was as much a principle of Cæsar as of Napoleon; the true theory of "big battalions" was in no sense a discovery of the eighteenth or nineteenth century. It has mattered little in history whether the opposing forces were armed with bows and arrows or with weapons of precision, whether the oar, the sail, or steam supplied the motive power of the fleets. Through all the ages, the factors making for success have varied little, if at all; it may be asserted with considerable confidence that it was not the armament of the rival fleets and armies which determined the result of Marathon or Salamis, of Trafalgar or Waterloo. Had Miltiades commanded a force armed with the weapons of 1815, and had the Persians been similarly equipped, it is at least probable that the result of the battle would have been as it was.

But when we pass from the domain of international war, and enter that of popular revolution, a totally different conclusion follows from the evidence of history. A sedition is governed by considerations, which do not affect the conflicts of nations; new factors make for victory or defeat. In general, the development of military science has been accompanied by an increase in the stability of an existing government, or at least by a change in the requisites for successful rebellion.

In the middle ages, almost anyone who possessed a defensible castle, and who could command the services of a body of men, was able to secure comparative independence. A Geoffrey de Mandeville could capture and hold to ransom the bride-elect of an heir apparent, and if this were in a time of anarchy, there are not

(ii.) The case of Barcelona, July, 1909, may be quoted as an apparent exception, but there the bombarding forts were peculiarly well placed, absolutely commanding the city from an easy range.

(iii.) That "air-power" will be the sole decisive factor in future wars would seem, indeed, to be the idea of Mr. H. G. Wells, whose speculations, however, are perhaps distinguished rather for their dazzling ingenuity, than for evidence of appreciation of the lessons of the history of the past or of the present.

wanting instances of more or less successful defiance of the authority of even a vigorous king. The barons of Aquitaine were sobered rather than subdued by the energetic government of Richard Coeur-de-Lion, "who," in the words of a not too friendly contemporary, "was hindered neither by precipices nor by towers of brass, though they had in all previous ages been impregnable, nor by the works of nature nor by those of man; who was not stayed by steep and lofty strongholds or by those which were hidden in the bowels of the earth." Even a petty baron, like Bertrand de Born, could boast that he had repelled with success the repeated attacks of his suzerain.

Instances of successful resistance during the succeeding period of English history may be multiplied, and in this respect the history of England is analagous to that of the continental states. No sooner had the prestige of John sustained the heavy blows of Bouvines and of the submission to Innocent III., than the barons realised their strength, and Magna Charta was signed. Hubert de Burgh created astonishment by daring to expel the alien Fulkes de Breauté; the independent attitude of Ranulf Blundeville went unpunished; and it was only the rivalry of De Montfort and De Clare which enabled the monarchy to secure the partial triumph embodied in the Dictum de Kenilworth. Bohun and Bigod raised an army on their estates sufficiently powerful to prevent the collection of taxes, and to force Edward I to assent to the Confirmatio Cartarum. Thomas of Lancaster overthrew Gaveston; the Lords Appellant drove De Vere into exile. The Lancastrian and Yorkist reigns are little more than a chronicle of the immunity of the "over-mighty subject," whose contempt for rule is illustrated in the Paston Letters, and whose greatest representative well merited the epithet of "Kingmaker." In general, throughout the middle ages, rebellion was easy, and the chances of success at least equal to those of failure.

When the Tudors ascended the throne, a sudden change occurred. That change, illustrated by the failure of Simnel at Stoke, by the execution of Buckingham, by the collapse of the Pilgrimage of Grace, and of the rebellions of Ket and Wyatt, cannot be attributed wholly to weariness of internal strife. Rather it was the result of a modification in the art of war, the introduction of musketry and the rise of the professional soldier. The central government, whether it created a standing army as in France, or relied on the ancient levies as in England, was always careful to accumulate in its own hands those military stores, without the possession of which the rebels were doomed to failure. In such cases as the Great Rebellion in England, where the executive was defeated, some part of those stores had fallen into the hands of the disloyal. Charles I. was excluded from Hull before he set up his standard at Nottingham, and he was additionally handicapped by the fact that the fleet sided with the Parliament and cut off possible supplies from the Continent. It may be added that the Great Rebellion was in no sense a movement organised by a class, and that the Parliamentarians, with all their advantages, were singularly unsuccessful until the genius of Cromwell organised victory. The history of the contemporary outbreak of the Fronde in France abundantly illustrates the fact that the days of baronial insurrection had passed. The introduction of the trained and disciplined soldier had changed the situation. (iv.)

(iv.) The "Brabancons" and the "Free Companies" of the middle ages were in a sense professional soldiers, but they were not trained and disciplined, as were the armies of the sixteenth century and later. And they were not under the control of a particular government, but sold themselves to the highest bidder, whoever he might be.

Indeed from the time of the establishment of centralised monarchies until the close of the eighteenth century, rebellion ceased to be easy. The motives for resistance to the central power remained, there was much stifled discontent, and there are not wanting instances of that discontent finding vent in popular outbreaks. (v.) But these risings failed, and failed because the executive controlled a disciplined fighting force. When the people rose against the government, it was merely a question of calling out the soldiery, whose permanent training in obedience led them to carry out any orders without questioning. At the command to fire, they fired; habit mastered any possible sympathy with their victims.

But in the last years of the eighteenth century, the populace of Paris discovered a way in which the forces of order might be met. The "personal element" was introduced; the soldiers learned that they were not automata. In other words, the Paris mob realised that an army is nothing more than a force of disciplined individuals, each of whom is personally actuated by all the motives which sway the mind of the

(v.) The rising of Masaniello and the sedition of Geneva may be noted especially.

civilian. It was merely the fact that they had been trained to act as a machine, which made it possible for a government to use troops against their fellow-countrymen, and if the sense of individuality could be restored to the soldiers, they would no longer give unhesitating obedience to any orders issued by the recognised authority. It was the discovery of the method by which this sense of individual responsibility could be restored to the army, which made the French Revolution successful, and the method was first the formation of crowds, and afterwards the erection of barricades.

Called upon to disperse a crowd of fellow-citizens, the first impulse of the army was to obey. But the mere appearance of a crowd, containing, in all probability, familiar faces, was enough to make the soldiers hesitate. It was one thing to shoot down a few isolated and unknown individuals; to fire into a multitude, including possibly relatives and friends, was another. And having hesitated, the soldiers allowed the natural to prevail over the artificial instinct; the sense of obedience was destroyed by the sense of ultimate community of interest, and fraternisation ensued. This, then, was the first lesson learned by the Parisian mob, that if troops were faced by a crowd and did not act at once, if they hesitated and so *thought*, then the victory of the crowd was assured.

Napoleon was equally alive to this fact. In the disturbances of Vendémiaire, 1795, he had recourse to artillery, that is, he enabled his men to operate from a distance, did not allow them, as it were, to look into the faces of their victims. The crowd was there, but it ceased to be an effective crowd; it became merely a target. Then the populace of Paris, the chief specialists in the art of popular revolution, made a second discovery. For the crowd to win, delay was essential; therefore, delay must be secured. The soldiers must be given time to hesitate, time to see their victims, time to think. Hence came the barricades, defences across streets, which had to be stormed, which compelled the troops to come face to face with the people, and which made hesitation inevitable. The result was seen in the revolution of July, 1830. Regiments fraternised with the rebels, whom they were ordered to disperse; once more the crowd became effective, once more the "personal element" was introduced. The soldiers became individuals by force of circumstances. (vi.)

At the present day, then, popular revolution will in all probability be successful, granted that a crowd can be collected and delay ensured. Such at least has been the case. But, as suggested already, every change in the art of war has affected the course of popular risings more than any other form of conflict, and on *à priori* grounds, it may be supposed that aviation will do so. This idea is supported by a consideration of the requisite for a successful popular rising, the introduction of the "personal element," by means of a crowd and of delay.

It can hardly be denied that the presence of a crowd would rather assist the operations of an airship than otherwise. For, in the first place, each bomb would be more likely to do damage; and in the second place, whereas an isolated rifleman would be at an advantage in a conflict with an airship, a crowd, when the taking of careful aim would be almost impossible, would be hopelessly handicapped. And the barricades would be equally valueless; they would not have to be stormed, and there would be no need for coming to close quarters. In a word, aviation would eliminate the "personal element."

The leaders of the popular revolution would thus be in a dilemma. They might organise their forces in large or small bands. If the former, they would be at the mercy of the airship; if the latter, they would be courting disaster by sacrificing the requisite for success. There is no historical instance in which the regular military forces of the state were not at first opposed to a popular rising. Their conversion, their fraternisation with the mob, has followed only as a result of contact with a crowd; isolated individuals, small bodies of men, have never been able to counteract the powerful influence of discipline, for they have not been able to make appeal to the "personal element." And the masses act neither with resolution nor with effect, except when operating in large bodies; the instinct of submission to the constituted authorities, however hateful those authorities may be, is only destroyed by union in multitudes. Thus, in either case, popular revolution would seem to be doomed to failure in the future, as a result of the introduction of airships,

(vi.) Cp. the recent instance of Barcelona, July, 1909. The barricades were built, but it was possible to shell the city from a fort near enough to make bombardment effective, far enough away to exclude the "personal element." The importance of broad streets, which cannot be barricaded easily, has been seen at Petersburg. If there were despots in the present day, the first care of each would be the benevolent widening of the streets of his capital.

unless, indeed, the aviators lead the rebellion.

Here, perhaps, lies the crux of the whole question. For the nature of the influence of aviation on political evolution must apparently depend upon the number and ownership of the dirigibles and aeroplanes. That the former can ever become even reasonably numerous and reasonably cheap appears to be quite unlikely, but, on the other hand, it does not seem to be an extravagant assumption that the aeroplanes will be comparatively common. It may be suggested that they will approximate in number and in cost to motor cars, and that they will tend to be owned by the same class of people. In other words, they will be the property of the rich or moderately rich; at any rate, they will not in general be owned by those who would organise or support a popular rising. It might, indeed, be that there would be some aviators on the side of the masses, but even so, the probability remains that the overwhelming majority would take sides with the constituted authorities.

That this fact is vital is clear enough, when it is remembered that the "popular" revolution depends for its success on the fraternisation of the opposing forces, as a result of the operation of the "personal element," introduced by the appearance of crowds and the compelling of hesitation of the forces of the executive. The probability that aeroplanes will not be owned by those who would naturally fraternise with the leaders of the "social revolution" is therefore of extreme importance. If airships were likely to be owned by the proletariat, then aviators might be moved by sympathy with the revolutionaries, as soldiers have been moved. But if they are the property of the more or less wealthy, of those "who have a stake in the country," it is more probable that devotion to the existing order, a sense of loyalty to government, and a realisation of divergence of interests, would be the motives which would operate upon them most powerfully. This is the more likely when it is remembered that there would be no enforced hesitation, no compulsory gazing into the faces of the victims. And finally, even one solitary airship would be sufficient to disperse a crowd; a fact which makes fraternisation even less probable.

Aviation, then, would seem to have impaired, if not to have destroyed, the chances of a successful popular revolution. It may be argued that the foregoing suggestions apply only to revolutions organised in large centres of population. But it may be answered that it is only such revolutions which would ever embarrass a modern government. This is essentially the age of the supremacy of large towns; a rebellion originating in a rural district is almost unthinkable, and in any case its success would depend on the degree to which it spread to and was supported in the cities. The age of manufacturers and commerce has for ever destroyed the political ascendency of the agrarian interest.

And it is not a valid objection that the aviator would need a base of operations. Nothing would be more easy than to find some place, inaccessible except by the air, to serve as the port, from which the airship would issue to do its work of destruction, and to which it could return.

But if it be true that aviation has thus given a new strength to the existing order, so far as resistance to forcible changes is concerned; if it be true that masses of people will no longer possess an inevitable supremacy, then we have indeed reached an epoch in the history of political development. The establishment in almost every country of representative institutions, of popular government in some shape or form, may fairly be attributed to the invincibility of "the Many." A crowd has been able to prevail, that is, the people, or the more noisy and more assertive section of the people. Popular government, like all other forms of government, rests ultimately upon the unanswerable argument of superior force. If that argument no longer support it. it may be asked whether the institution will itself endure. Visions of a despotism may appear to be no longer mere wild imaginings, of a depotism of aviators, who will have the one final argument on their side.

But whether or no such visions are more than idle dreams, there would seem to be some grounds for asking whether it is not in the sphere of political evolution that aviation will have the profoundest influence. That the art of international war, commerce and locomotion will be in some degree affected may be admitted, and yet it may be doubted whether any of these will be affected to the same degree as the development of government. It may be suggested that the forcible carrying through of the "social revolution" is no longer possible; that "Days of July" are for ever a thing of the past.

AN AERIAL POSSIBILITY,

BY RÉNÉ DECLARGES.

It is generally accepted that Air Craft will shortly be of great service as scouts both by sea and land. But it is almost universally held that this is likely to be their only function until in the distant future the airship of schoolboy romance becomes a reality. At present there is no sign of this. There is indeed some talk on the one hand of existing types being able to inflict much damage by dropping bombs, and on the other hand of the construction of monster aeroplanes able to keep the air for weeks, and to carry such guns as will enable them to compete in open and straightforward combat with the strongest forts and battleships. But in spite of these suggestions, it is not too much to say that most thinking men assign very definite limits to the usefulness of air craft in warfare, and hold that we are as far as ever from the day when armies and fleets will merge into one common flying service.

The author of this paper would not think of disputing these conclusions when put forward merely as generalities. But as in most concerns of life, so here there is a special case affecting England most of all, in which these conclusions cease to hold.

It is possible that this special case may never arise, or it is possible that the next five years will see its full development.

Let it be presupposed that during the next two years Germany succeeds in building airships which have a speed of 80 miles an hour, and which carry fuel for a week at this speed.

It may be said that this is an extreme assumption, but in view of the immense progress marked by the *Schütte* type as compared with the *Zeppelins*, and in view of the fact that never before in the march of progress has there been a science in which greater results could be obtained so simply and with so few drawbacks by merely increasing the dimensions, it must be considered a quite possible one. It may be that Germany will not consider the "game to be worth the candle," in which case these airships will not be built; but otherwise they are already as absolutely certain of construction and successful completion as are the *Nassaus*, the German *Dreadnoughts*, of her latest naval programme.

Suppose Germany to have a fleet of such ships, this fleet would still be almost useless against armies or warships: but it will not have been for use against *Dreadnoughts* that they will have been created. When they see a battleship, they will gently and discreetly move away.

Come! let us try to foresee the great day of war.

What will it avail England if her Navy is all powerful, if the German Fleet is pent (not unwillingly) in its harbours, if the great Territorial Army is ready and efficient to its last gaiter button, so long as the German Air Fleet has spread around the coasts and swept outwards along all the trade routes?

A paltry bomb on the deck of a *Dreadnought* would pass almost unnoticed, but not so on the deck of a liner or a tramp. There can be no easier task for an efficient airship, of whatever type, than to sweep from the Channel outwards along one of the established routes, deflecting all the English commerce to some distant port, under threat of destruction, and to sink in a few minutes any that disobeys.

Cruisers would be useless against these craft, which can avoid them as a cat could avoid a tortoise. The blockade and conquest of England would be accomplished with a total loss perhaps of only a few tramps and a liner or two. Such is the picture! So long as Germany sees a reasonable chance of competing with England on the sea, it is doubtful if she will attempt to turn the position in such a way as this; but once it is seen that the "Two Keels to One" policy is one by which England means to stand at whatever cost, there can, be little doubt as to Germany's course of action. It has too long been believed that the beating of an enemy's fleet or army is the object of a war. A moment's reflection shows that such occurrences are merely incidents which have hitherto contributed to the desired end, namely the compelling of an enemy to agree to certain terms. If this can be attained by Germany without anything but a mocking laugh being accorded to the mighty useless British Fleet, then most certainly and without the slightest doubt she will take the necessary steps. Consequently it may be soon, possibly in 1911, that England will find herself compelled to take preventive measures. The one correct way of doing this is to build an aerial fleet, and meet the enemy in the air.

If Germany's fleet is intended solely against England, and if she changes her policy and diverts her resources from waterships to airships, can England neglect to do the same?

She would be faced with the alternative of risking her national existence, or of following suit, allowing the navy to dwindle to the requirements of a police force in foreign waters, and only maintaining a 2-power policy in home waters, while straining every nerve to gain the mastery in the air. It is one thing to hold an acquired superiority, but quite another to make up lost time and gain the lead, and this is what England is or will be called upon to do.

Briefly this paper attempts to draw attention to the great error of supposing that navies and armies are necessary to war, or that airships must be judged by their ability to cope with these. Airships, in a war with England, can afford to mock at her fleets, and ignore her troops. Therefore England's enemies will assuredly take to the air. Are they not doing so? Yet it will be a sad day when the grand Fleet of England is no more her "All in All."

RÉNE DÉCLARGES.

AERIAL ENGINEERING,

BY CHARLES DE GRAVE SELLS, M. INST. C.E.

OF all methods of progression it is the idea of "flying" that has ever exercised the greatest fascination on the human race, and from the earliest ages the flight of birds and other winged creatures has been studied with a view to discover the secret of their swift and apparently easy movement through the air. And yet, even to-day, we seem as far as ever from any real knowledge of the subject, or indeed of what gravity is, or how it acts, so that we might best learn how to overcome it.

At the same time, the ingenuity of man is such, that for generations past, a flying machine has become a possibility, and the airships of to-day have been designed on lines very similar to those projected in early days. The sketches of Leonardo da Vinci, show that he had a true conception of the problem to be solved, and there is no reason why his flying machine should not have been successful, had he had the material which to-day is disposable.

But the great step in advance, that has brought us to the very threshold of the day when a flying machine may be considered an ordinary and generally available means of conveyance has been in the introduction of machinery which is capable of developing a very considerable power in proportion to its weight, and the recent ratio of progress in this direction has been so rapid, that but few realise that only nineteen years ago, a steam engine and boiler was still looked upon as the only possible means of obtaining the desired power.

It was not until 1780 that the Conquest of the Air began seriously to occupy the thoughts of engineers, and the natural commencement was with machines lighter than air.

Once these had become actualities it was not very long before a desire of navigating the air at will and in a definite direction took the place of the original idea of merely rising from the earth and being carried whither the wind permitted.

One of the earliest to perceive the analogy between water and air navigation, was Lavoisier, who in December, 1783, gave before the French Academy, a list of conditions which would have to be fulfilled by vessels for the navigation of the air, and the last of these is as follows:—

"By employing the force of man it appears certain that it will be possible to cause the direction of the balloon to vary from the direction of the wind, under an angle of several degrees.

This idea was discussed and considered by Montgolfier, who had made his first public experiments with a balloon six months before, and it was decided to try and adapt oars to his balloon.

Several experiments were made in this direction but none of them were successful, so that gradually the great things which were hoped for, consequent on the success of the first balloon ascents, were looked upon as not feasible, and all faith in air locomotion was abandoned.

But the real difficulty was in finding an adequate means of applying power, just as later on the difficulty has been that of obtaining a sufficiency of power to attain the end desired.

It was not until half a century afterwards that the introduction of the screw propeller for water navigation provided a means of propulsion for aerial navigation which was entirely suitable for complete immersion, for which the apparatus of partial immersion, such as paddle wheels, feathering fins and oars, up to then the only means available, had not been suitable.

The great advantage of the new apparatus was that it was of the simplest character, it could be easily worked, and was capable of applying in the most effectual way any amount of power obtainable.

The first person to make a serious attempt to utilize the screw propeller for aerial navigation was M. Henri Giffard, the inventor of the injector, but he soon found that the application of such an apparatus to a circular balloon was not likely to prove a success. So he followed out the analogy of water navigation and came to the conclusion that the shape of the vessel for navigating the air must resemble the form found most suitable for propulsion in water and be of elongated form with a pointed bow and stern; that it should also be provided with a keel and a sufficiently powerful rudder placed in a similar position to that of a water-bourne vessel.

He accordingly constructed a balloon 39 feet diameter and 144 feet long. From this was suspended a keel extending for rather over two-thirds of the total length of the vessel and at the stern of which was a triangular sail serving as the rudder. From the keel was hung the car carrying the propeller and its machinery which is shown in Fig. 1.

The steam engine was of three h.p., and was directly attached to a screw propeller of 11 feet diameter, having two blades, and which had a normal speed of 110 revolutions per minute.

The great danger to be guarded against was fire, so he adopted the ingenious arrangement shown in the illustration, and turned the funnel downwards, and the draught was produced by a steam blast.

M. Giffard made the first ascent with his dirigible at Paris, on the 24th of September, 1852, and he found that the balloon could be propelled in the direction desired and could be readily steered. The screw gave a velocity through the air of from 2 to 3 metres a second, or $4\frac{1}{2}$ to $6\frac{3}{4}$ miles an hour.

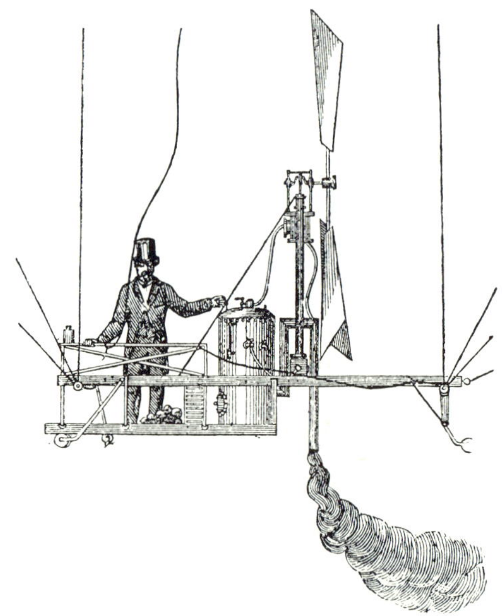

Fig. 1.
Engine of Mons. H. Giffard's Dirigible, 1852.

But little else was done in this direction for some years, and it was not until the siege of Paris, in 1870, that the necessity of dirigibility becoming urgent, attention was again given to the matter. Balloons were then being used to get despatches out of the city, but something that would take them in was greatly desired, and remembering the success obtained by M. Giffard, the French Government had experiments carried out under the direction of M. Dupuy de Lome, then Naval Architect to the French Government. He constructed a balloon of elongated form, 49 feet diameter and 118 feet long. From this was suspended a car which carried a screw propeller of 29 feet 6 inch diameter, and having two blades. This was arranged to be worked by man-power, normally with four men, and the lifting power of the balloon was arranged so that a relay of four other men could be taken up to relieve them.

The experiments were interrupted by the Communist Insurrection but were eventually completed, and trials made in February, 1872, proved that the balloon had a motion independent of the wind and could be directed where desired. When the two gangs of men were working together a full speed of $27\frac{1}{2}$ revs. per minute was obtained, which gave the airship a velocity of 9·2 feet per second, or about $6\frac{1}{4}$ miles per hour.

But the urgent need of such a vessel having come to an end, the matter again dropped, and it was not until sometime after that dirigibility was again been attempted.

Following these experimenters were those who adopted batteries of electric accumulators as the motive power. In 1881, Gaston Tissandier applied them to drive the screw-propeller of his dirigible which was 30 feet diameter and 91 feet long.

This electric motor made 1,800 revs. per minute and was connected by spur gearing to the propeller shaft which had a normal speed of 180 revs. The propeller was 9 feet in diameter and with it a maximum speed of 8 miles per hour was obtained.

This was followed by the dirigible of Captains Renard and Krebs, whose balloon was 27 feet diameter, and 163 feet long, the propeller being 23 feet in diameter. The weights of the machinery were as follows:

Battery of 32 cells	958 lbs.
Electric Motor	216 ,,
Gearing and shaft	170 ,,
Propeller	90 ,,

The battery was designed to give 9,000 watts to the motor for four hours. The speed of the motor was 3,000 r.p.m. giving the propeller about 40 revs. per minute, at which velocity a speed of over 15 miles per hour was realised. It was in connection with this dirigible that an aerial log was first adopted, constructed on the lines of an ordinary log at sea. A small balloon made of gold-beaters skin and filled with gas was attached to the end of a silk thread wound upon a reel. On being let go it would recede from the dirigible and as soon as the full length of the thread had been run out the time elapsed would be noted and the speed calculated, and the thread would be rolled up again.

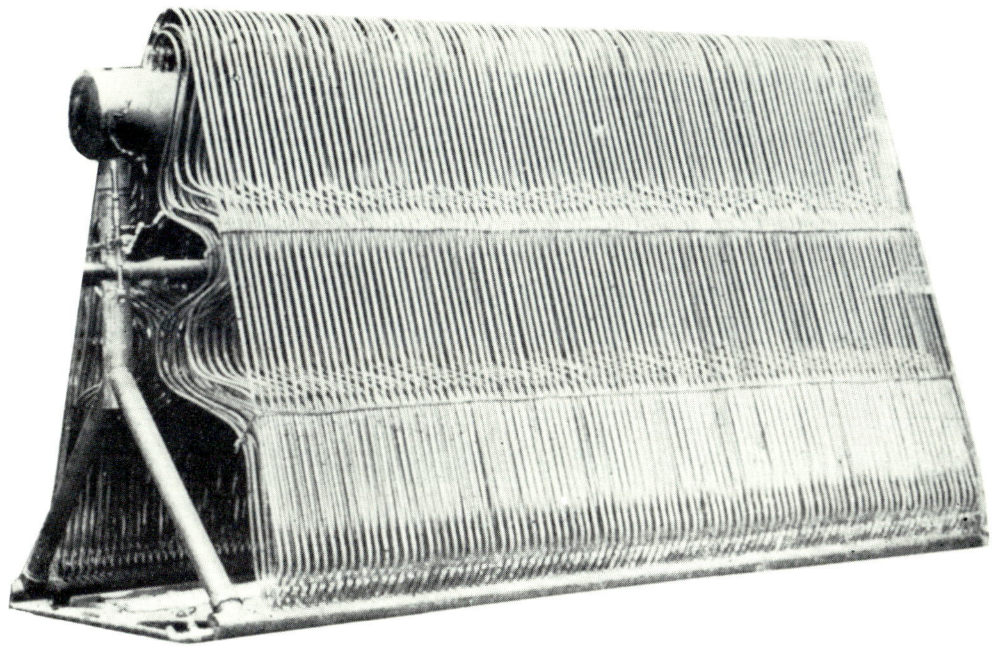

Fig. 3.
WATER-TUBE BOILER OF SIR HIRAM MAXIM, 1889.

The distance that this dirigible could travel was of course very limited, and as soon as the internal combustion engine became possible it was adopted as the motor power, and since that time there has been a steady increase in France, and lately in Germany and the United States also, both in the speed and in the power of such Air-ships.

Fig. 2 shows the ideas that prevailed in 1836, of the possibilities of a steam driven flyer, and when years afterwards Sir Hiram Maxim constructed his flying machine, it was steam that he adopted as being the only practical motive power available.

The engine was a two cylinder compound, the high pressure piston having an area of 20 square inches and the low pressure piston an area of 50·26 square inches, the stroke being 12 inches.

With a boiler pressure of 320 lbs. per square inch the pressure on the low pressure piston was 125 lbs. per square inch and this abnormally high pressure was due to the fact that there was a very large clearance in the high pressure cylinder to prevent damage in case water should go over direct from the boiler when the machine tilted or pitched. The engines were made as light as possible, the cylinders being only $\frac{3}{32}$ of an inch thick and the crank shaft hollow. When first put together they weighed 300 lbs. each, but after completion with all their fittings the weight was 320 lbs., or a total of 640 lbs. for the two engines which developed 362 h.p. at full power.

The steam generator constructed for these engines was a distinct advance on anything of the sort built up to then. It was 4 feet 6 inches wide at the bottom, 8 feet long and 6 feet high and formed of tubes of $\frac{3}{8}$ inch external diameter and $\frac{1}{30}$ of an inch thick, arranged as shewn in Fig. 3 (page 335). Over this again was another series of tubes $\frac{1}{4}$ inch diameter and $\frac{1}{60}$ of an inch thick forming a feed water heater which utilized the heat of the products of combustion after they had passed through the boiler and greatly reduced their temperature, whilst it raised the temperature of the feed water so that it entered the boiler at 250° Fahr. The total heating surface of the boiler including the feed heater was 800 square feet, and the total weight including the casing, the funnel and all its fittings was rather less than 1,000 lbs.

A most ingenious design was adopted for the condenser, after a number of experiments had been carried out which showed that in the case of air cooling the tubes of the condensing surface must be widely distributed, so that a very large amount of air is encountered, and the air which has struck one tube and become heated in extracting the heat from the steam must on no account come in contact with a second tube.

The condenser tubes were not only designed so as to give the maximum cooling effect, but they were to act also as sustainers and gave a decided lifting effect. They were made of very thin copper and in the form of a small aeroplane as shown in Fig. 4, A being a soldered

Fig. 4.
MAXIM CONDENSING TUBES.

joint, and the steam passing through the space B. These were arranged as shown in Fig. 5 so that they passed freely through the air, without driving forward or compressing it.

The experiments showed that a condenser made in this form sustained considerably more than its own weight and the weight of its contents in the air, and that all the steam was condensed into water sufficiently cooled to be pumped with certainty.

Since the construction of Sir Hiram Maxim's steam engine the progress in the design and construction of internal combustion engines using benzine as the fuel has been very considerable and they are now generally adopted for the propulsion of air craft of every description.

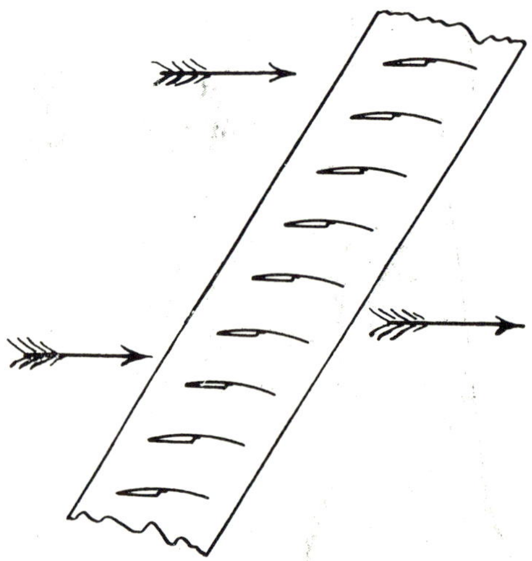

Fig. 5.
ARRANGEMENT OF MAXIM CONDENSING TUBES.

The first engines of this description were fairly heavy but as experience was gained with them the weights were gradually reduced, until at last it was evident, that an engine was actually procurable which it was possible to utilize for flying machines heavier than air. Unfortunately a craze set in for reducing the weight at all costs and regardless of the consequences, with the inevitable result that the reduction was very much overdone, and the engine on which the flight so largely depends became unreliable and developed the bad habit of stopping when it was not intended to and without notice, with the inevitable result that machine and aviator came quickly to the ground.

The extraordinary thing is, that flying at all being considered so wonderful, these enforced descents are looked upon as more or less as what might be expected, and lives are placed in peril and much property destroyed entirely without reason. Take for instance the records of a short recent period.

M. Bleriot, at Douai, the motor "commenced missfiring," and the day after there was "a derangement of the motor"; M. Paulhan, at Issy, "trouble with the motor and the machine would only make short jumps"; Orville Wright, at Fort Myer, U.S.A., "after nine circuits the engine suddenly stopped"; M. Sommer, at Chalons, "after being in the air for a few minutes, the motor suddenly stopped"; Mr. McCurdy, at Petewawa, Canada,

"The motor stopped suddenly"; and the way that Mr. Latham's motor behaved on his two trips across the English Channel is known to everyone.

Besides these defects due to what are termed "unknown causes," there are the mishaps directly traceable to defects of design and construction, such as breakage of the main shaft or other working part, bursting of the radiator, etc., which are far too frequent.

It is not too much to say that the one thing now wanted to make flying a definite certainty is

RELIABILITY

on the part of the motor and there is no reason whatever why this should not be attained.

On land a mishap with the propelling machinery, or the breakdown of any part, is hardly likely to produce more inconvenience than leaving one stranded at a considerable distance from help or shelter, and possibly the ignominy of having to have the machine propelled vehicle towed to the nearest place where the necessary repairs can be effected, by one of the more-or-less displaced quadrupeds.

In any case there is but little chance of a breakage producing danger to the machine, or to the passenger. And it has now become the rule to carry spare pieces of such working parts of the machinery as may possibly give way, and so reduce to a minimum even such an inconvenience.

At sea, such a mishap may be productive of more inconvenience and possible danger, should it occur away from the vicinity or track of other vessels, but such is but seldom likely to occur in the case of small craft, whilst for passenger vessels the adoption of twin screws has now become so universal that it is only in very rare cases that a breakdown is likely to destroy entirely the propelling power of a vessel. Even then the sea-going engineers of the day are such excellent mechanics and so eminently resourceful, that in the case of anything short of an almost entire collapse of the engine they can generally manage to effect such temporary repairs as to get sufficient way on the vessel. Many interesting cases are on record of the main propelling shaft being doctored up with some hastily made keyways cut under circumstances of great difficulty with the vessel plunging in heavy seas, a few fathoms of chain, some wooden wedges, and some spare bolts cut up to form the keys, serving as the means of repair.

But in the case of any other working part having to be changed, such an elaborate set of spare gear is usually carried in the case of both large and small engines that any mishap is well provided for, and the temporary disablement of the machinery can be comparatively quickly remedied.

In the case of sky-craft the conditions are very different indeed; it is not inconvenience, delay, or mere damage to the machinery that may be caused by a mishap or the stoppage of the engine, but for all such vessels other than aerostats such an incident means the probable collapse of the vehicle with its machinery, and absolute peril to the conductors and passengers; and the more so as the danger is ever present and instantaneous.

Given an engine well designed, well constructed of the best materials, and well fitted, there are two causes of stoppage which should be specially guarded against, and to one or the other of which, most, if not all of the undesired stoppages of the engine are due:—

1st.—Lubrication Troubles.
2nd.—Ignition Troubles

both of which are equally reprehensible, and equally preventible.

With regard to the first of these, it may be taken that defects due to lubrication troubles are partly caused by the foolish desire to cut down weight, so as to be able to proclaim that such and such an engine weighs per h.p. a few ounces less than the engines of rival makers, and partly by the want of thorough experience on the part of the constructing and driving staff of the engine builder. Both of these causes can be and should be quickly remedied. It is an absurd idea to suppose that the extra weight of a few pounds can in any case whatever be of more importance than rendering the lubricating arrangements so perfect and so efficient that if they are properly looked after, they may be thoroughly relied upon. Each working part and each joint that requires lubrication should have its needs studied, and this be duly provided for, nothing being left to chance, as it certainly is in the wasteful and comparatively inefficient system termed "splash lubrication."

Matters in this respect will be quickly rectified when sky-craft are no longer being made by owners who also act as their own engineers, and do not thoroughly understand such matters. As soon as such craft come into more general use, contemplating purchasers will quickly learn to discriminate between a reliable engine and one that has an unpleasant habit of coming to a standstill undesired and unlooked for. And the insistance by purchasers of a definite guarantee on the part of the vendor, for not only the damage to the engine and machine, but also full responsibility for possible injury to life and limb due to defective design or construction of the engine will soon bring about a better state of things in this respect.

The want of sufficiently thorough experience on the part of the staff of the constructor is largely due to the speedy development of the motor car industry in a relatively short time, and the fact that a motor car may come to an unlooked for standstill without danger to the car or its occupants has led to disregard of the needful attention to details that is necessary if such incidents are to be rendered improbable.

Although matters have really improved of late in this respect, even to-day there is still truth in the saying, that the only certainty of a motor car journey is the time of departure, and no one can guarantee the time of arrival. Such a condition of things may be considered tolerable for motor-car work, but is no longer admissable for flying machines.

With regard to the second point, Ignition Troubles, here again the craving for reduction of weight would appear to be largely the cause of them, and should no longer be allowed to stand in the way of a complete elimination of them by the invariable fitting of a duplicate system.

Recurring again to the analogy of marine navigation, when it was recognised that certain fittings of the marine boiler were subject to failure which might be due to either human carelessness or defects of the inanimate substances, and that such a failure might be the cause of danger to both life and limb, and destruction of property, the authorities (in this case the Board of Trade) stepped in, and insisted on such fittings being duplicated, so that there are *two* safety valves, *two* water gauges, *two* pressure gauges fitted to each boiler.

And it is highly desirable that the leading authority in the matter of sky-craft should at once take steps to rectify this matter and promptly ordain that no such craft should ascend with a passenger or with anyone other than the owner, which is not fitted with two independent arrangements of ignition.

In the case of a boiler it is not sufficient that there should be merely two pressure gauges, but these two gauges have to be entirely independent of one another, with their separate lines of piping and each with its own outtake and cock on the boiler.

With the water gauges also, in good naval practice it is not allowable to have one column with one attachment only to the boiler, both the water gauges being attached to it, but each water gauge must have its own independent connecting pipe to the boiler.

In like manner, it is not sufficient that the system of ignition should be duplicated, but these systems should be entirely independent so that there can be no possibility of a failure.

In naval engines there has been a constant endeavour to reduce weights as much as possible, and in the case of a vessel with 20 boilers, the weight of these duplicate fittings mounts up to a considerable total, but it is necessary to provide them to ensure *reliability* as far as possible, and since this duplication has been insisted on, dangerous explosions of boilers have been reduced to a minimum.

In the case of an engine for an aerial vessel, the ignition arrangements form an even more important part of the machinery than such fittings of a boiler in the case of the machinery for marine navigation. It is therefore still more important that their duplication should be insisted upon.

In the case of marine navigation, even with the duplicate system of water gauges and their pipes, mishaps sometimes still occur due to human carelessness or neglect, cocks being left closed when they should be open, or pipes being choked which should be kept clear, and the matter being overlooked by those in charge. In like manner occasional mishaps may be looked for also in aerial navigation, human nature being what it is, and troubles will arise from defective contacts or similar preventible causes, but it should be made a punishable offence to send or take a man up in the air without having taken all reasonable and due precautions to guard against failure in such a very important part of the machine as the propelling machinery.

Of course such a step will be speedily taken by Government authorities as soon as such craft are used for any purposes other than sporting ones, but it is greatly to be desired that the leading aerial authorities should take such steps as may lead to such a condition being voluntarily complied with, and without waiting for compulsion being necessary.

* * * *

The following are the best types of engines which have been adopted for air craft up to the present:—

THE ADAMS-FARWELL ENGINE.

One of the most novel of aeroplane motors is that designed by Mr. F. O. Farwell, and known as the Adams-Farwell Motor. It is constructed by the Adams Co., at Dubuque, Iowa, U.S.A., and a similar type has been in use for their motor cars for the past ten years, and with good results.

This engine is not a rotary but a revolving motor, and it operates much in the same way as the ordinary internal combustion engine, except that the cylinders revolve instead of the crank shaft which in this case is keyed to a stationary base. The force of the explosion, being confined between two objects, moves the one offering the least resistance, and as the crank shaft is secured to the base the remainder of the motor revolves.

This renders a fly wheel unnecessary, and the cylinders moving rapidly through the air like the spokes of a wheel are automatically cooled without water jackets, radiator or fan. It is stated that this feature was the incentive that led to the invention of this motor, but in working out the details it was found that centrifugal force and the placing of the cylinders radially about the crank shaft made it possible greatly to simplify the valve mechanism, ignition, crank shaft, crank case, etc., and do away entirely with exhaust pipes and muffler, besides reducing frictional losses and weight in many ways.

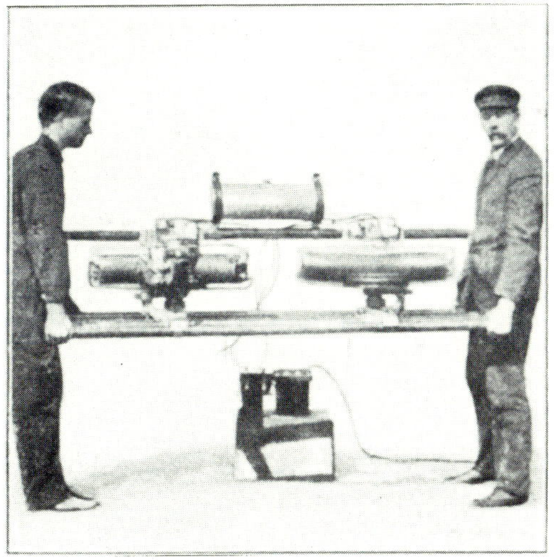

Fig. 6.
TWO 36 H.P. ADAMS-FARWELL MOTORS, ONE OF WHICH IS RUNNING.

The absence of reciprocating parts makes it possible to mount the motor on a light frame-work that would not stand the vibration set up by an ordinary motor, while the heavy revolving element (about 85 per cent. of the entire weight of the motor) insures a very steady motion. This is well shewn in Fig 6, where the right hand motor is revolving at 1,000 revs. per minute, and the clearness of the photograph illustrates the absence of vibration.

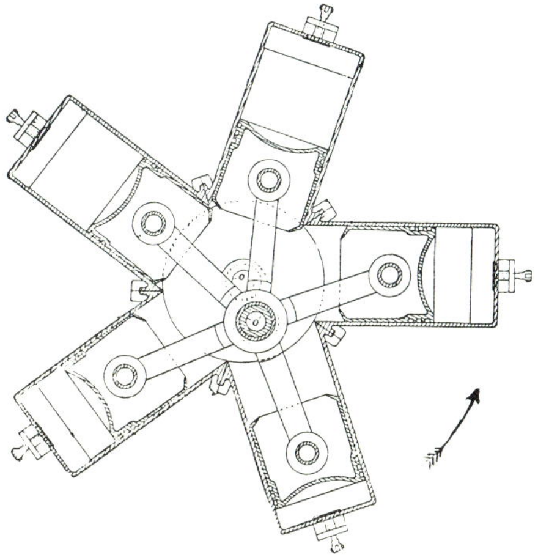

Fig. 7.
ADAMS-FARWELL MOTOR.

The tubes on which the motors are shewn fixed, form part of the frame-work of an air-ship, and have nothing to do with the motor itself.

A section of the engine is shewn in Fig 7, and it will be seen that all five pistons are connected to the same crank pin which is part of the stationary crank shaft, and therefore, the pistons travel in a perfect circle about their common centre.

The cylinders being bolted together revolve as one piece around the centre of the crank shaft. The circle described by the pistons being eccentric to that of the cylinders, the pistons approach the heads of the cylinders at one point and the bases at another point but each part moves in a perfect circle.

The cooling is a most important feature and when the motor revolves, the system becomes operative at once as the centrifugal force removes the air in contact with the cylinders, and atmospheric pressure supplies fresh air.

The circulation is equal for all the cylinders and varies directly as the speed of the motor, thus insuring positive results under all conditions.

It is further claimed as a special advantage that the circulation of air is equally rapid on all sides of the cylinders, and as the cylinder walls are of equal thickness on all sides, the expansion is equal and the pistons can be very closely fitted with no danger of sticking or scoring.

In the ordinary type of engine, air blown upon the cylinders by a fan strikes only one side, and even if several fans were provided eddy currents would result, leaving hot spots which would distort the cylinders and cause trouble.

The arrangement of the cylinders permits the use of a single throw crank, and, as the compression and explosion come on opposite sides of the same crank pin, the only strain upon the shaft is the net driving torque which is always in the same direction and nearly constant.

The method of controlling the motor consists in regulating the closing of the inlet valve so as to retain only the required charge in the cylinder, allowing the balance of the gas to be drawn in by the next cylinder. As no throttle is used the pistons are not retarded on the suction stroke, and another advantage claimed is that the gas is forced into the cylinders under the pressure resulting from the action of centrifugal force upon the gas in the radial intake channels.

The inlet and outlet valves are both placed in the head of the cylinder where they are actuated by a pull and push lever worked by a single cam which is common to both the inlet and the outlet valves on all five cylinders. The valves closing outwardly are held shut by centrifugal force, which varies directly as the motor speed and insures proper valve closing at high speeds without the aid of springs.

It is claimed that the valves never require regrinding, and that this is due to the perfect valve operation and the efficiency of the cooling system which supplies an extra volume of air to the cylinder heads and valves. The hottest part of the exhaust passes out through the auxiliary ports which on this motor require no valves. As the valves themselves are closed positively this ensures that none of the explosion can escape past them and makes this motor free from pitted or leaky valves.

In this engine the system of ignition is very simple, the wiring consisting of one primary wire to the timer, one short wire to the battery, one secondary wire and a ground wire. A single pole timer, a three pole coil with a single vibrator, and a storage battery complete the equipment. As the cylinders revolve they pick up the spark as they pass a certain point, and there being but a single contact point in the timer, any wear affects all cylinders alike and they can not get out of time.

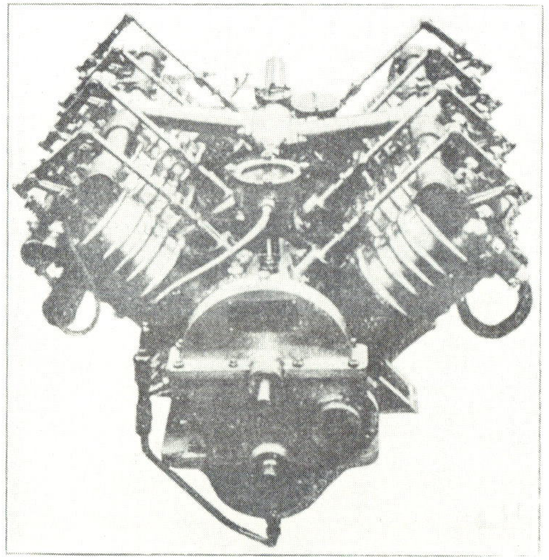

Fig. 8.
The "Aeroplane" Engine.

Two sizes of this motor are made, the particulars being as follows:—

Horse power	35	63
Diameter of cylinders	$4\frac{3}{16}''$	$5\frac{5}{8}''$
Length of stroke	$3\frac{1}{2}''$	$5''$
Normal revolutions	1,000	800
Max. revolutions	1,500	1,200
Total weight	97 lbs. (44 kilos)	250 lbs. (113 kilos)

The weight given is the total for the complete engine including carburettor, timer, secondary distributor, wiring, oil tank, and oil pump. And it is claimed that the low weight is due to the simplicity of the motor and its construction rather than to reducing the weight of its parts. In the larger size longitudinal ribs are cast on the cylinders for their entire length and still further assist in the cooling.

"AEROPLANE" ENGINE.

This engine was designed by Mr. W. L. Adams, and is constructed by the Aeroplane Engine Co., of Redbridge, Harts., whose engines for motor boats and hydroplanes have obtained a good reputation. The eight-cylinder engine of 88 b.h.p. shown in Fig. 8 is of the V type with the cylinders set at an angle of 90 degrees. The cylinders are 4 ins. (102 m/m) diameter and $4\frac{3}{4}$ ins. (121 m/m) stroke, and have a water cooled auxiliary exhaust which relieves the exhaust valves of practically all pressure. When the piston is $\frac{5}{8}$ in. from the end of the stroke, it uncovers a number of small holes drilled through the walls of the cylinders, and the exhaust gases rush through these holes into a deep annular passage, which is entirely surrounded by a considerable body of water, so that the gases are cooled and instantly reduced to atmospheric pressure. The piston is designed of a greater length than the stroke, so that the exhaust parts are never uncovered from below, and it is impossible for the oil in the crank chamber to be blown out.

The valves are mechanically operated by a single cam shaft and overhead rocking lever, the cam shaft running in ball bearings.

The crank shaft is hollow, of nickel steel, and also runs in ball bearings.

"ANTOINETTE."

The "Antoinette" engine constructed by the Antoinette Co. at Puteaux (Seine) has to its credit more flights than any other aerial engine in existence, and was generally used for all the earlier experiments in Europe.

The size generally used has eight cylinders arranged as a V in two lines of four, the centres being inclined at an angle of 45 degrees from the vertical, as shown in Fig. 9.

The main shaft of the engine has four cranks and is carried in five bearings, the centres of one row of the cylinders being forward of the centres of the other row, so that the two connecting rods of opposite cylinders are connected up to the same crank pin. The use of eight cylinders allows a fly wheel to be dispensed with, and great evenness in running is obtained.

Fig. 9.
"Antoinette" Engine.

The cylinders are water jacketted, and the cooling arrangements are so effectual that but little water is required. It is quickly converted into steam, and then passes to a very effective radiator condenser, the tubes of which are disposed along the framing of the aircraft. For a 50 h.p. engine only 8 litres of water are required, this quantity being partly in the cylinder jackets and the surplus in a small tank. The entire cooling equipment on the aeroplane used by Mr. Latham for crossing the English Channel weighed only 43 lbs. (22 kilos), inclusive of the water.

The valve gear consists of a pinion fitted on the crank shaft, which drives an intermediate pinion of double the diameter carried on the cam shaft.

The lubrication is effected by a small gear-driven pump, which forces the oil under pressure through a copper pipe pierced with a number of small holes, throwing it in all directions.

The water is circulated by means of a pump which drives it through the water jackets of the cylinders and then through two pipes situated at their lower end, each pipe feeding four cylinders. After issuing from the upper ends of the cylinders it passes through two pipes to the radiator, which is always placed at a higher level than the cylinders, and renders impossible any formation of steam in the water jackets which might impede the circulation.

A special feature of this engine is the absence of a carburettor, the carburation being effected by direct supply. A small pump driven by the motor draws the essence from the tank and forces it into eight small distributors, placed on the eight suction valves of the motor. These regulate the quantity necessary for each cylinder and store it, awaiting the suction, during which it is pulverized and vaporized.

The quantity of essence delivered can be readily regulated by changing the stroke of the piston, whilst retaining the automatic feature of the supply due to its being driven by the motor.

Whatever the temperature of the air may be it is claimed that the best possible carburation can be obtained by this means, and a very economical working of the engine.

Another advantage of this system of carburation is that it allows the suppression of all heavy and cumbersome piping.

In order to ensure perfect working of this system, strainers of very fine wire gauze are fitted in the essence pipes, so as to prevent any foreign matter entering from the supply tanks, and beside this four other strainers are fitted in the motor itself, one for each group of two cylinders, which still further purify the essence already cleared by the other strainers, so that no essence can enter the cylinders that is not perfectly freed from impurities.

It is claimed for this system that the carburation is more reliable under all conditions.

ANZANI.

The Anzani Motor brought into special notice by Bleriot's successful flight across the English Channel in one of the simplest of engines, with as little complication as possible and very strong for its work, and for these reasons has found great favour for motor-boat work, when reliability is considered very desirable.

It is made in three sizes, 25, 35 and 45 h.p., the smaller of these being the one used for the Channel flight. All the sizes have three cylinders, the central one vertical, and the other two on either side at an angle of about 60 degrees with its centre line, as shewn in Fig. 10.

This engine relies entirely upon air cooling, the cylinders being fitted with radiating fins, no fan is required, and water jackets, tank, piping, radiator and cooling water are dispensed with. The ignition is by a simple induction coil and make-and-break on the engine, the current being obtained from accumulators. The main shaft has but a single crank to which the connecting rods of all three cylinders are attached.

As will be seen the design is simpler than that of any other type

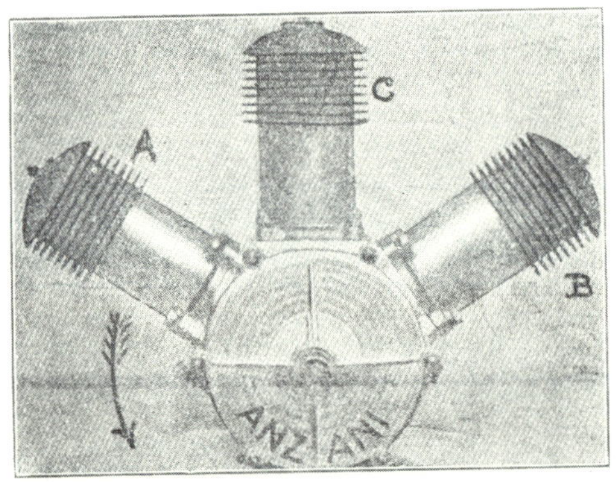

Fig. 10.
ANZANI ENGINE.

of engine adopted for airships, and it is claimed that the material is of the very best, and very great attention is paid to the workmanship, it being of the finest.

The arrangement of the ignition of the charges with an engine of this description is as follows:—Starting from the ignition in cylinder A, with the engine making a left-handed turn, the crank travels through 240 degrees, and the charge is fired in cylinder B; it then travels through 420 degrees and C is fired, and then after a travel of 60 degrees it begins the cycle again with ignition in cylinder A. The lack of regularity is compensated for by balancing the fly wheel, which constitutes in large proportion the weight of the entire engine, as will be seen from the following:—

	A	B	C
Rating in h.p.	25	35	45
Diameter of cylinders ...	105 m/m	120 m/m	135 m/m
Length of stroke ...	130 m/m	130 m/m	150 m/m
Revs. per min...	1600	1500	1400
Weight of fly-wheel in lbs...	57	70	80
Total weight of engine in lbs...	145	176	231

When fitted with water cooling apparatus the increase in weight is about 10 per cent.

CLÉMENT BAYARD.

Special interest attaches itself to Fig. 10a, which shows the most powerful engine yet constructed for aerial navigation. It is of 200 h.p., and built by Mons. A. Clément at the works at Levallois-Perret, and is one of the two to propel the new dirigible "Clément-Bayard No. 2." It is especially interesting, inasmuch as it is so far the only one arranged with reversing gear, and of which the speed can be regulated as desired. The two motors are to be so arranged

Fig. 10a.
THE CLEMENT BAYARD ENGINE.

that they can be used together, or one only, for driving both propellers.

This engine has four vertical cylinders, cast in pairs, with copper water jackets. The valves are arranged in the cylinder head, on opposite sides, and operated by an overhead cam shaft. The diameter of the cylinders is $7\frac{1}{2}$ ins. (190 m/m), and the stroke of the engine 9 ins. (230 m/m). The total weight of the engine is about 1,100 lbs. (500 kilos).

CURTISS.

Fig. 11 shows the Curtiss engine, which has given excellent results, and is compact and well built.

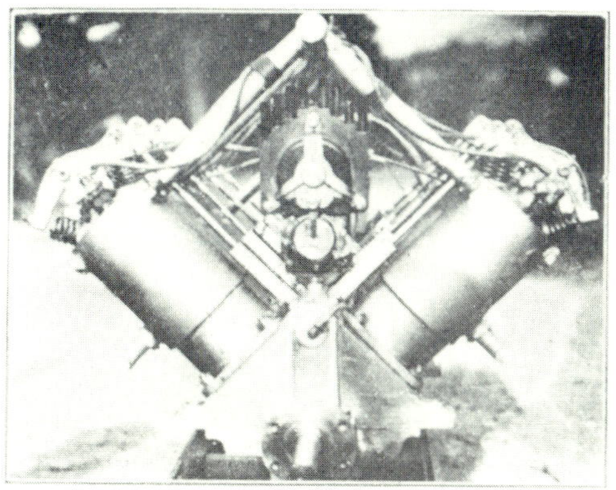

Fig. 11.
The Curtiss Engine.

The four-cylinder type of this engine resembles half the engine illustrated, the four cylinders being placed in a row, and the engine being rated at 30 h.p., with a speed at full power of 1200 revs. per minute.

The cylinders are cast separately, and have copper water jackets. The inlet and outlet valves are in the head of the cylinder, and operated by means of pivotted rocking levers actuated by tappet rods worked from the car shaft.

The same shaft drives a water pump by means of gearing at one end, and at the other end a high tension magneto, whilst in the centre there is an oil pump for the lubrication.

The radiator for this engine carries two gallons of water, and weighs 4 lbs. The total weight of the engine and fittings is 192 lbs. (87 kgs.), of which the engine itself accounts for 85 lbs. ($38\frac{1}{2}$ kgs.), and the magneto $12\frac{1}{2}$ lbs. The eight cylinder engine is rather lighter in proportion to the power developed.

E.N.V.

The E.N.V. motor built at Courbevoie (Seine), is an eight-cylinder engine of the V type, as shown in Fig. 12, the cylinders being set at an angle of 90 degrees with one another. The cylinders are water jacketted, the casings being formed of copper electrically deposited; a practice which ensures absolutely perfect joints, besides being the lightest method by which the desired result can be obtained.

The connecting rods work on a shaft with four cranks carried in three long bearings, the cranks being arranged two and two at an angle of 180 degrees in relation to one another. The webs are formed of discs with an outside rim, which adds considerably to their rigidity.

The cam shaft for operating the valves is machined from a solid piece, and hollow for its entire length. The magneto is placed towards the front of the motor, and is worked from the crank shaft by means of set of inclined gear. The pinion on the crank shaft can be moved longitudinally, and produces a common movement of the cams and the distributor of the magneto thereby varying the moment of firing.

The cooling arrangement consists of a small turbine pump fixed in front of the motor, and between the radiator and the water jackets of the cylinders.

Special attention has been given to the lubrication, which is provided for by a pump driven by a cam turned on the cam shaft, and which forces the oil through a circulating system, the quantity being regulated by a byepass, and the level of the oil being governed by a float. The oil is forced into the three main bearings, it next passes through the crank shaft, and then through a short pipe into the cam shaft. From the crank pins, it is forced up the

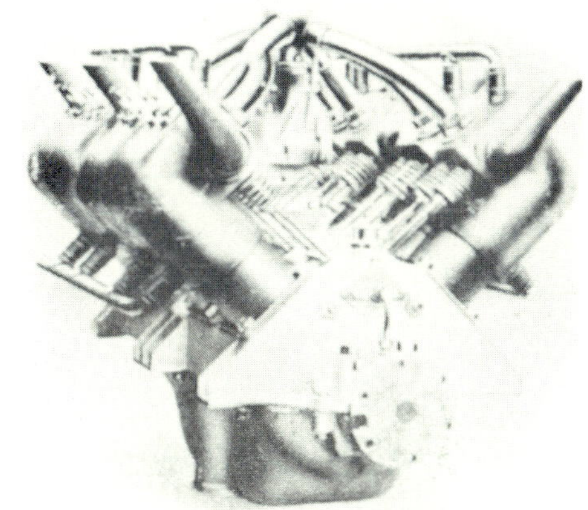

Fig. 12.
The E.N.V. Engine.

hollow connecting rods, and through the crosshead pins and eventually to the cylinder walls.

When properly looked after the lubrication is therefore practically certain, and waste of oil is avoided.

The larger size of this engine is rated at 80 h.p., the cylinders being 100 m/m diameter, with 130 m/m stroke. On a trial run at 1180 revs. per minute the consumption of petrol worked out at ·53 lbs. (·345 litres) per H.P. per hour.

The weight of the engine, complete with carburettor, magneto, water pipes and pump, and including a thrust-block with ball-bearings is 360 lbs. (163 kgs.)

This engine, both as regards design and construction, is one of the best yet designed for ærial use, and has given good results where it has been adopted.

FARCOT.

The Farcot motor is an engine of entirely distinctive type, with eight cylinders arranged in two planes. The cylinders are distributed at equal distances around the entire circumference of the engine, the four of the second plane coming in between the four of the first plane. The crank shaft has two crank pins at 180 degrees with one another, to one of which is attached the connecting rods of four cylinders in one plane, whilst the other four are attached to the opposite crank pin.

The cylinders are placed horizontally and work a vertical shaft, at the end of which is fixed a fan for cooling the cylinders by means of the current of air it throws on them. The power is transmitted to the horizontal propeller shaft by means of gearing. A feature of this engine is the Farcot double valve, which serves for both supply and exhaust.

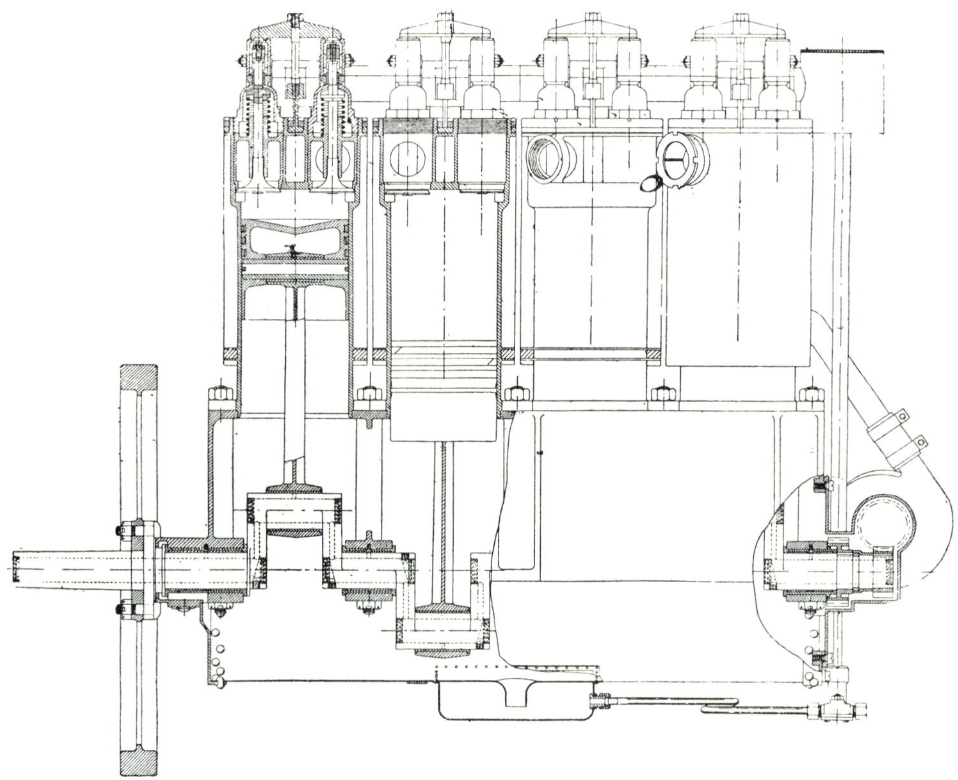

Fig. 13.
GREEN'S AERIAL ENGINE.

The 50 h.p. engine has eight cylinders of 100 m/m diameter, and 125 m/m stroke, the full speed being 1600 rvs. per minute and the weight is stated to be 125 lbs. (56·70 kgs.)

GNOME.

The Gnome engine is very similar to the Adams-Farwell motor already described, having a fixed crank shaft, with the cylinders revolving round it, but in this case there are seven cylinders instead of five. They are fitted with radiating fins, and are self-cooling as they revolve. In this engine, however, the valve is in the piston itself, the charge passing through it from the crank case to the cylinder.

As all the cylinders are in the same plane, all seven connecting rods work on the one crank pin. The rod serving as the main connecting rod has its crank end made with two steel disc rings which surround the crank pin, and in this there are six eyes, through which pass the pins at the extremities of the six auxiliary rods. The cylinders of this engine are 110 m/m diameter and 120 m/m stroke, and the engine complete has a total weight of 165 lbs. (75 kgs.)

GOBRON.

This engine. built by the Gobron-Brillié Co. at Boulogne, is of a novel type, and has eight cylinders disposed in a X in two vertical planes, each cylinder containing two pistons which, on the charge being fired, work in opposite directions. In this engine the cranks are exactly opposite one another, and as all the reciprocating parts are opposed to one another, the running of the motor is very even.

The valve mechanism is very simple, there being no gearing, cam shafts or cams. The inlet valves are automatic, and the exhaust valves are actuated by a very simple arrangement.

The total weight of the engine is 330 lbs. (150 kilos), and it develops 55 h.p. at 1,150 revolutions per minute.

GREEN.

Great simplicity and considerable reduction in weight are claimed as the special advantages of Green's Aerial Motor, manufactured for Green's Motor Patents Syndicate by the Aster Engineering Co.

This engine which is shown in Fig. 13 (page 344) works on the Otto, or four-stroke cycle of operations, and has four (or more) separately mounted cylinders. Each barrel is of uniform thickness, is cast with its crown and vertical valve-chambers in high grade steel, and is machined both on its exterior as well as interior surfaces. The valves are situated in the head, so that there are no passages or pockets in which unburnt gases might accumulate, and the combustion-spaces are exactly equal, giving perfectly synchronized compression in all the cylinders. The water-jacket consists of a fluted copper casing, which is pressed in a die from a single piece of sheet metal. The lower end of the casing is attached to the cylinder in a special way so as to provide for free expansion and contraction, a grey rubber rectangular ring being inserted in a groove running round the cylinder. The casing is slightly bell-mouthed at its open end, and merely by pressing it gently over the groove in the belt containing the rubber ring referred to, a perfectly water-tight joint is made.

The valves are of the mushroom spring-closed type, and are actuated by the end of a tappet pin, as shown in Fig. 14.

The cam shaft is rotated by the vertical shaft shown at the end of the engine, and driven through a pair of worm-wheels by a half-speed shaft. Inside the camshaft case are pivoted the rocking levers, the striking ends of which are provided with adjusting screws, and the ends operated by the cams are provided with rollers.

The cam case is so arranged that the undoing of two clamping screws allows of it being rotated on the axis of the cam-shaft within its brackets to a sufficient degree to obtain easy access to the valves and cages, so that the cam-shaft is not disturbed in any way, nor the timing of the motor interfered with.

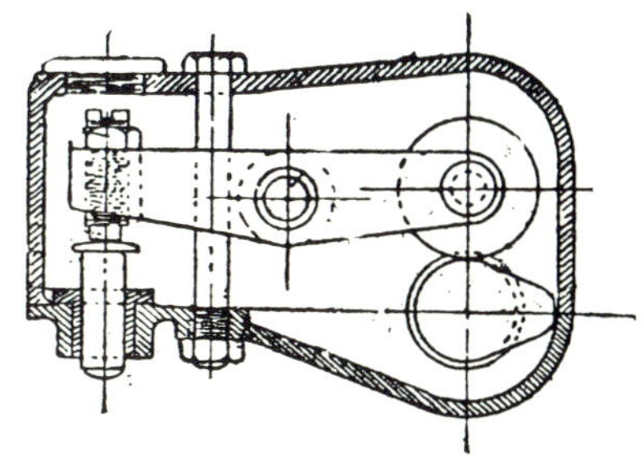

Fig. 14.
Valve Gear of Green's Engine.

For lubrication, oil is forced through the bearings by means of a pump driven from the lower end of vertical spindle. Ignition is effected by a magneto fixed on a bracket on the crank chamber, and direct driven by a short transverse shaft passing between the middle pair of cylinders in the top half of the crank case, and by the vertical spindle through a pair of worm wheels. At its opposite end the transverse shaft drives the water circulating pump. When running the only part of the engine which can be seen revolving is the flywheel, as every other working part is enclosed within oil-tight casings; whilst it is claimed that every part is as easy of approach as though it were fully exposed.

The engine illustrated is rated at 30 to 40 h.p., and its cylinders are 105 m/m, with a stroke of 120 m/m, and on trial at 1,200 revs. it gives an actual h.p. of 38.

The total weight of this engine complete, but without flywheel is 142 lbs. (64½ kgs.) The larger size also with four cylinders arranged as in the illustration is of 50 to 60 h.p., having cylinders of 140 m/m diameter, with 146 m/m stroke, the total weight under the same conditions being 236 lbs. (107 kgs.)

The dirigible of the British War Office of 1909 is being fitted with an engine of this description, but of the V type, having eight cylinders, and of about 100 h.p., and a six cylinder engine of the same type is in course of construction which should develop 82 h.p.

NEW ENGINE Co.'s MOTOR.

This engine works on the 2-cycle principle, that is an explosion occurs every time the piston is at the top of its stroke. Consequently there are double the number of impulses as compared with a 4-cycle engine, the idea being that the torque will be far better, and the engine considerably smaller and lighter for the power produced.

At the lower end of the cylinder is a ring of ports—on the one side for admission; on the other for the exhaust. As the piston descends it uncovers the exhaust ports and the exhaust rushes out, reducing the pressure inside the cylinder to that of the atmosphere. As the piston moves further down inlet ports are opened, to which air is supplied under pressure. The moment the inlet ports open this air commences to rush through the cylinder driving the exhaust before it, and it is thus completely filled with fresh air. As the piston rises on the compression stroke it commences to close the ports. Just before the inlet is closed the necessary gas is introduced into the cylinder. Immediately afterwards the exhaust port is closed and the gas compressed. At the top of the stroke the mixture is fired in the usual manner.

It will be seen there are no valves, cams, springs, or valve gear, and the engine is therefore very simple and reliable, and works very silently.

The ports are large, and there is, therefore, no throttling of the mixture at high speeds, such as sometimes occurs with poppet valves, and the power of the engine increases in proportion to the speed. The exhaust is driven out of the cylinder immediately after its work is done, whilst on the 4-cycle engine the exhaust gas remains in during the exhaust stroke and passes a great deal of heat to the cylinder walls, so that as this is avoided the engine requires considerably less cooling.

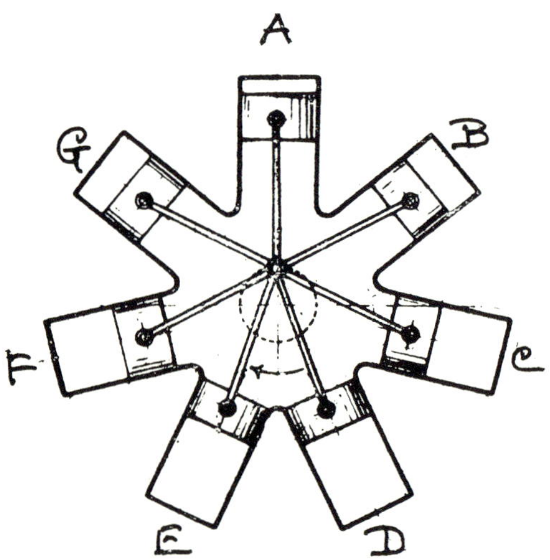

Fig. 16.
7-CYLINDER ENGINE.

The special advantages of the 2-cycle system have always been admitted, but the difficulty has been in the method adopted for driving the air into the cylinder. To do this it has hitherto been necessary to compress the air into the crank chamber, causing great trouble with the lubrication, or to compress the air in a separate cylinder, adding greatly to the complications and weight. With this type of engine it is claimed that these disadvantages are avoided, while the advantages of the 2-cycle system are retained.

These engines are built in three sizes, the smallest has two cylinders and develops 12 h.p. at 1,200 revs. per minute.

The next size is fitted with four cylinders and develops 25 h.p. at 1,200 revs. per minute, whilst the third size is fitted with six cylinders and develops 35 h.p. at that speed.

It claimed that a 2-cycle engine is the equal in smoothness of running and evenness of torque of a 4-cycle engine, with double the number of cylinders, and that to obtain the same power from the same number of cylinders the force of each explosion is little more than half what it must be in the 4 cycle engine, so that the strains on the engine and on all the driving gear are thus greatly reduced.

R.E.P.

The first engine built of this type was designed by its inventor, M. Robert Esnault-Pelterie for his own experiments in aviation, and seven cylinders were adopted as being the number best adapted for a multiple cylinder engine, as it gives much smoother running than can be obtained with any other number, the explosions following one another at exactly equal distances. This may be clearly seen on reference to Fig. 16, in which the ignition in the cylinders occurs in the following order, A, C, E, G, B, D, F, A, and so on. But it was desired to keep all the cylinders on the upper portion of the circumference, so the fan-shaped arrangement, shewn in Fig. 17, was adopted, and as it was impossible to get such a large number of cylinders arranged on the upper half of a base of moderate dimensions, they were disposed in two planes, four cylinders being in one plane and three in the other.

The explosions are arranged to follow one another in the same order as shown in Fig. 16, so that the engine has $3\frac{1}{2}$ power-strokes per revolution, or seven in each two revolutions of the crank shaft. As will be seen from the illustration, the two cranks are arranged opposite each other, and four of the connecting rods are coupled up to one crank

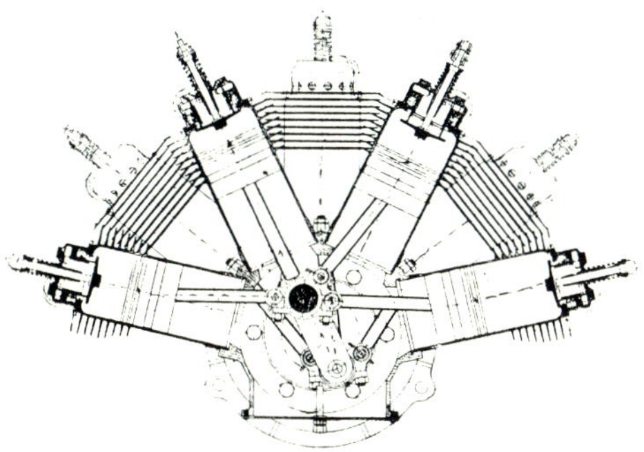

Fig. 17.
The R.E.P. Engine.

bearing, and three to the other. In each group one of the crank heads is solid, and carries in one case three, and in the other, two eyes to which the other connecting rods are attached. The cylinders have a number of radiating ribs cast on their upper portion and are air cooled.

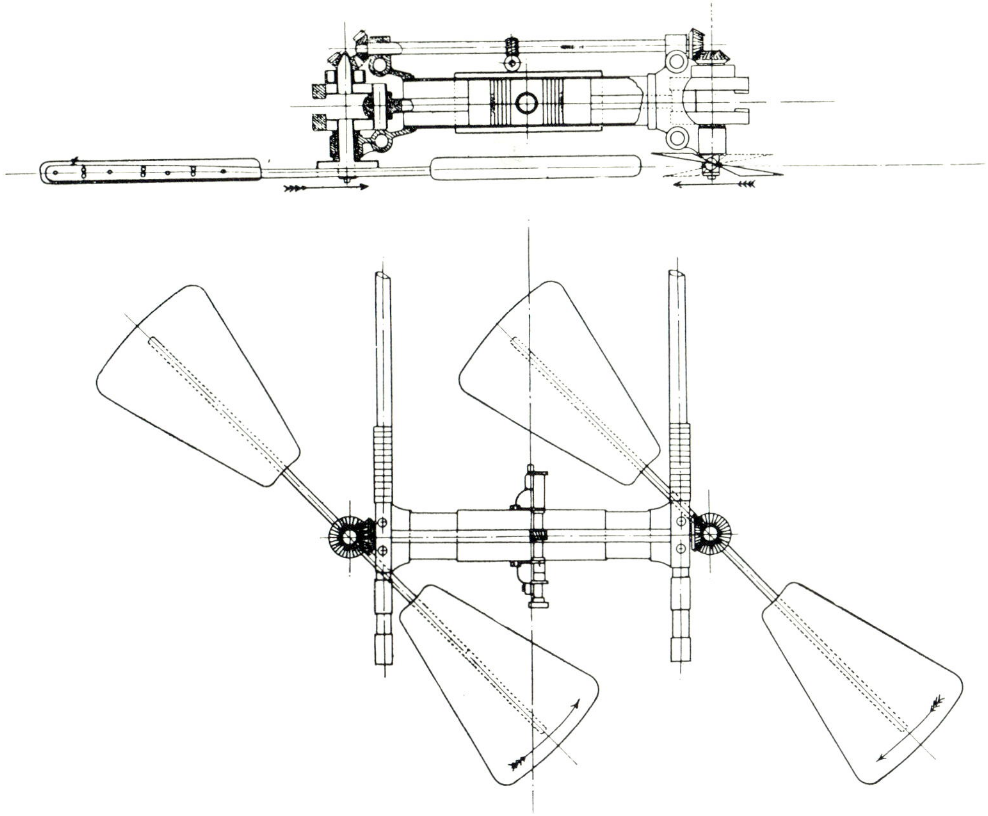

Fig. 18.
S.P.A. Engine.

A special feature of this engine is that one valve serves both for the intake and for the exhaust. It is shown in the head of the cylinder, and it will be seen that it is of specially large diameter. It is actuated by a rocker arm operated by the cam shaft. For admission the valve itself is lifted off its seat, allowing the intake of the mixture in the usual manner, whilst for exhaust the steel collar surrounding the valve stem uncovers the holes round the circumference at the top of the dome, and thus allows the spent gases to be discharged.

The total weight of the 30 h.p. engine in working order is 115 lbs. (52 kilos.), and the double engine with 14 cylinders and rated at 60 H.P. has a weight of 216 lbs. (98 kilos.) in like condition.

S.P.A.

The most novel type of all aerial engines is the S.P.A. engine, built by the Società Ligure Piemontese Automobile at Turin, and shewn in Fig. 18. It has the speciality that it is not an ordinary motor car engine, lightened as far as possible and adapted to aerial conditions, but is a well thought out attempt to deal with the requisites for such conditions, and has been specially designed to that end.

It will be seen that in each cylinder there are two pistons, and the explosion takes place in the centre between them, causing them each to travel outwards at the same time. The inlet and ignition arrangements are all in the centre of the engine. The pistons are of trunk form, with connecting rods which are attached to the crank pins of the perfectly balanced cranks of the main shaft. The screws are fitted directly on these shafts, which are attached to one another by means of gearing, so that the pistons and the screws always occupy the same relative position to one another. The arrangement shown represents the ordinary engine, with the screws fairly close together, but the distance between them can of course be varied as desired and increased to any amount by lengthening the connecting rods. The motor illustrated is the smallest size. The large ones have four cylinders, each with two pistons, and in the latest type there are several modifications in the minor details, but the general arrangement of the engine is identical with that shewn.

The advantages claimed for this engine are as follows:—

1—Absence of vibrations, inasmuch as the motor is completely symetrical, and therefore well balanced in its working.

2—It does not disturb in any way the equilibrium of the aeroplane, as it is fixed exactly on the line of the centre of gravity and of the centre of resistance. With aeroplanes having one screw only and one engine, it is generally found necessary to have a counterweight to balance the turning effort of the motor, so that the propulsive force of the screw, the centre of mass, and the centre of resistance are in three different lines.

3—With the two screws it is possible to make them smaller and more convenient both as regards pitch and diameter.

4—As the screws are placed upon the main shaft the drive is direct, weight saved, and the arrangement more reliable.

5—Although each cylinder gives the equivalent of two cylinders in an ordinary engine, there is but one admission valve, one exhaust, and one point of ignition.

The engine is very stoutly built, as reliability has been looked upon as the chief desideratum, and very little lubrication is required, as the main shaft runs in ball bearings throughout. Water cooling is adopted, and a double set of ignition gear is provided.

The engine is made in three sizes. the particulars being as follows:—

	A	B	C
H.P. rating of engine ...	20	40	100
No. of cylinders	1	4	4
Diameter of cylinder in ins....	$3\frac{7}{8}$	$2\frac{7}{8}$	$4\frac{3}{8}$
,, ,, m/m...	100	75	112
Stroke in ins.	$5\frac{1}{8}$	$5\frac{7}{8}$	$5\frac{7}{8}$
,, m/m	130	150	150
Weight of engine in lbs. ...	106	220	530
,, ,, kilos ...	48	100	240

In the aeroplanes to which these motors have been fitted, for the A type the distance apart of the main shafts has been 2 ft. 9 ins. (840 m/m), the diameter of the screws 4 ft. $3\frac{1}{8}$ ins. (met. 1,300), and their pitch 3 ft. $3\frac{3}{8}$ ins. (met. 1,000), the revolutions of the engine being 1,200. In the C type the distance apart of the shafts is 3 ft. $7\frac{3}{4}$ ins. (met. 1,110), the diameter of the screws 6 ft. $6\frac{3}{4}$ ins. (met. 2,000), and their pitch 4ft. 11 ins. (met. 1,500), the revolutions of the engine being 1,200 as in the other case.

WOLSELY.

The Wolsely Co., has introduced a light engine of their well known type for aerial use, which is shewn in Fig. 19, and it is constructed to suit two different types of drive.

1st.—Coupling the propeller direct to the crank shaft.

2nd.—Driving the propeller at half the speed of the main shaft through gearing by means of the cam shaft and a third arrangement has now been introduced by which two propellers are driven in opposite directions by means of gears and chain wheels.

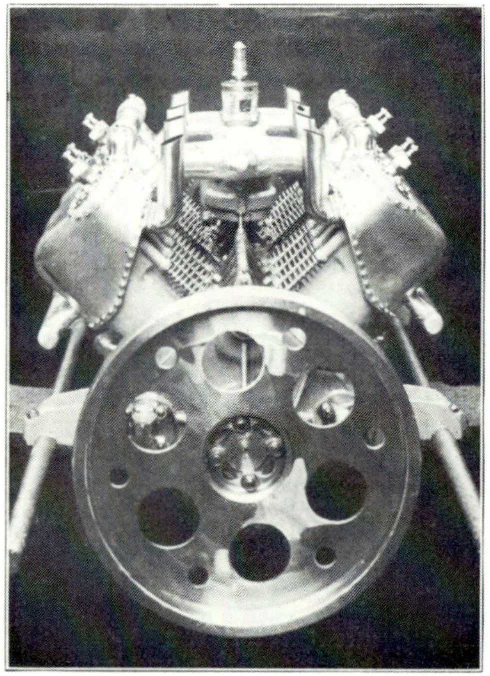

Fig. 19.
The Wolsely Engine.

The standard engine is of the V type with eight cylinders having a diameter of 3¾ ins. (95 m/m) and of 5 ins. (127 m/m) stroke. They are set at an angle of 90° to each other and fixed to an aluminium crank case. A special feature of these cylinders is that the jackets are made of planished sheet aluminium. All the valves are underneath and on the inside and are operated from a central cam shaft by means of rocker levers. The crank shaft is carried in three bearings of ample proportions and the cranks are set so as to give an even turning movement.

The power of the engine is 50 b.h.p. at 1,350 revolutions and can be worked up to a maximum of 70 b.h.p. In working order the total weight is 340 lbs. (154 kgs.) or if the propellers are coupled direct, the flywheel is dispensed with, and the weight is reduced to 320 lbs. (145 kgs.) It is claimed that in no case has the weight been reduced below the safe working limits and that the engine may be used for long periods at full load without any trouble manifesting itself.

A very thorough method of testing is adopted with this engine. It is mounted on the back of an ordinary motor car chassis complete with its radiator, and a fan brake is coupled to the end of the main shaft, the brake being so arranged that it can be set to absorb any horse power at any speed of the engine which may be desired.

The car is then driven over rough ground at a speed of between 20 and 30 miles per hour, and it is claimed that such conditions fairly represent the usage an engine would receive in actual flight. The test is made with the aerial engine running at full power and for a period of about three to four hours.

WRIGHT.

The aerial motor of the Wright Brothers is very similar to ordinary standard types. It is a four cylinder engine of the two-cycle type, the cylinders of which are made separately and fitted with water jackets, and are mounted on a built-up crank case made of aluminium, all the valves are in the cylinder heads, and the valve gearing is of the usual type. The engine has a free exhaust and auxiliary exhaust ports are provided at the end of the cylinders. The one special feature of this engine is that there is no carburettor, the liquid fuel being forced directly through a mixing valve into the cylinders by a pump worked by worm gear from the cam shaft.

From the same shaft is driven also the oil pump for lubricating, and the base of the crank case forms an oil reservoir from which the oil is pumped up to the oil circulation system. Ignition is effected by means of a high tension magneto worked from the end of the cam shaft. A small rotary pump for circulating the cooling water is worked off the end of the crank shaft, the radiator consisting of a number of flattened brass tubes connected at the top and bottom into suitable healers. The total weight of the engine is 170 lbs. (77 kgs.) and it develops from 25 to 30 h.p.

SCREW PROPELLERS.

In the early days of the introduction of the screw propeller for marine navigation it was a very hard matter to persuade the then Authorities of its superiority to the paddle-wheel, then the almost universally used means of propulsion of steam vessels. And the most effectual way to convince the unbelievers, amongst whom was nearly every Admiral in every Navy in the world, was to attach the stern of a paddle vessel to that of a screw vessel of similar form and displacement and of exactly the same power by means of a tow-rope, and let each do its best in a tug-of-war.

This was first done in 1845 with the screw vessel *Rattler* and the paddle vessel *Alecto*, each of 1,140 tons displacement and of about 300 i.h.p. The two ships were attached stern and stern, and then the engines of the *Alecto* were started, and the speed gradually increased until the *Rattler* was being towed at the rate of two knots. The engines of the *Rattler* were then started and their speed in both vessels increased to full power. In a very short time the sternward movement of the *Rattler* ceased, and she towed the *Alecto* astern at a rate of 2·8 knots. Since that time the superiority of the screw propeller has been considered to be proved beyond question.

There are even now some inventors who claim that a kind of stern paddle wheel will give better results for aerial propulsion than a screw propeller, but for aeroplanes, at least in the form they are at present, there can be no question as to the superiority of the latter.

Even taking this for granted, there is still a splendid field open for experiments: — the advantages of one type of screw over another—the best number of screws to adopt—whether they are most effectual as tractors or as propellers or whether a combination of the two systems should be adopted—the best number of blades and the best form to give them—the best material and the best method of construction—all these matters are as much open to discussion as they were when flying first became a possibility.

And there is yet another point open for decision.

In marine navigation the diameter has to be kept within certain limits imposed by the draught of water of the vessel—that is to say the available limits of the medium in which it is to work. But in aerial propulsion there are no such limits. However the general tendency is to make the screw of a handy size and suit the revolutions to it, rather than to make the screw as large as it is possible to get on the machine.

As a result of years of practice in marine navigation, some surprising results have been obtained with screw propellers in recent years, but conditions vary so largely that it is still quite impossible to lay down any definite law and to assert that a screw propeller of a certain design and certain dimensions will be the best and most efficient for any given case. In this instance, moreover, one is dealing with vessels all more or less of the same form, and with the propeller always in comparatively the same place relative to that vessel, whilst for aerial navigation, also these conditions vary.

There is however growing up a general feeling that the most efficient lines to be followed for aerial propulsion correspond largely with those found to be most efficient in marine propulsion and this especially as regards the position of the screw and the form of its blades.

With regard to the former a great authority is Sir Hiram Maxim, who from 1889 onwards, has carried out an infinitude of experiments and he deals with the matter as follows:*

"Many experimenters have imagined that a screw is just as efficient placed in front of a machine as at the rear, and it is quite probable that, in the early days of steamships a similar state of things existed. For several years there were steamboats running on the Hudson River, New York, with screws at their bows instead of at their stern. Inventors of, and experimenters with, flying machines are not at all agreed by any means in regard to the best position for the screw. It would appear that many, having noticed that a horse-propelled carriage is drawn instead of pushed, have come to the conclusion that, in a flying machine, the screw ought, in the very nature of things, to be attached to the front of the machine, so as to draw it through the air. Railway trains have their propelling power in front, and why should it not be the same with flying machines? But this is very bad reasoning. There is but one place for the screw, and that is in the immediate wake, and in the centre of the greatest atmospheric disturbance. While a machine is running, although there is a marked difference between water and air as far as skin friction is concerned, still the conditions are the same as far as the position of screw is concerned."

*Artificial and Natural Flight. Sir Hiram Maxim, 1909, page 47.

After explaining why the stern is the best place for the screw in ships, Sir Hiram goes on to say

"What is true of ships is true of flying machines. Good results can never be obtained by placing the screw in front instead of in the rear of the machine. If the screw is in front, the backwash strikes the machine and certainly has a decidedly retarding action. The framework, motor, etc., offer a good deal of resistance to the passage of the air, and if the air has already had imparted to it a backward motion, the resistance is greatly increased. The framework will always require a considerable amount of energy to drive it through the air, and the whole of this energy is spent in imparting a forward motion to the air, so if we place the propelling screw at the rear of the machine in the centre of the greatest atmospheric resistance, it will recover a portion of the lost energy, as in the steamship referred to. It will therefore be seen that when the screw is at the rear, it is running in air which is already moving forward with a considerable velocity, which reduces the slip of the screw in a corresponding degree."

But practice is very varied on this point, some aircraft having one screw in front, others have it at the rear, the Weiss monoplane has two screws at the rear, whilst the Howard Wright biplane has two screws, arranged tandem-wise and revolving in opposite directions.

M. Bleriot, places his screw at the front for structural and mechanical reasons, and it is also claimed that the air driven astern with increased velocity helps to support the machine, but it is more probable however that an increased resistance is set up.

The Wright Brothers have studied the question very fully and their experiments have enabled them to get together a large amount of data in the matter, and they claim that two screws placed at the rear give the best efficiency. They claim also that their experiments have shown that a propeller which would give sufficient thrust to start from a position at rest is not the most efficient when the aeroplane is travelling through the air, and this is the reason why they use an auxiliary starting device.

After the first introduction of the screw propeller for marine navigation, a large number of experiments were carried out at Lambeth in 1856, by the late Mr. Charles Sells, to determine the relative values of propellers with different forms of blades, at varying pitches, and with varied numbers of blades, and amongst them are the following, which will be found equally enlightening in respect to aerial navigation.

These trials were carried out with model screws, six inches in diameter and of nine inches pitch, and were made

a.—at a constant speed of screw of 250 feet per minute, and

b.—at a constant thrust of 2·27 lbs.

The first series gives the relative efficiency of screws having the same diameter, pitch and total area of blades and varying only in the number of blades. It will be seen that when the total area of the screw was divided up into several blades, there was a distinct improvement in its performance. This, provided that the blades are of the same form in each case and that the narrowest of them is of reasonable width.

The second series given shews the relative efficiency of screws having the same number of blades and corresponding in every particular except in the form of the blades, and it will be seen that the generally adopted form in every case gives the best results.

The third series gives the results of shrouding the blades on one, and on both sides, and it will be seen that there is a distinct loss in both cases.

For air propellers this plan has been recommended, at least for the following edge of the driving side, as giving a better "grip" of the air, but if the speed is at all considerable, it is certain to set up a churning action, and so reduce the efficiency of the screw.

The adoption of only two blades for so many of the screws for aerial navigation is a curious example of the force of custom.

Reference has already been made to Sir Hiram Maxim's illustration of the horse pulling the carriage and the locomotive drawing the train as giving the idea that has led to the screw being placed in front in some aircraft, and here we have another instance of the way aerial navigation is copying the older methods of marine navigation.

From the very beginning, there has been no reason whatever for limiting the number of blades of the screw to two, so far as efficiency and facility of construction is concerned. But it will be remembered that when the screw propeller was first introduced, steam vessels were still largely of the nature of a compromise. In many vessels the steam engine was only fitted as an auxiliary and in those cases where it was designed also to give full speed to the vessel there was still the desire to retain sails as an auxiliary, generally with the idea that the sails should propel the vessel when the wind was favour-

SCREW PROPELLER EXPERIMENTS.—MR. CHARLES SELLS, 1856.

A

Number of Experiments	Number of Blades	Coll. Area of Blades	Pitch of Screw	Working at a uniform velocity of 250 ft. per minute: Thrust of Screw	Working at a uniform velocity of 250 ft. per minute: Relative power employed	Working at a uniform thrust of 2·27 lbs.: Velocity of Screw	Working at a uniform thrust of 2·27 lbs.: Relative power employed	Relative Efficiency
			ins.	lbs.		feet		
10	2	·286	9	2·27	471	250	474	1·00
9	3	·286	9	2·22	435	253	450	1·05
7	4	·286	9	2·22	429	253	445	1·06
8	6	·286	9	2·33	417	247	400	1·18

B

		Number of Experiments	Number of Blades	Coll. Area of Blades	Pitch of Screw	Working at a uniform velocity of 250 ft. per minute: Thrust of Screw	Working at a uniform velocity of 250 ft. per minute: Relative power employed	Working at a uniform thrust of 2·27 lbs.: Velocity of Screw	Working at a uniform thrust of 2·27 lbs.: Relative power employed	Relative Efficiency
					ins.	lbs.		feet		
	A	11	2	·190	9	1·79	345	281	490	·97
	B	13	2	·190	9	1·73	365	286	550	·86
	A	13	2	·143	9	1·52	275	306	500	·95
	B	17	2	·143	9	1·47	270	310	575	·92
	A	7	4	·286	9	2·22	429	253	445	1·06
	B	2	4	·286	9	1·98	378	268	468	1·01
	A	14	2	·238	9	2·01	412	263	480	·99
	B	6	2	·238	9	1·37	265	322	565	·84
Series XXXVII		14	2	·238	9	2·04	412	263	480	.99
" XXXVIII		9	2	·238	9	1·99	440	267	540	.88
" XXXIX		13	2	·238	9	2·19	530	256	570	.85

able and served sufficiently well, and also that they might serve as a standby in case the steam machinery, still looked upon with a certain amount of doubt, should prove unreliable.

Two methods were adopted to obviate the drag of the propeller at the stern when the sails were being used and the engines stopped. One was to provide a well at the stern of the vessel just above the propeller and into which it was lifted clear of the water, the other was to make the stern-shaft and propeller-boss hollow and to fit gear in them by which the blades of the propeller could be turned so that they stood fore and aft, and with the propeller itself standing in line with the stern post it offered but little resistance when the vessel was under sail.

But in both these cases it will be seen that only a two bladed propeller was permissible and so it came about that even when sails were done away with altogether, the two-bladed propeller was still adopted as the standard practice. Since then this has been modified and the general rule to-day is four-bladed propellers for ordinary mercantile vessels and three-bladed propellers for war vessels and express passenger services.

But for aircraft, there seems no good reason for thus limiting the number of blades and it seems probable that for screws running at a moderate speed improved results could be obtained by the adoption of six or even more blades.

Fig. 20.

Propelling Machinery of Dirigible "Belgique."

Ingénieur Louis Godard.

But here we come to the question of cavitation and to the influence it may have on the efficiency of the screw of an aerial vessel. In marine navigation, when the screw has a number of blades and revolves at a very high speed, cavitation is said to be the formation of cavities of air in the seawater due to such cause. It would then appear that it would be impossible for such a phenomenon to occur in the air itself. But recent researches tend to show that it is not always air which is carried round with the blades of a propeller, but that sometimes there is something of the nature of a vacuum produced in such circumstances. Should this be so, it is possible that some such phenomenon may also be found to exist with screws revolving in air at a very high velocity when the form and proportions of the blade are not suitable.

It is now well known what a considerable effect the depth of water under the bottom of a large vessel has on the speed, the resistance increasing considerably as the hull of the vessel approaches the bottom of the sea.

On the other hand with small fast craft, such as torpedo boats, there are instances when, after passing the critical point of extreme resistance, a distinct gain in speed has manifested itself on getting into shallow water.

It will be interesting to note if anything of the sort occurs with sky craft on their approaching the earth; and in the case of similar results, it should be studied in order to ascertain how best it can be taken advantage of.

When circumstances require the diameter of the screw to be made as small as possible and it is desirable to get greater area of blade, it is probable that the adoption of more than one screw on the same shaft will give better results than have been realized in marine propulsion, if care is taken to place the screws sufficiently far apart.

But when skycraft commences to be used for a regular passenger service, it will be found indispensable to adopt two engines driving two independent screws. Fig. 20 shows the admirable arrangement adopted by those veteran aeronauts, Messrs. Godard, in their dirigible "Belgique." There are two independent engines, each of 60 h.p., and running at 1,200 to 1,400 revolutions. This speed is reduced to 250 to 280 at the main shaft, and this number of revolutions of the screw should give the dirigible a speed of 25 miles per hour.

The screws have a diameter of 16 ft. 4 in. (5 metres), and a mean pitch of 11 ft. 6 in. (3·5 metres).

In the cases where a wider framing is adopted, the best arrangement will be to have the two independent engines placed side by side in the centre, carrying the main shafts fore and aft, with a propeller at each end, thus giving a dirigible the turning and manœuvring power of a twin-screw vessel, even in the case of the rudder or steering arrangement being carried away or disabled. Provision should also be made for driving both shafts from one engine in case of need, sprocket wheels being provided on the main shafts, and the necessary driving chain being carried as spare gear. This would allow of a temporary disablement being easily overcome without having to descend to *terra firma* for repairs.

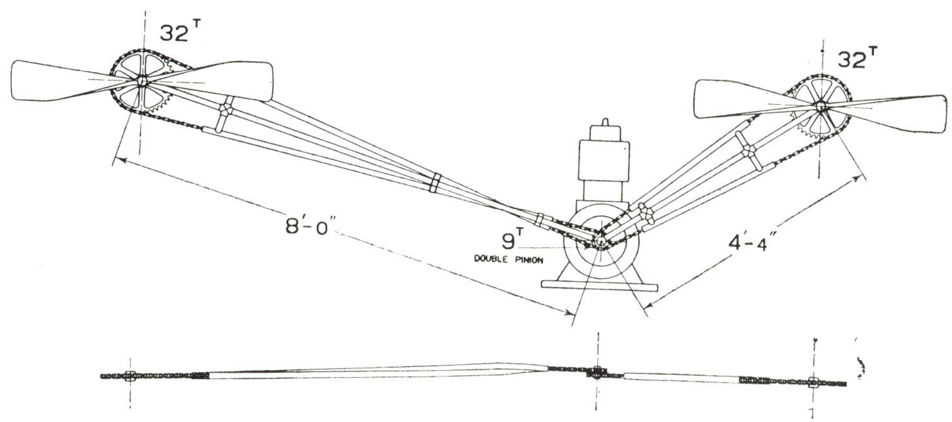

Fig. 21.
Chain Drive of Wright Bros. Aeroplane.

In the case of aerostats, the wind has a very considerable effect both in assisting the speed when the direction is favourable and in retarding it when the airship has to oppose it: and although advantage can be taken of the former, there is no means of dispensing with the latter. With the heavier-than-air machine of the future, however, there can be no doubt that some provision will be made for taking advantage of a favouring wind, something being provided in the nature of sails that can be set, or planes that can be trimmed, so as to derive all possible help from the air currents found to be constant and favourable. Here again is a large field available for research when the new "territory" of the air can be thoroughly taken possession of.

In the case of aeroplanes of box form, there is no doubt that the arrangement adopted by Messrs. Wright, and shown in Fig. 21, is theoretically correct, and that the driving by two screws gives considerably more stability to the machine than the use of one only. The screws are driven by means of the bush roller chain of Messrs. Renold, which quite overcomes the difficulties of having to drive the two propellers in opposite directions from the one engine, as with this type of chain it is possible to run with a crossed chain as shown, for driving the screw, seen on the left of the illustration. The chains run through guide tubes for the greater part of their length, so that they run very smoothly, and whatever the movements of the machine may be, there is no tendency for them to vibrate or jump off the sprocket wheels on the engine or screw shafts. The engine is of 25 to 30 h.p., and the screws revolve at from 350 to 400 revolutions per minute. The chains are of one inch pitch, and have a breaking load of 4,000 lbs.

We are so used to regarding the stability of a vehicle or a vessel as such an established fact, that the possibility of it being overthrown by a brusque manœuvre of the propelling force is almost unthinkable.

But in the case of an aeroplane with only one large and very powerful propeller, it is quite possible that a sudden change in the speed of the motor, or an overquick reversal of its action, may result in the propeller trying to turn the machine, and the equilibrium of the aeroplane be so upset, that unless the navigator is very skilful, it may be difficult to regain it.

So long as the engines of aeroplanes run in one direction only, and this without any variation in speed, there is no danger of this kind, but as soon as they begin to be used for the carriage of passengers, and are fitted with high-powered reversible engines, the use of one screw only will have to be abandoned, and either the Wright method be adopted, driving the two screws from one engine, or independent engines be provided for each of the screws.

Both as regards the material of the blades and their shape, present practice is very varied indeed, but there is no doubt that in these two respects the integral screws of Mons. L. Chauviére, and constructed at his works in Paris, are by far the best. Each single screw is specially designed for the work it has to do, and certainly they have been wonderfully successful. The form of these screws is shown in Fig 22, and the form is designed to follow the correct screw line. The blades are so constructed that the centre of pressure coincides with the centre of gravity, and it is claimed that this ensures uniform stability of the blades and avoids all vibration of the edges. The pitch of these propellers varies as a rule from the boss to the periphery, in order to throw the air next to the boss to the outside, whereas the extremities of the blades suck in the surrounding air and drive it straight astern. This is especially the case with tractor screws, and this variation in pitch is invariably carried out in such cases, as it would be useless to force the air near the boss straight astern because of the immediate vicinity of the masses of the engine and aeroplane, and the air striking these parts much power would be wasted. It is claimed that this method of design accounts for the high efficiency of these screws, which, when tested under stationary conditions, show actual efficiencies of from 90 to 97 per cent.

These screws are built up in layers of the finest selected French walnut in such a way as to ensure the grain of the wood extending from tip to tip of the propeller. The single layers are joined by an insoluble glue, and this method has the advantage that an absolute uniformity can be obtained in the structure of the material, as it is next to impossible to obtain perfect blocks of wood in the required lengths of from 7 to 20 feet, and as a rule in such long pieces the density of the wood varies to an extent which would throw a screw thus made completely out of balance.

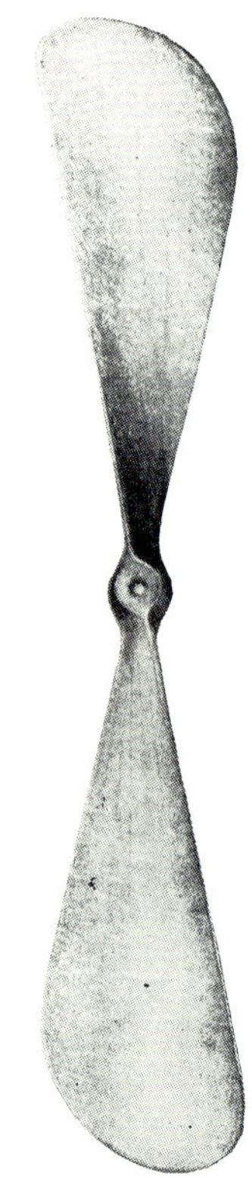

Fig. 22.
"Integral" Propelller.
Ingénieur L. Chauvière.

In the case of two-bladed screws, where great lightness is required, this type presents considerable advantages over those made of metal, being about half the weight, besides being much safer.

In the case of the screw also, it is necessary to lay great stress on the absolute necessity for reliability.

This is such an important part of the apparatus that any mishap with it may cause a disaster, and in some cases the strength has been so reduced with a view to cutting down the weight, that should the engine run away for any cause, the blades would probably be twisted off. It must also be remembered that metal screws, when coupled direct to the engine shaft, are liable to internal crystallization due to the incessant hammering of the engine, which in course of time reduces the strength of the material and may lead to failure. In every case ample strength must be provided whatever material the screw is made of.

It is claimed that with the Chauviére method of construction the screws are perfectly safe with a speed of periphery of 650 feet (200 metres) per second, and that at this speed they run without vibration or any change of form.

The Chauviére screw used by Mr. Bleriot on his famous cross-channel flight had a diameter of 6 ft. 10½ ins. (met. 2,100), and was of variable pitch, increasing from 2 ft. 5½ ins. (met. 0.750) at the periphery to 4 ft. 11 ins. (met. 1,500) at the boss. The normal number of revolutions was 1,600, and the slip of the screw at this velocity when the aeroplane was in flight amounted to only 15 per cent.

The screw shewn in Fig. 23 is a French model which has been extensively used for aeroplanes, but whilst it is acceptable for experimental work, it is of the worst possible form for use as a permanent screw. The blades are made of a sheet of metal, rivetted to a bar, which passes through a clamp fixed on the end of the main shaft, and which can be readily loosened so as to permit of the blades being twisted to modify the pitch, or

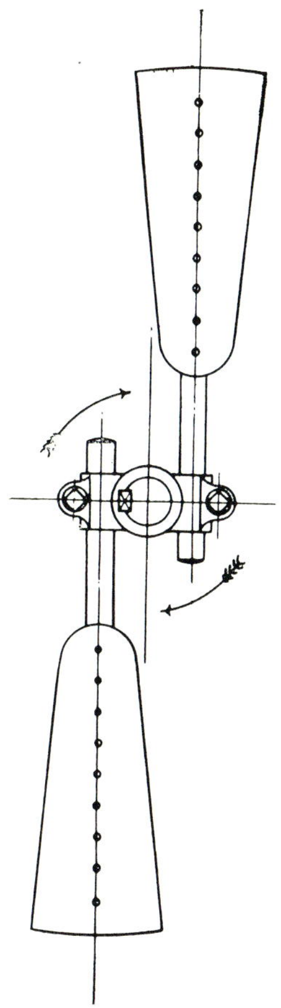

Fig. 23.
Metal Propeller of Aeroplane.

to be shifted in the direction of its length, and so change the diameter. It will be seen that it will serve to give an idea of the pitch and diameter best suited to any given case. But beyond this it has no advantages but many disadvantages. It is very heavy in relation to its efficiency, it is very wasteful as to power, and the projections on the back of the blades considerably reduce its efficiency.

Fig. 24 shows the R.E.P. screw, which has been used considerably in France. The boss is of steel, the arms are of tubes of a specially tough steel, and the blades are of aluminium. They are very light and supple, and it is claimed that when working under normal conditions the efficiency obtained amounts to 80 per cent.

Screws have also been made of a steel frame covered with woven fabric, but it is very difficult to keep the fabric taut, nor can it be made smooth, so that the results are very poor, and the efficiency is less than half that of a well made screw of wood.

The efficiency of the screw is measured by comparing the useful work done by it with the actual power imparted to it, and everything possible should be done to reduce the difference between these two items to the lowest possible figure.

Frictional resistance always causes a considerable loss of energy, and the surface of the blades should have no projections whatever, so as to offer as little resistance as possible. They should also be as smooth as

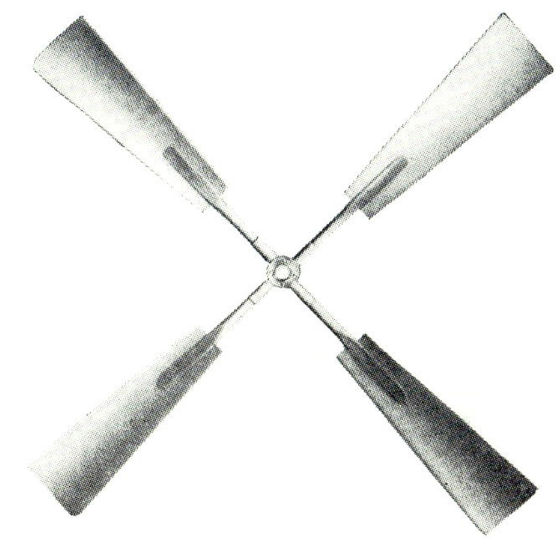

Fig. 24.
The R.E.P. Propeller.

possible, and, when made of wood, be highly polished.

A large number of the screws used to-day for skycraft are ill suited for their work and very wasteful of power.

When proper attention is given to this matter it will be found that the flying capacities of most machines can be largely augmented by thoroughly utilizing the power developed by the engine.

CHAIN DRIVING.

The practical success that has been obtained with chain driving since the

Fig. 25.
Renold Bush Roller Chain.

Renold chain was introduced has been of great service for airships, and this chain is being largely adopted for nearly every class of aerial motors. The type found most useful for this purpose is shown in Fig. 25, and is known as the Renold Bush Roller Chain. Each outside link is composed of two light steel plates and two shouldered rivets, whilst each inside link is formed of two larger side plates, two bushes drilled and turned from solid bar, and two rollers also machined from the solid. For chains above $1\frac{3}{4}$ inch in pitch the side plates are composed of two or more pieces.

In the construction of the chain the rivets are forced into the outside plates, and slightly rivetted over, whilst the bushes are forced into the inside plates, and form a bearing for the rivet on the inside and the roller on the outside.

The rivets, bushes, and rollers are all case-hardened, thus ensuring good hard wearing surfaces.

The large addition to the wearing surface by the Renold method of construction has largely added to the life of the chain. It is commonly supposed that a chain stretches in use, but it is very seldom indeed that such occurs with any type of chain, the truth being that there is only an increase in the length of the pitch due to wear, so that the larger surface it is possible to obtain, the less trouble there will be from this cause.

But to maintain the correct tension it is always necessary to provide for taking up slack which is bound to come eventually from wear, and especially so in the case of an aeroplane with only one screw, where the drive is generally short and in a vertical direction.

A hard wood packing under the motor is the best arrangement, as this can be easily thinned down as required, and makes a simple and efficient adjustment. This is sufficient where the line of force of the chain is at right angles to the base of the engine, but where the chain drive is parallel with the engine base the bolt holes should be elongated inwards towards the shaft centre, the exact amount of regulation of position being provided for either by a couple of set screws at the inner end, or a double wedge.

When the chain is a long one the wear can be taken up by shortening it by one link, in which case two links must be removed, and be substituted by one cranked link, but if possible it is preferable to reduce the length by two links.

Fig. 26 shows the new type of joint that has been introduced, and which has entirely overcome the trouble found with the old methods. The special feature is the T headed bolt, which is machined from the solid, the head fitting into the slot milled in the outside of the link, which is thickened in order to maintain the necessary strength. The end of this pin is secured by a castle nut, which is prevented from turning by a split pin. The trouble with the old type of joint was the tendency of the bolt to work back, which loosened the nut, and eventually sheared the split pin. In the new type of joint the bolt is not screwed into the link but passes right through it, and cannot turn, as it is keyed to the link by the T head, so that all danger of the split pin shearing is now eliminated.

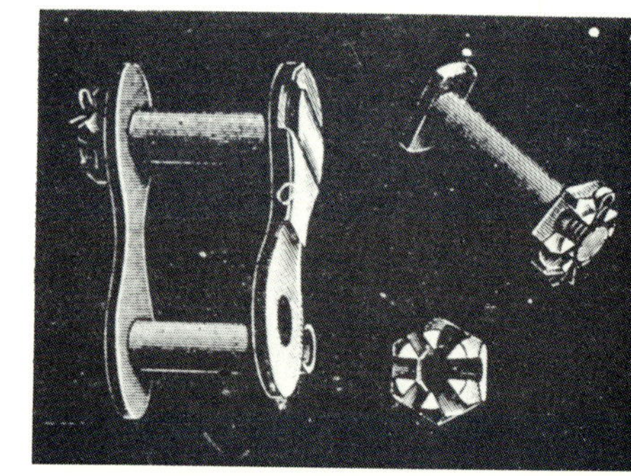

Fig. 26.
Joint of Renold Chain.

These chains are used on nearly all types of dirigibles and aeroplanes both for direct driving and for crossed chains. A good example of this latter is the drive of the starboard propeller of the aeroplane of the Wright brothers, as already described and shown in Fig. 21.

CARBURETTORS.

Next in importance to the propelling engine and machinery comes the carburettor, and as flights increase in length and height, modifications in adjustment will become indispensable.

The function of this important part of the apparatus is to vaporize the fuel and to mix with it the necessary amount of air to form an explosive mixture; when this mixture is in precisely the correct proportions the whole of the vapour is consumed and no matter is deposited in the cylinders or passages, nor is any smoke produced. The limits between which such perfect combustion is obtainable are comparatively narrow, so that the carburettor should preferably have some means of adjustment so as to be able to correct the proportions of the supply should such be necessary.

The popular idea is that the piston sucks the mixture into the cylinder, but of course this is not so. In reality it is merely that "nature abhorring a vacuum" the pressure of the atmosphere forces it in to occupy the space left as the piston recedes. Stated in these terms it is evident that no greater pressure can be obtained under ordinary circumstances than that due to the atmosphere itself. But as an airship rises in the air the atmospheric pressure naturally becomes less, with the result that less of the mixture is forced into the cylinder, and a consequent reduction in the amount of the charge and in the pressure resulting from the explosion, so that if steady running at full power is required for different altitudes it becomes necessary to have the means of readily modifying the relative amounts of air and fuel in the charge.

The ordinary type of carburettor is shewn in Fig. 29, and the inlet from the supply tank of the fuel is at the bottom of the float chamber on the left. When the carburettor is dry the float shown sinks to the bottom of this chamber, the two weights at the top of the float follow it, and lift the spindle of the admission valve, thus allowing the liquid to flow into the chamber. The float is of course so adjusted that it rises when the proper amount has passed in, and it then lifts the weights and the valve is closed. In actual work the amount of fuel in the float chamber is maintained constant at the same level.

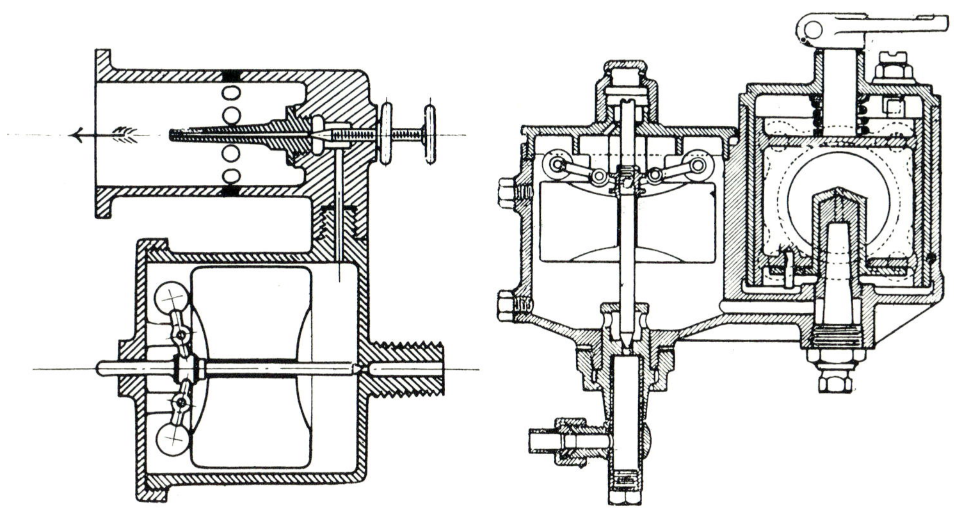

Fig. 29.
CARBURETTOR.

Fig. 30 shews the carburettor with a variable jet, made by Messrs. White & Poppe, of Coventry. It will be seen that the ordinary type of float chamber is used, but rollers are fitted to the little weights so as to reduce the friction as much as possible. On passing into the carburetting chamber the fluid flows up through a hollow nipple, which has a conical top, in which the spray hole is drilled eccentrically. Over this nipple is another, which turns freely upon it, and which has also a similar hole in it, which hole in one certain position is exactly in line with the hole in the lower nipple. This upper nipple is attached to a cylindrical plug, termed the throttle, and forms part of it. This throttle is a working fit in the circular sleeve, which again can turn freely in the outside shell. At right angles to the supply passage from the float chamber are two large ports, one being for the supply of air, and the other being the passage to the cylinder for the mixture.

The joint at the apex of the conical-headed nipple is ground and petrol-tight, a spring at the top of the throttle being employed to keep the surfaces in close contact.

When the throttle plug is rotated the opening for the supply of air is restricted, as also the opening for the supply of fuel through the holes in the nozzle, since the holes will no longer be opposite one another, and on the engine side of the carburettor the supply of gas will also be restricted.

The opening and closing movement of the carburettor is controlled by a lever which is clamped on the throttle spindle, and the travel of this lever is limited by two eccentric screw stops fitted with lock nuts, which project into the inner side, with a stop stud on the inside of the chamber. The cover of the throttle chamber is in one piece with the moveable sleeve.

Once the correct mixture has been obtained, at every position of the throttle the proportion of air and petrol taken into the engine in the form of gas will be precisely similar.

With the throttle full open there is practically no resistance in the carburettor, and the perfect mixture which results ensures good combustion and smooth running.

Another apparatus designed with same end in view is the Capel Carburettor, manufactured by Messrs. Capel & Co., London.

It was designed primarily to meet the want of a reliable and simple carburettor which was suitable for the heavier oils such as paraffin, but has been extensively used for all classes of fuel with satisfactory results. It is shewn in Fig. 31, and it will be seen that it is very simple in construction, and that in this case the float chamber is entirely done away with.

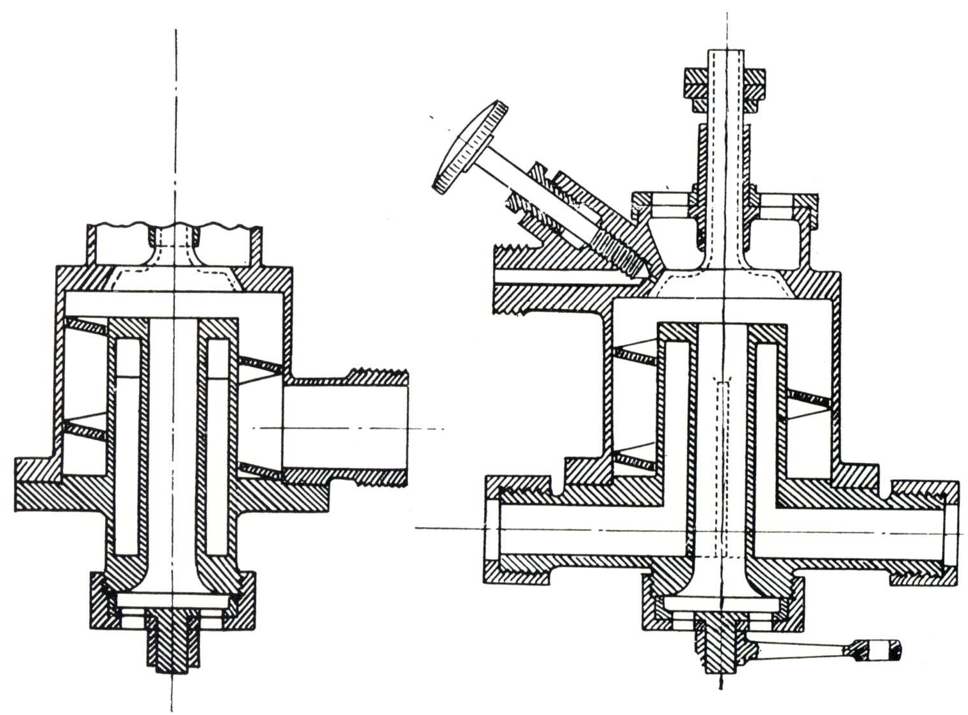

Fig. 31.
The Capel Carburettor.

At the top of the chamber there is a spring-loaded lift valve which is opened by the suction of the engine. The valve has a broad face, to the centre of which the petrol supply is brought, so that when the valve is closed both fuel and air are simultaneously shut off.

Upon the valve being opened by the suction of the engine, the fuel is drawn in through a small jet in the face of the valve seat. The mixture enters the chamber below the seat, and is made to take a spiral course through the body of the carburettor. The spiral in the interior of the carburettor serves to atomize the fuel as it impinges upon same, so that the air and fuel are thoroughly mixed before being admitted to the engine. In this way a perfect mixture is obtained, which is evident by the invisible exhaust obtained from the engine.

Provision is made for taking in additional air through the diaphragm valve in the bottom of the carburettor. The air passes through the central tube and enters the chamber below the valve, being thus thoroughly mixed with the essence, which is admitted by the carburettor valve. A spiral groove surrounds the central tube, through which part of the exhaust gases may be led if desired, and in cases where the engine is working in exposed places this is desirable. This, it is stated, prevents all tendency to freeze, and not only heats the mixture but also the additional air.

In cases where heavier fuels are used it is essential, and it will be found that the carburettor will work well on any fuel having a flash point up to 84° Fahr.

The carburettor automatically regulates the proportion of fuel and air to the speed of the motor, consequently when the carburettor has once been adjusted for the engine on which it is intended to run, it requires no further regulation.

The lift of the main valve varies with different engines, but is usually about $\frac{3}{16}$ to $\frac{1}{4}$ of an inch. The exact regulation of this lift is important, as it determines the proportion of air and fuel admitted.

A thumb screw is provided for adjusting the size of jet, and the position giving the best results may be readily found while the engine is running, after which it is not necessary or desirable to alter it, as long as the fuel and the circumstances of the engine remain the same.

The louvre plate at top of carburettor is usually kept full open, and the proper mixture found by adjusting the additional air valve at the bottom of the carburettor. A greater suction on the jet may be obtained by partly closing the louvre plate, but if this is found necessary it is usually due to dirt lodging in the jet of the carburettor and preventing the free flow of fuel.

AIRSHIP STAYS.

In every kind of airship wire stays are a necessity, and a satisfactory method of tightening them is an important feature.

The type of tightener generally adopted is a small and light form of the rigging screw, or "turn buckle" as it is called in the United States, consisting of two eye screws, one having a right-handed and the other a left-handed screw, with a long nut open in the centre, having corresponding threads to the two screws, and which unites them.

To avoid the screws slacking back when the wire is inclined to twist keep nuts are generally added, but when the twisting effect is pronounced it sometimes happens that when slightly loosened by vibration both the eye screw and the keep nut slack back together.

To obviate this the Eyquem's Patents Co. have introduced the new type shown in Fig. 27. It is very similar to the fitting described above with keep nuts, but instead of these being placed on the eye screws, they are screwed on the outside of the centre piece, the screwed part being slightly tapered. When the correct length of the stay is obtained, the keep nuts are screwed up, and the eye screws are thus firmly gripped, and cannot slack back however much vibration they may be subjected to.

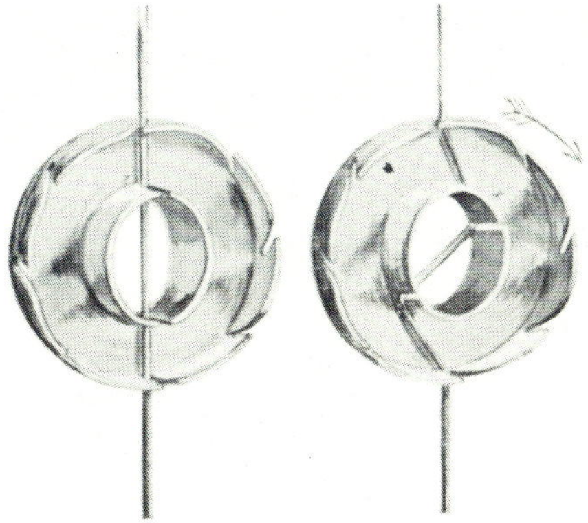

Fig. 28.
Tightening Appliance for Stays.

But for aeroplanes especially there are many short lengths of wire which may in course of time require to be tightened up, and this without cutting them and remaking the joint, and for this purpose a very ingenious device has been introduced by Mr. Alfred Portway, and is being manufactured by Messrs. Handley Page. It is shewn in Fig. 28, and consists of a saucer-shaped steel stamping, one inch in diameter, with a central boss. Round the rim are four or more inclined slots, arranged in pairs, diametrically opposite to one another. The centre boss has two slots only, but not inclined, and which are in line with one pair of the rim slots.

When it is required to slightly tighten a wire the device is placed upon it, with the wire running through the two centre slots and the two rim slots, which are in line with it. It can then be rotated by means of a key or screw-driver placed in the slots (the rotation being in the direction of the inclined slots), and the wire slips from the slots it is resting in, and as soon as the next pair of slots come beneath it, it falls into them, and is automatically locked by its own tension. Should a considerable amount of wire require to be taken up to save cutting it, the surplus length is of course wound on the centre boss. When in place the device is absolutely secure, and it will be seen that it cannot possibly slack back.

THE PERSONAL ELEMENT.

Recent events have clearly demonstrated how very largely the personal element enters into the success or otherwise of skycraft, so that most countries are already acquirjng a certain number of them in order to train a staff capable of using them.

In France a Higher School of Aeronautics has been established, which is to be principally concerned with practical work, and the students who succeed in passing the examination at the end of the course will receive the title of Aeronautical Engineer.

Germany is equipping a distinct corps of airship engineers, and is insisting on their having a thorough practical training as mechanics. They will form a body similar to that of the seagoing engineer officers of the German navy.

In Italy so far the matter has been left to the existing officers of the army and navy, the former having command of the dirigibles and the latter that of the aeroplanes, but officers of the navy have now been nominated to assist also with the dirigibles.

The official world in England has at last awakened to the possibilities of aerial navigation, and tardily seems inclined to make up for lost time. In March the British Government appointed an Advisory Committee for Aeronautics, which in August presented an *interim* report, from which it appears that steps were first taken to obtain the fullest possible particulars of all that had been done up to that time, and that reports had been obtained from experts dealing (amongst other matters) with the screw propeller, petrol motors for aeronautical purposes, and light alloys of metals. As these contain matters of general interest, it has been arranged to publish them, together with abstracts of published papers and memoirs dealing with such subjects.

A list of the experimental work to be undertaken at the National Physical Laboratory includes the following:—

PROPELLER EXPERIMENTS.

(*a*) Efficiency, and the effect on the efficiency of variations in blade area, pitch and slip.
(*b*) Position of the propeller relative to the machine.

MOTORS.

(*a*) Efficiency.
(*b*) Reliability and steadiness.
(*c*) Materials of construction.
(*d*) Design.

And experiments were commenced at once with different kinds of motors and propellers.

Special equipments are being installed, so that the work may be carried out as thoroughly as possible, and comprise the following:—

(*a*) A wind channel, 4 ft. square and about 20 feet long, with a fan giving a draught of 40 feet per second, special arrangements being made to obtain a uniform flow.

(*b*) A whirling table of 70 feet diameter, for which a special building is being erected. This will be employed for a repetition of Dines' and Langley's experiments, as well as for the propeller tests.

(*c*) Two wind towers for experiments in the open, which it is hoped will afford valuable information as to the varying conditions which obtain in practice.

(*d*) Apparatus for efficiency tests on high-speed motors up to 50 h.p.

It is considered that valuable information may be obtained by towing models under water, and for this purpose the tank now under construction for experiments on ship models will be available. This will be provided with a carriage able to attain a velocity of 25 feet per second over a run of 500 feet, and will, it is hoped, also prove of value for air pressure and propeller experiments, whilst a lighter carriage will be designed for special experiments at higher speeds.

A very important matter to inventors has been the position taken up by the Committee with regard to the inventions and ideas which their authors might desire to be treated as confidential, and on this point the Committee have decided that—

Inventors wishing to bring their ideas before them must previously take the steps necessary to fully protect their inventions, unless they are prepared to allow the Committee to use them freely in their work.

DEFENCE AND OFFENCE.

The tremendous change in conditions which will be brought about when aviation becomes possible for all is as yet hardly realized, but the many and great problems which will have to be faced by the various Governments, and especially those of European countries, are already giving serious thought to those concerned.

From the earliest times when anyone wished to ensure security or privacy it was sufficient to build a high, strong wall, and the same held good of communities, the only difference being of size and degree of strength. And up to this very day the same rule has applied in spite of all the increase in capacity of modern means of attack; and the siege of Port Arthur shows how very difficult it is to take a well equipped and well defended stronghold. The principle has ever been the same, and up to to-day might be relied upon.

But the possibility of freely navigating the air has brought about such a complete change of conditions that in the course of the next war we shall assuredly see the most tremendous scares in large defenceless towns at the bare appearance of a single airship, in case it should prove to be an enemy's airship, whilst the strongest and most thoroughly fortified strongholds, according to present ideas, will be as much open to attack as absolutely undefended places. At the same time such attack is by no means the easy matter that it is imagined to be, and the dropping of packets of explosive on a given point is a very uncertain possibility. To keep well out of danger it would be necessary for the airship to maintain a considerable height, and it is just as probable that a packet of explosive intended for Woolwich Arsenal would make a harmless hole in the Thames or in the fields inland as that it would do any harm to the Arsenal itself. To overcome such difficulties, a kind of vertical cannon, constructed of aluminium, is being fitted in some of the aerial warships, and the packets of explosives are made up somewhat in the shape of a modern projectile, with a very long point. But the problem to be solved in obtaining accuracy of aim in dropping a cylinder on a given spot is so very different to that involved in ensuring correctness in the horizontal flight of a mechanically propelled body that under present conditions the only effectual way would be to try and induce sufficient terrorism by dropping quantities of such explosives promiscuously, in the hope that they would do such an amount of destruction as to enforce submission. But this involves the carrying of a large amount of explosive, and at present no airship can afford to devote very much of its lifting power for this purpose.

A clause in the Hague Convention of 1899 prohibited the bombardment of all undefended towns, villages or buildings, but this arrangement was only binding for five years. In the Hague Convention of 1907 it was proposed to continue this agreement, with the addition of the words "by any means whatever," the phrase being specially designed to deal with the dropping of explosives on unprotected places and destroying inoffensive civilians and their property. But very considerable differences of opinion manifested themselves on this occasion. Twenty-nine of the Powers, including Great Britain, were in favour of affirming the prohibition, whilst eight, of which Germany was the principal, were altogether opposed to any such limitations, and this lead was followed by France, Russia and Spain. The Japanese delegates pointed out that it was useless to vote on the subject, seeing the impossibility of the Powers coming to any amicable and unanimous decision on the matter, and eventually they and six other Powers refrained from voting. So that at present there is nothing to prevent belligerents from dropping explosives how and when they like, and seeing that it is those Powers that consider they have much to lose by such an arrangement that are the principal opponents of such a measure, there does not seem the slightest possibility of any agreement being come to in the future. So that most of the Powers are taking steps to perfect means of offence against attack from above, and some of them have already attained a considerable measure of success in this direction.

Italy has for a long time had her hail cannon, which would certainly be effective enough against a dirigible in the immediate vicinity, but would be useless against an aeroplane or flying machine at any distance.

France is certainly the foremost so far in aeronautics, and is also well to the fore in engines of destruction for invaders coming from above, and has already made considerable progress with quick-firing guns, specially designed for their annihilation.

Britain, however, contents herself with trying to obtain knowledge of what other nations are doing, but experiments are to be carried out on Salisbury Plain to test the effect of the new 18-pounder shell against captive balloons.

It had been laid down as a practical rule that if the balloon were at a greater distance than three miles, and at an elevation greater than 1,000 feet, it was perfectly safe from any projectile that might be discharged at it. But the progress made by other nations in arms for balloon attack renders it very doubtful if this is so any longer.

Germany, with her characteristic thoroughness, is already far in advance of other nations in this direction, and the famous Krupp firm has devoted considerable attention to the matter, with the result that two types of gun with high angle fire have been constructed for specially combating airships.

The smallest is of about 2½ inches diameter (actually 65 millimetres) and has a length of 35 calibres. It may be elevated to 60 degrees, and will propel a projectile of about 9 lbs. at a velocity of over 2,000 feet per second. This gun is mounted on a carriage which at first sight appears of the usual type, but the wheel axles are attached to the frame of the carriage with a hinged joint, and so arranged that the wheels can be advanced until they nearly meet, as shewn in Fig. 32. The carriage may then be run round the central pin at the rear fixed in the earth.

The sighting of course has to be done by means of a telescope, and the gun is also fitted with a range-finder to give the elevation necessary for any given range and height.

Fig. 33.
75 m/m gun on automobile carriage.

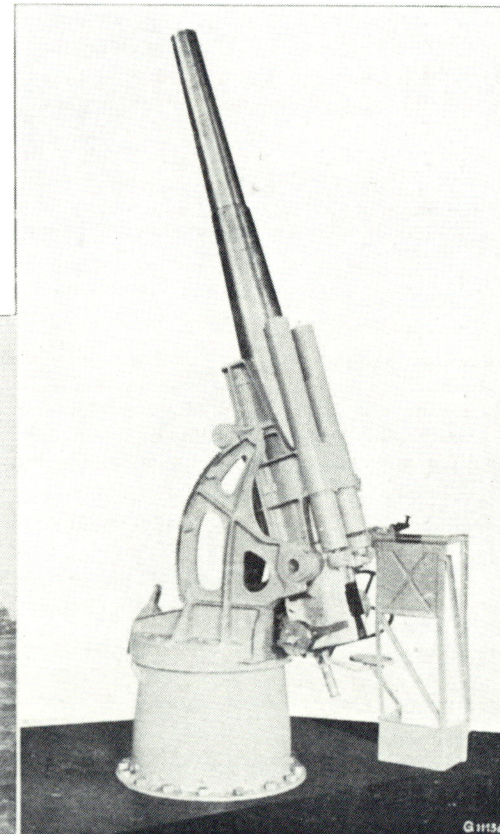

Fig. 34.
75 m/m gun for warships and fortifications.

Fig. 32.
65 m/m gun with field carriage.

KRUPP GUNS FOR ATTACKING AIR-SHIPS.

The second gun is of 3 inches bore (75 millimetres), also 35 calibres in length, and fires a 14 lb. projectile at a velocity of over 2,100 feet per second. Fig. 33 shows the gun mounted on a Krupp carriage driven by a petrol engine and specially designed for rapid mobility and for following an airship at a high speed.

Another type of this gun is shown in Fig. 34, specially arranged with a pivot mounting, for use on board ship or for fortress work. The possible elevation is 75 degrees. This gun is fitted with an air recoil apparatus, shown above the barrel, and which is always under pressure ready for firing. When fired the trigger releases the gun, which immediately runs out by the pressure of the compressed air, and on reaching the end of the stroke, the shot is fired automatically, and the energy of the recoil of the gun due to the explosion, recharges the run out cylinder. For quick firing there is an automatic arrangement for releasing the trigger, or for single rounds this is done by hand. There is also provided means of recharging the air cylinder with compressed air, on commencing to fire, or in case of a miss fire.

It is sometimes difficult enough to pick up the range on land or at sea, but it is found still more so when firing at an object in the air, when distances are very misleading.

The opportunity was taken at Rheims to carry out some interesting experiments in this direction, and they clearly manifested the special difficulties of correctly estimating the distance or elevation of machines when flying.

The amount of accuracy obtained may be realized on taking the result attained when the Curtiss machine was in the air, and which may be considered a fair example of the difficulty in arriving at a correct result, even when every possible care has been taken to obtain it.

The officer taking the sights gave the distance as 7,220 feet, and the elevation as 50 feet, whereas the actual distance was found to be 6,150 feet, and the height 80 feet.

Another point brought out by these experiments was the greatly increased difficulty of correctly estimating the position of a flying machine in motion, when starting from a point the distance of which is known, over what is met with in the case of a vessel at sea or a vehicle on land. The vessel or the vehicle moving in the one plane has two directions to move in with regard to the observer, or a compound of them, and by noting the direction, and the time taken, a fairly correct judgment can be made of the new position.

But the aerial machine, having three directions to move in, or a direction compounded of any proportion of these three, completely upsets all calculations, and observers are completely at fault almost immediately after the machine has left the position of which the distance and elevation are known.

In the German official trials of July, 1909, a captive balloon was anchored at a height of 1,200 ft. (365 metres), and in the car two dummies were placed to represent men.

It was first attacked by a body of infantry from a distance of 4,000 ft. (1,220 metres), and 5,000 cartridges were fired at it in volleys, but without any perceptible effect.

The second trial consisted of an attack by two machine guns from the same distance, and 4,000 cartridges were expended, but the balloon seemed none the worse.

Finally an attack was made by field howitzers at 6,560 ft. (2,000 metres) distance, and one of the shot managed to score a hit, and made a considerable rent in the balloon. However, it sank so slowly that the occupants of the car would not have been injured in the descent. On examination afterwards it was found that of the 9,000 shots fired of small calibre only 20 had penetrated the fabric, and the holes were so small that the edges of the material had closed together again and prevented the gas from escaping. The only damage done to the dummy men was that one had got a bullet in his leg.

It will, therefore, seem that it is a by no means an easy matter to hit aircraft of any description even with the best of weapons, so that in the event of a fortunate hit being secured, it is very desirable that the projectile should be of such a type as to at least disable the airship.

Many types of projectiles have been suggested, but there seems no actual type that will specially meet the needs for airship attack. A shrapnel is the one which would seem to promise the greatest amount of execution, but with an airship divided into many separate compartments, or unless an aeroplane is struck in a vital part of the machinery, even this would not do great damage.

The United States Government is adopting projectiles with holes in the base, in which is placed a material which is ignited as the shell leaves the gun and burns with a vivid flame, so that if it struck an airship depending on gas for its flotation powers, or using benzine as fuel, ignition might be caused and bring down the airship. But this, of course, presupposes that the airship will be hit, a rather doubtful matter in the present state of affairs.

The most promising would appear to be a projectile, containing a very high explosive, combined with an appliance of the type just mentioned, and which, on bursting anywhere in the neighbourhood of any skycraft, would create such a concussion as to destroy a dirigible, or so disturb the equilibrium of an aeroplane or flying machine as to overthrow it.

For such purposes percussion fuses would be comparatively useless, and everything would depend upon the time fuse. But there is no time fuse yet in the possession of any nation that can be absolutely relied upon in such a case, even if the distance is absolutely known.

The German shell which has now been adopted for this purpose is a flaring shell, filled with a mixture of black powder, saltpetre, magnesium and resin, which emits a great sheet of flame. It is also fitted with a smoke tracer, the smoke escaping through holes in the shell, so that the course of the projectile can be traced, and it is so arranged that the emission of the smoke can be retarded until the projectile has already completed a considerable portion of its flight.

In the branches of engineering dealing with projectiles and guns there is now an immense and untouched field presenting itself, and besides weapons for the use of terrestial forces against aerial ones, there is the further, and eventually still more important, question of weapons for the use of one airship against another.

For here again the analogy of marine navigation holds good, and the only really satisfactory and truly efficient way of disposing of an enemy's fleet of airships will be to go in search of them with better and faster airships, and make an end of them wherever they may be.

It therefore behoves any nation that desires its voice should be heard in the Council of the World to see to the matter in time. We are living in days of progress, and those who rely on merely copying what other nations are doing, in the fond idea that airships are so readily made, quickly constructed, and easily handled, will assuredly have a rude awakening when the evil day comes that there is need of them.

CHAS. DE GRAVE SELLS.

Owing to a decision to produce this book earlier than had been anticipated, one or two illustrations referred to in the text are missing.

ADDENDA.

BRITISH.

Howard Wright. Biplane.

British Dirigible, Lebaudy type. Length, 328 feet (100 m.) Capacity, *about* 282,500 c. feet (8,000m³.) 2 engines of 135 h.p. each. Propellers are of wood.

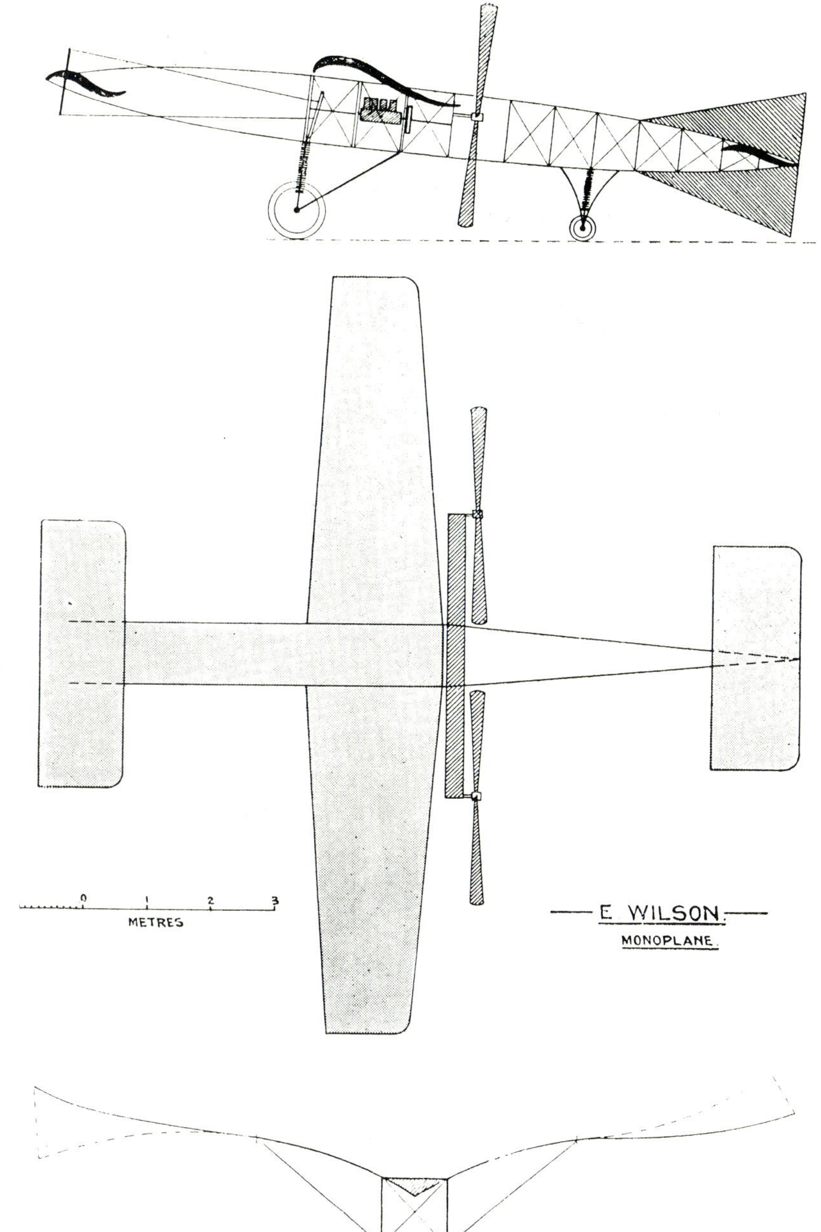

E. WILSON.
MONOPLANE.

BRITISH.

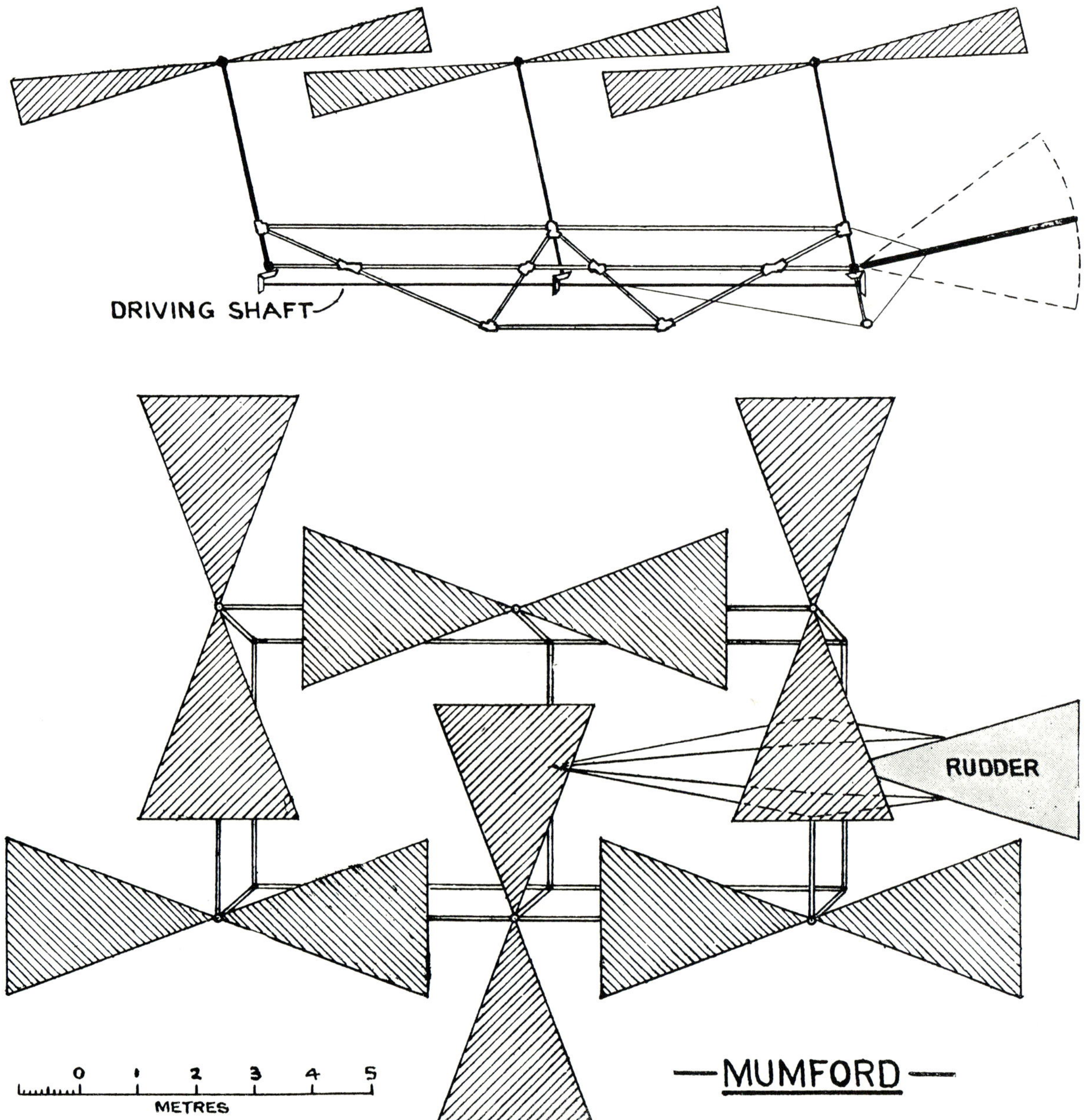

—MUMFORD—

FRENCH.

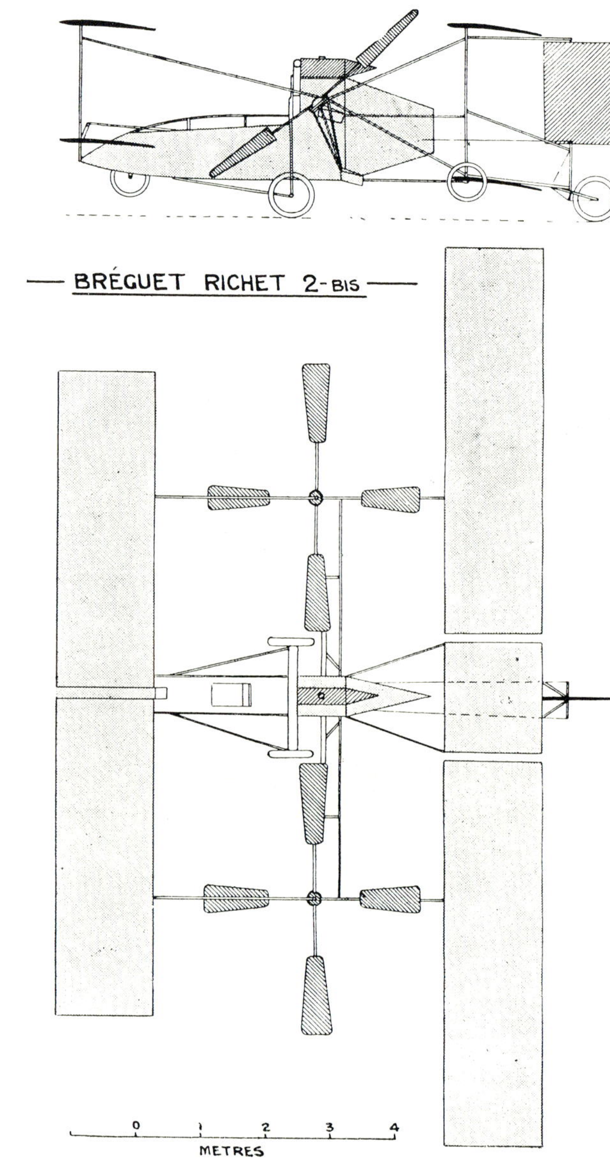

—BRÉGUET RICHET 2-BIS—

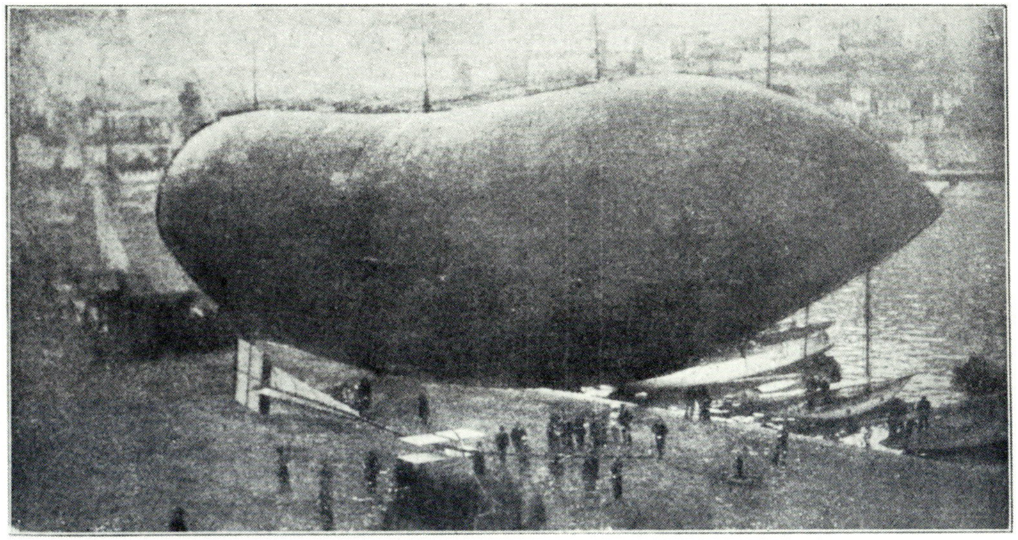

Faure.

Republique.

M. M. Clement-Bayard are building two patterns of the *Santos Dumont XX*. The spans are respectively 18 feet (5·60 m.) and 21⅓ feet (6·60 m.)

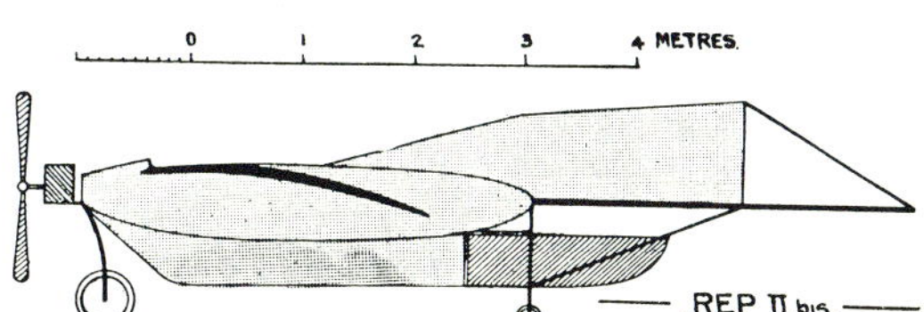

FRENCH.

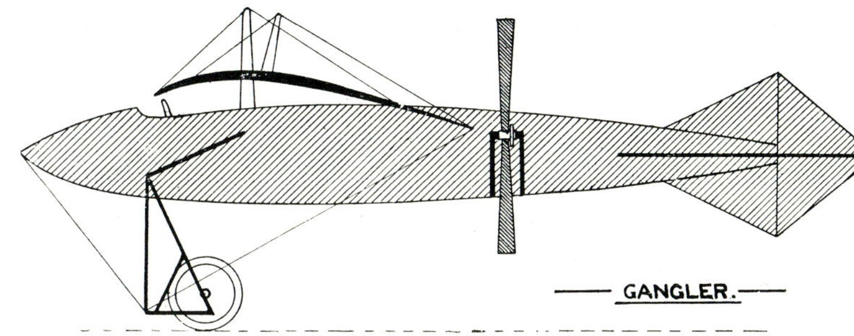

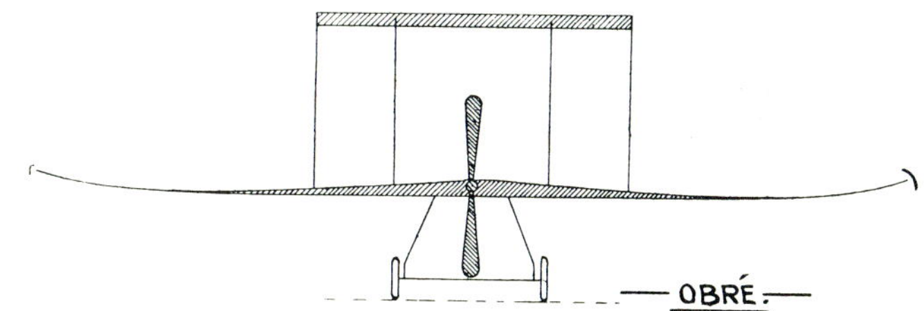

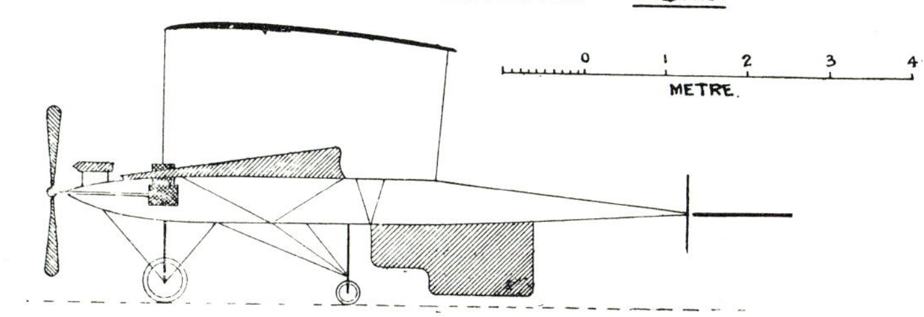

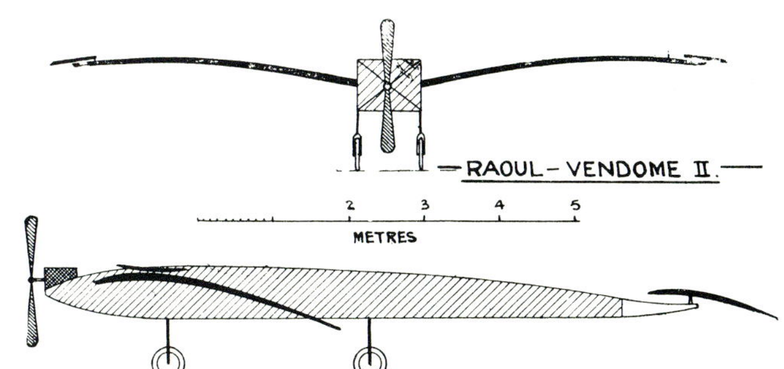

GERMAN.

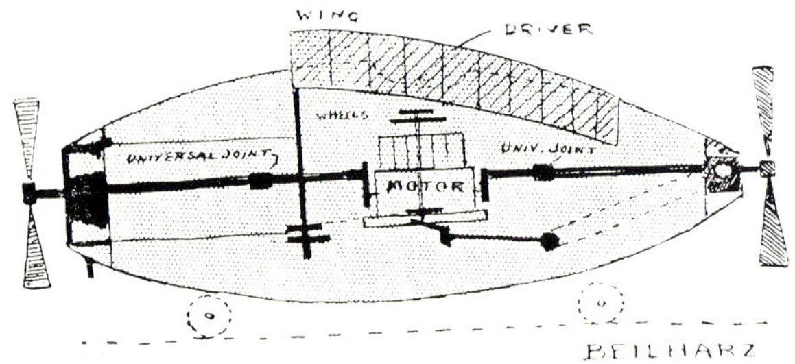

ITALIAN.

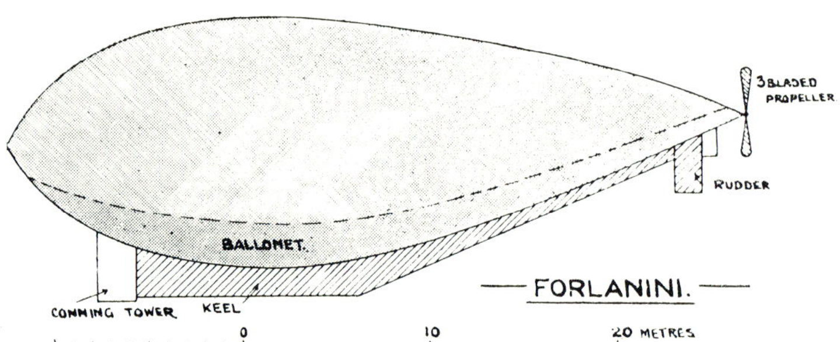

U. S. A.

Call (H. L.) Biplane.

Luyties Otto.

KIMBALL.

WALDEN (DR. H. G.) Biplane.

BABCOCK.

SHNEIDER. Biplane.

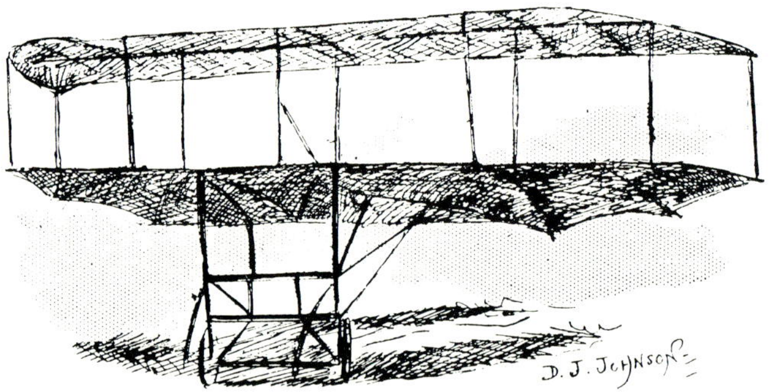
D. J. JOHNSON

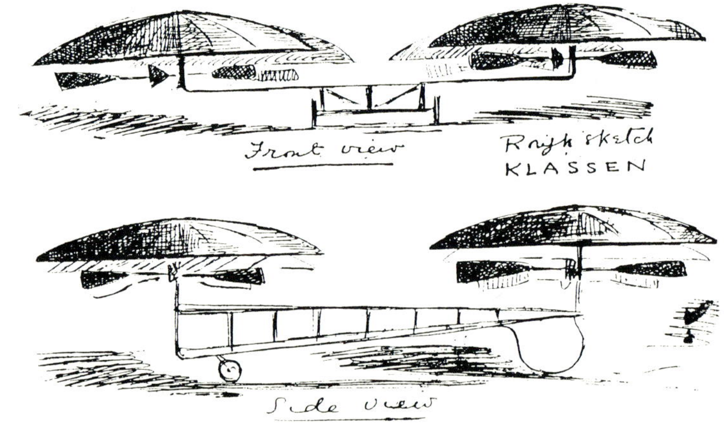
Front view
Rough sketch
KLASSEN
Side view

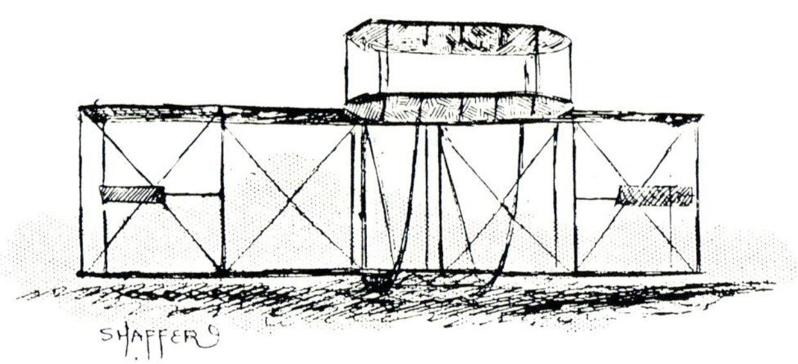
SHAFFER

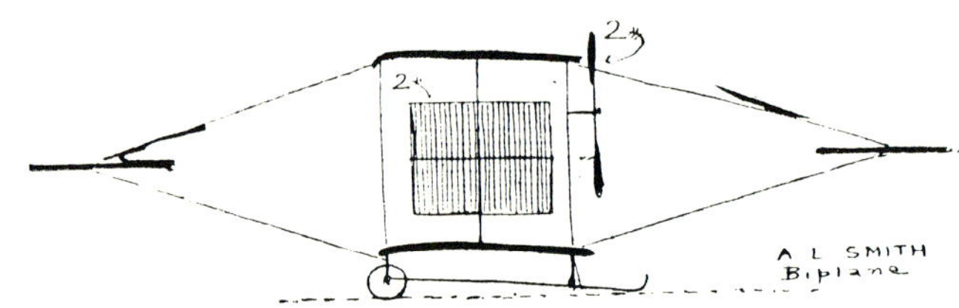
A L SMITH
Biplane

NOTE.

Owing to the rapid progress that is being made in Aviation, it has been decided to produce Quarterly Supplements to this Annual for the next year or so. The price of this Supplement will be **5**/-, and they can be obtained through any bookseller.

NATIONAL INDEX OF AEROPLANES AND AVIATORS.

NOTE.—It has proved impossible, for this edition, to obtain the names of all owners of standard machines, but where possible they are included in this index.

Austrian.

PAGE

Burchard ... 20
Etrich ... 18
Farman triplane ... 22
Hippsich ... 17
Henri Farman I ... 22
Nemethy ... 22
Stolfa ... 21
Urbanek ... 21
Wright Bros. ... 23
Wels ... 18
Zolyi ... 19

Balkan States.

Goliescu ... 25

Belgian.

De Caters ... 27
De la Hault ... 29
Louis de Laminne ... 28
Lepousse ... 28
Marchand ... 27
Vleminckz ... 30
Vandenberg ... 30
Voisin Freres ... 27

British.

PAGE

Antoinette type ... 35
Baddeck ... 75
Baden-Powell ... 44
Barber ... 39
Barnwell ... 45
Bell, Dr. Graham ... 74
Bleriot Ltd. ... 35
Bull's Eye ... 61
Cockburn ... 49
Cochrane ... 55
Cody ... 43
Cox ... 35
Cygnet II ... 76
Dunne I ... 42
Dunne II ... 42
Ellis ... 77
Farman machines ... 49
Fay Wilkinson ... 48
Feary ... 38
Friswell ... 52
Gibson ... 77
Gross ... 38
Hammond ... 51
Hinde ... 35, 49
Howard Wright ... 65
Humphrey ... 50
Lamplough ... 66
Ling ... 37
MacFie ... 37
McClean ... 70
McCurd ... 72
Moore Brabazon ... 57
Mumford ... 67
Moering ... 57
Neale ... 38
Nyborg ... 69
Ogilvy ... 60

PAGE

Page, Handley ... 36
Parkinson ... 35
Phillips, A. ... 52
Phillips, H. ... 68
Piffard ... 51
Pirie ... 53
Pischoff Koechlin type ... 71
Pitman ... 57
Porte ... 53
Pride ... 64
Richardson ... 35
Roe ... 61, 62
Rolls ... 73
Rolls ... 60
Rucker ... 73
Sandars ... 56
Saul ... 63, 64
Short ... 54
Silver Dart ... 74
Simms ... 57
Simplex M. Co. ... 35
Stocks ... 55
Smith and Mateka ... 70
Tinline ... 56
Voisin Freres ... 57
Weiss ... 40
White ... 35
Wilkinson ... 48
Wilson ... 40, 58
Wilkes ... 58
Windham I ... 46
Windham II ... 47
Windham ... 35, 59
Worswick ... 41
Wright Bros. ... 60
Wright (Howard) ... 65
At Yarmouth ... 71
Wokingham Flyer ... 72

Chinese.

Chatley ... 85

Danish.

PAGE

Delagrange ... 89
Ellehammer ... 87
Farman type ... 88
Hansen ... 88
Wright Bros. ... 89

Dutch.

Küller ... 91
Lefevre ... 91
Wright Bros. ... 91

French.

Aérienne ... 93, 99, 146
Aerodrome ... 93, 147
Abris Callas ... 130
Aeriel Soc. ... 147
Antoinette ... 93-97
Anzani ... 146
Auffm Ordt ... 98
Aimé-Salmson ... 144
Barratoux ... 147
Bayard-Clement ... 103
Baylac ... 155
Bertin ... 155
Bertrand ... 133
Beaufeist ... 144c
Bello ... 146
Bibesco ... 146
Blanc ... 104
Bleriot ... 99
Bleriot IX ... 100
Bleriet X ... 135
Bleriot XI ... 101
Bleriot XII ... 102
Bolotoff ... 146
Bonnet-Labranche ... 132
Borgnis-Desbordes ... 106
Borgnis de Savignon ... 149
Bourdariat ... 105

PAGE

Bousson-Borgnis ... 151
Braniff ... 146
Brequet Richet 120, 144*a*; *also* 154
Brissand ... 131
Bulot ... 104
Bunau-Varilla ... 146
Burgeat ... 93
Capferer ... 114
Caters ... 146
Charpentier ... 124
Charry ... 93
Chedville ... 125
Collomb ... 148
Colliex ... 146
Cornu ... 156
Cox ... 99
Crini ... 133
Dadé ... 146
Delagrange ... 99
Demanest ... 93
Dennisell ... 156
De Dion Bouton ... 130
D'Equivelly ... 113
De Rue (*see* Ferber) ... 146
Deschamps and Blondeau ... 108
Doutre ... 140
Dufaux ... 149
Dumont (Santos) ... 107
Duray Matthys ... 136
Dutheil and Chalmers ... 136
Euler ... 146
Farman ... 146
Farman, Henri ... 137, 138
Farman, Maurice ... 109
Ferber, Capitaine ... 146
Fernandez ... 131, 135
Florio ... 146
Fournier ... 146
Foquet ... 117
Frettes ... 146
Gangler ... 120
Gasnier ... 144*b*, 147
Gastambide and Mengin ... 110
George ... 146
Gaudart ... 146
Gilly ... 150
Girardville ... 147
Givaudan ... 158

PAGE

Gobron ... 146
Godville ... 156
Goupy ... 139, 146
Grahame-White ... 146
Gregoire Gyp ... 110
Gron ... 112
Guillebaud ... 111
Guyot ... 112
Hansen ... 146
Hedburg ... 146
Hinde ... 99
Henriot ... 106
Jacquelin ... 144*e*
Jerme ... 140
Juge and Rolland ... 157
Juvissy ... 147
Kapferer ... 114, 147
Koch ... 146
Koechlin ... 119, 143
Lambert ... 147
La Demoiselle XIX and XX ... 107
Lande ... 144*d*
Lastenas ... 141
Latham ... 93
Leblanc ... 147
Lefebvre ... 147
Legagneux ... 146
Legrand ... 146
Lejeune ... 142
Levy Caillat ... 115
Mangin ... 116
Matthys ... 136
Military Aeroplane ... 115
Moering ... 146
Monin ... 117
Moore-Brabazon ... 146
Odessa Aero Club ... 146
Odier ... 118
Obré ... 157
Ouviere ... 153
Parkinson ... 99
Pasquier ... 116
Paulhan ... 146
Penteado ... 141
Piquerez ... 153
Pischoff Koechlin ... 119, 123, 143

NATIONAL INDEX OF AEROPLANES AND AVIATORS—*Continued.*

PAGE

French.—*Continued.*

Raoul Vendome ... 127
R.E P. ... 121, 122
Richet ... 144*a*
Robart ... 144
Rolland ... 157
Rouchonnet ... 93
Rougier ... 146

Salvert ... 146
Sanchis Besa ... 134, 146
Schrek ... 147
Seymour ... 146
Simms ... 146
Singer ... 146
Soiron ... 118
Stourdza ... 146

Taris-Busseron ... 123
Theroul ... 150
Tissandier ... 147

Vaniman ... 151
Vendome ... 125, 126
Vermorel ... 129, 144*d*
Voisin Fréres ... 145, 146
Vosgien ... 129
Vuia ... 128
Vuitton Huber ... 151

Wehrle ... 105
White ... 99, 146
Windham ... 93, 99
Witzig-Liore Duttikuil ... 127, 152
Wright Bros. ... 147

Zens ... 148
Zipfel ... 146

German.

Beilharz ... 179

Degn ... 184

Euler ... 188
Erhard ... 183

Fabrice ... 188
Farman, H. ... 187
Flugel ... 182
Fritzche ... 178

Ganz Fabrice ... 190
Grade ... 184

Huth ... 181

Jatho ... 185
Jorch ... 183
Jospe ... 181

Lorenzen ... 186

Meschner ... 190

Nolte ... 189

Parseval ... 178

Ruthenburg ... 182
Reissner ... 189

Schnell ... 186
Scholz ... 180
Schuler ... 190
Schulke ... 180
Staechel ... 179

Voisin type ... 188

Wright type ... 188

Italian.

Anzani ... 215

Bertelli ... 212*g*
Bossi ... 212*e*

Cadore ... 215
Cagno ... 215
Calderara ... 215

Facilli ... 212
Florio ... 215
Fuseri ... 212*h*
Gemma ... 212
Gloria ... 215

Howard Wright ... 213, 214

Miller ... 212*b*
Miller ... 212*c*
Moncher ... 212*a*
Moncher ... 212*f*

Radice ... 212*d*

Voisin Fréres ... 215

Wright Bros. ... 215

Japanese.

Harrison ... 220

Yamada ... 221

Russian.

Baronorski ... 226
Bolatoff ... 233

Farman type ... 230

Lebedeff ... 229
Loutzky ... 228

Malynski ... 231

Odessa ... 231

Pischoff-Koechlin type ... 230

Tatouroff ... 232

Voisin Fréres ... 231

Wright Bros. ... 231

Zatoff ... 227

Spanish.

Ama ... 240

Barker ... 241

Canella ... 240

Salamanca ... 243
Sanchez ... 242

Voisin Fréres ... 241

Wright Bros. ... 243

Swedish.

Hansen ... 247

Voisin Fréres ... 247

Swiss.

Comisetti ... 249

Geneve ... 249
Guignet ... 249

Plazzeriaud ... 249

U.S.A.

Adams ... 285
Algar ... 277
Andræ ... 299

Babcock ... 259
Baptiste ... 290
Barnes ... 277
Bates ... 286
Beach ... 260
Beach, Willard ... 254
Berliner and Williams ... 252
Bishop, Courtland F. ... 262
Bokor ... 280
Boland ... 253
Bourdin ... 253
Bradley ... 270
Bulask-Hidalgo ... 289
Bultzingslowen ... 291

C. A. Mfg. Co. ... 285
Call ... 288
Clarke ... 284
Colby ... 277
Culver ... 293
Curtis ... 261
Curtis ... 262
Cuthbertson ... 258

Dayton ... 277
De Haven and Watkins ... 258
Dixon ... 287
Dorland ... 291
Downer ... 281

Eichenfeldt ... 266
English ... 302

Gillespie ... 292
Gould ... 264
Greene ... 263
Greene ... 277

Hendrickson ... 263
Hendrick ... 255
Herring-Curtis ... 262
Hezog ... 265
Hudson and O'Brien ... 274

Jean ... 269
Johnson ... 266
June Bug ... 256

Kerrison ... 265
Kimball ... 267
Klassen ... 293
Kunow ... 296

Lake ... 300
Lane ... 294
Lawrence ... 268
Lindsay ... 269
Loose ... 259
Luyties Otto ... 295
Metcalf ... 302
Myers ... 283

Newman ... 270

Ochua ... 255
Orme ... 271

Peach ... 260
Pitman ... 277
Posadas ... 287
Potts ... 290
Purcell ... 304

Raiche ... 273
Randolph ... 277
Red Wing ... 256
Richardson ... 257
Rickman ... 272
Robinson ... 297
Rogers ... 278

Shneider ... 273
Scott ... 281
Shaffer ... 292
Smith ... 275
Smith ... 303
Smidley ... 298
Snell ... 295
Stadtler ... 281
Steffan ... 282

Telfer ... 274
Thomas ... 275
Thompson ... 296
Triaca ... 283
Twining ... 294

Wade ... 301
Walden ... 278
Wilson ... 288
Willard ... 262
Willoughby ... 289
White Wing ... 256
Wright Bros. ... 276, 277

Zornes ... 279

DIRIGIBLES

ALPHABETICALLY ARRANGED IN NATIONALITIES.

Austrian.

	PAGE
Parseval	21

Belgian.

	PAGE
Belgique	31
Flandre	33
Lambert	33
St. Marcq	32

British.

	PAGE
Naval I.	81
Army II.	81
Army III.	82
Clement Bayard	79
Lebaudy	80
Willows I.	83
Willows II.	84

French.

	PAGE
Astra type	163
Capazza	167
Colonel Renard	166
Faure	173
La Flandre	176
Lebaudy II.	160
Liberté	162
Malecot	174
Marcay	175
Patrie	159
Petit Journal I.	169
Petit Journal II.	175
Republique	161
Ville de Bordeaux	165
Ville de Nancy	165
Ville de Paris	164
Zodiac I.	169
Zodiac II.	170
Zodiac III.	171

German.

	PAGE
Clouth	206
Dürkopp	209
Gross I.	198
Gross II.	199
Gross III.	199
Gross IV.	200
Gross V.	200
Harburg	209
Hildebrandt	209
Parseval I.	202
Parseval II.	203
Parseval III.	204
Parseval IV.	204
Parseval V.	205
Parseval VI.	205
Ruthenberg	207
R. W. M. G.	208
Schütte	196
Siemens	209
Unger	206
Zeppelin type	192
Zeppelin I.	193
Zeppelin II.	194
Zeppelin III.	195

Italian.

	PAGE
Forlanini	218
Italia (Da Schio)	216
Ricaldoni I.	217
Ricaldoni II.	217

Russian.

	PAGE
Bayard Clement	235
Lebed (Lebaudy's "Russie")	236
La Flandre	234
Outchebny	237

Spanish.

	PAGE
Astra type	215
Torres Quevedo	214

Swiss.

	PAGE
Genève	219

U.S.A.

	PAGE
America	307
Anderson	308
Baldwin	305
Charlemagne Sirch	308
East St. Louis	309
Goodale	309
Hildebrand	302
Knabenshue	306
Lincoln Beachy	310
Myers	310
Purcell	304
Rekar	311
Rigg & Rice	311
Schutt	312
Wellman	304